GEOLOGY
and the Environment

GEOLOGY
and the Environment

BERNARD W. PIPKIN
University of Southern California

■

with contributions by
D. D. TRENT
Citrus College

WEST PUBLISHING COMPANY
Minneapolis/St. Paul ■ New York ■ Los Angeles ■ San Francisco

PRODUCTION CREDITS

Artwork: Carto-Graphics, Precision Graphics
Individual credits follow the index
Composition: Parkwood Composition Services, Inc.
Copyediting: Carole Bailey Reagle
Text design: K. M. Weber
Index: Terry Casey
Cover Image: Lofoten Islands, Norway, © Charlie Crangle, JLM Visuals
The author gratefully acknowledges Reed Monroe and James S. Wicander for providing a number of illustrations from their books, *Physical Geology: Exploring the Earth* and *Historical Geology: Evolution of the Earth and Life through Time,* both published by West Publishing Company.

Production, Prepress, Printing, and Binding by West Publishing Company

WEST'S COMMITMENT TO THE ENVIRONMENT
In 1906, West Publishing Company began recycling materials left over from the production of books. This began a tradition of efficient and responsible use of resources. Today, up to 95 percent of our legal books and 70 percent of our college texts are printed on recycled, acid-free stock. West also recycles nearly 22 million pounds of scrap paper annually—the equivalent of 181,717 trees. Since the 1960s, West has devised ways to capture and recycle waste inks, solvents, oils, and vapors created in the printing process. We also recycle plastics of all kinds, wood, glass, corrugated cardboard, and batteries and have eliminated the use of Styrofoam book packaging. We at West are proud of the longevity and the scope of our commitment to the environment.

610 Opperman Drive
P.O. Box 64526
St. Paul, MN 55164-0526

Printed in the United States of America

01 00 99 98 97 96 95 8 7 6 5 4 3 2

Library of Congress Cataloging-in-Publication Data

Pipkin, Bernard W.
Geology and the environment / Bernard W. Pipkin, with contributions by D. D. Trent.
p. cm.
Includes index.
ISBN 0-314-02834-X
1. Environmental geology. I. Trent, D. D. II, Title.
QE38.P48 1994
550—dc20

93-40808

CIP

Contents in Brief

Case Studies

Contents

CHAPTER

8

Subsidence and Collapse 188

CHAPTER

9

Water on Land 208

CHAPTER

10

Water Under The Ground 242

CHAPTER

11

Oceans and Coasts 266

CHAPTER

12

Arid Regions, Desertification, and Glaciation 296

CHAPTER

13

Energy 334

CHAPTER

1

ENVIRONMENTAL GEOLOGY AND HUMAN EXISTENCE

The power of population is infinitely greater than the power in the earth to produce subsistence for man.

THOMAS MALTHUS (1766–1834)

Geology is the study of the earth, and **environmental geology** is the subdiscipline that focuses on the relationship between humans and their earth environment. A moment's reflection reveals that environmental geology encompasses many essentials of our daily lives: the resources used to manufacture our material goods, the water we drink, the soil that grows our food, and the very ground on which we stand and travel. Many of our life choices have geological ramifications. The decision of where to live, for example, should involve geological questions. Is the home site stable against landslides? What are the risks of disastrous floods? Could an earthquake damage this dwelling?

Yes, environmental geology is the study of human interaction with the land, with all its sociological, economic, and political ramifications. It is the application of the science of geology to solve the problems that arise from the complex interactions of the earth's five systems: water, ice, air, the solid earth, and life. As such, environ-

The Statue of Liberty with New York City and Manhattan Island in the background. The towers of the World Trade Center can be seen in the center of the photograph. Underneath the tall skyscrapers is a maze of subway and sewer tunnels. This is possible because the city is built upon competent metamorphic bedrock that is capable of supporting enormous loads. New York City has one of the largest concentrations of people in the world and, as a result, is beset with problems. Many of the problems are geologic in nature, such as safe disposal of solid and liquid wastes, adequate supply of fresh water, and pollution of the Hudson River Estuary and Long Island Sound.

mental geology encompasses all of the traditional branches of geological study: mineralogy and economic geology (mineral resources), petrology (the study of rocks), engineering geology (the building of structures on land), sedimentology (the origin and interpretation of sediments), structural geology (how rocks deform), geophysics and geochemistry (the physics and chemistry of the earth), and hydrology (the study of water). ➢ Figure 1.1 illustrates the components of environmental geology. Environmental geology applies all that is known about geology to the problems of human existence, particularly to the problems of geologic hazards and how the earth's resources can best be utilized.

Although the world's peoples have been applying the general principles of environmental geology without thinking very much about them for several thousand years, the designation "environmental" is a very recent addition to the formal study of geology. In Europe and elsewhere, many old monasteries, cathedrals, castles, and fortresses are sited on the soundest locations for miles around. Until about 1000 B.C., large Lake Texcoco occupied the Valley of Mexico, the home of the Aztec civilization. The Aztecs drained the lake in order to improve foundation conditions and placed impermeable walls into the ground to impede sulfur-contaminated water from mixing with fresh underground water. The dry bed of ancient Lake Texcoco is now occupied by Mexico City, one of the largest cities in the world (➢ Figure 1.2). In ancient Persia, present-day Iran, underground water was collected for daily use by digging tunnels, known as *kanats,* many kilometers into hillsides (➢ Figure 1.3). Although the practice started in Persia, it spread rapidly, and by about 500 B.C. kanats were being used in Afghanistan, China, and Egypt. Many of them are still operative, and ten new ones were constructed in the People's Republic of China in 1949. Even years ago, however, some unfavorable or hazardous geological conditions were ignored or were not recognized. This explains, for example, why some towers in Italy now tilt due to the underlying soil conditions (see Chapter 8). Many ancient structures that were built in concert with the environment are still serving human needs today; others, built to overwhelm or to command the environment, are in ruins.

A major goal of those who work in environmental geology is to identify adverse geologic conditions in advance by collecting surface data from rock outcrops and subsurface data from drilled boreholes and by using geophysical techniques. This need to identify potential geological problems was recognized in the 1950s in the United States because of burgeoning urban growth. Marginally suitable lands that are subject to landslides, earthquake damage, and flooding were being developed for

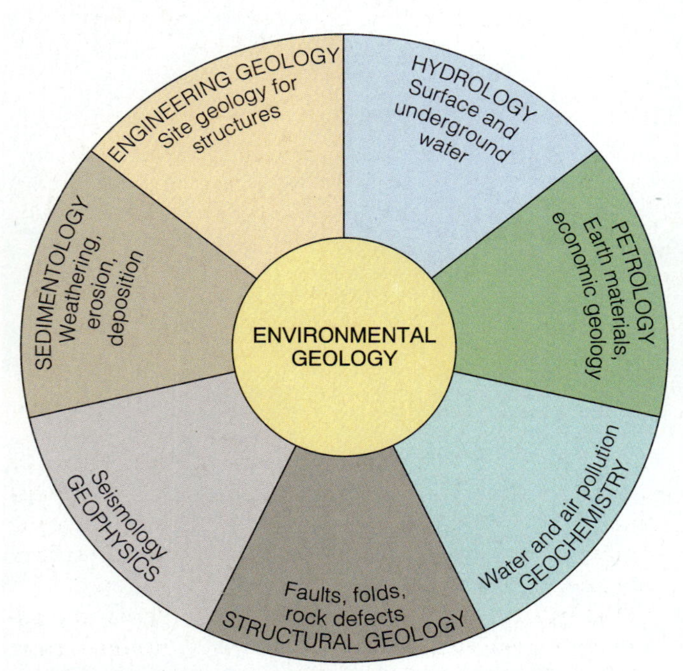

➢ **FIGURE 1.1 The relationship between the geological sciences and environmental geology. Environmental geology incorporates many skills and specialties of the discipline of geology.**

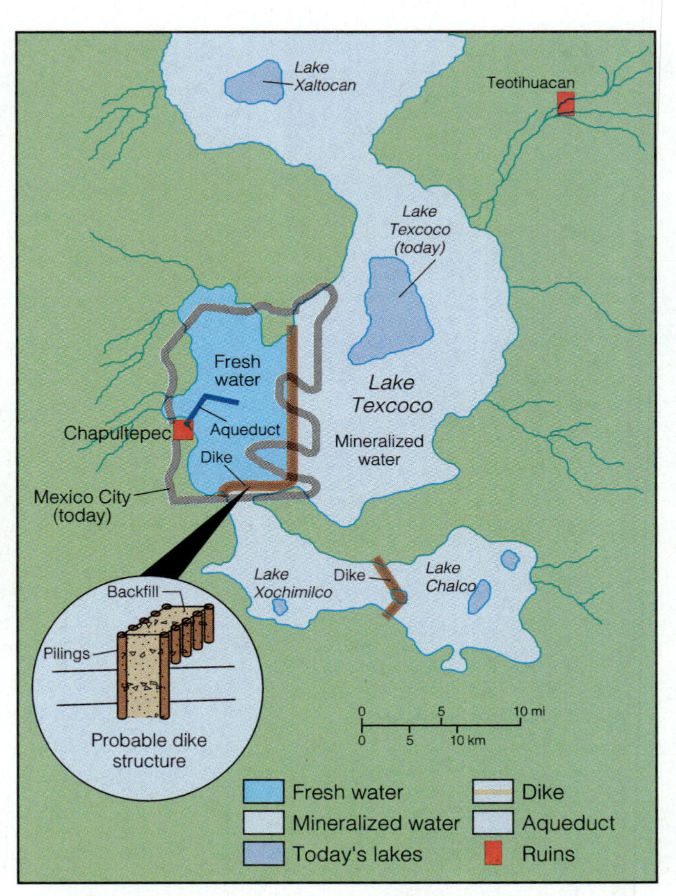

➢ **FIGURE 1.2 An ancient Aztec solution to ground water problems. About 1000 B.C. ancient Lake Texcoco was drained, leaving Lakes Xochimilco and Chalco as remnants in the Mexico City area.**

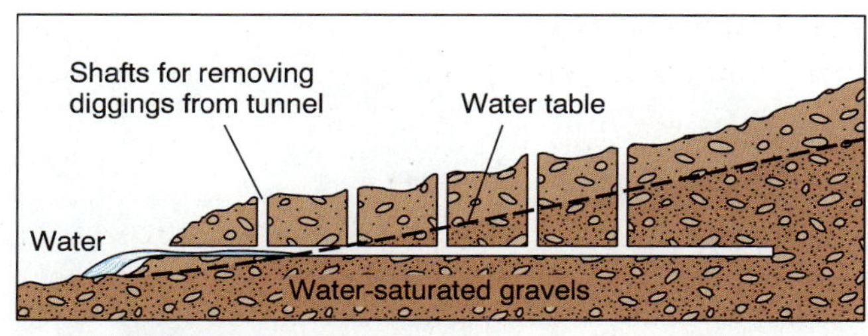

➢ **FIGURE 1.3 An ancient water-collecting tunnel, or *kanat*, driven into a gravel deposit. Kanats are still used today.**

human use. In addition, the demand for resources increased as cities grew, which required advance planning for local sourcing of building materials, providing adequate drinking water of good quality, and deciding what would be the best use for given parcels of land. Most of the planning problems then, as now, ultimately became engineering or geological ones. Not surprisingly, many city planners are engineers, and these civil engineers were among the first to realize the need for input from professional geologists. Thus the field of engineering geology expanded into "urban geology," and as communities became larger and more complex, "environmental geology" was born. Although the original definition of **engineering geology** was restricted to "the application of geological data and principles to the design and construction of engineering works," it has been expanded to include "the development, protection, and remediation (cleanup) of underground water resources." Such environmental applications of geology as ridding waters and soils of toxic substances remaining from past industrial and commercial activities are now major activities of geological firms that once dealt only in heavy construction and land development.

Multipurpose maps of areas being considered for development are essential to effective planning, and these maps must be available to planners long before the bulldozers arrive. Maps (such as the ones in ➢ Figure 1.4) can be generated that show areas of unstable ground, locations of active earthquake faults, areas with high flood potential, the depth to underground water of a particular quality, and possible mineral resources. Such maps can also indicate areas that are unsuitable for residential or commercial development but which could be used for parks and recreation. The generation of a multipurpose map usually begins with the drawing of a geologic map that shows the distribution of rocks and their spatial relationships at the surface and below, and drainage patterns. The National Geologic Mapping Act of 1992 authorized the federal government to spend $184 million over four years for geologic mapping and support activities. The act recognizes that "geologic mapping is a fundamental part of most earth-science studies, with applications in such fields as applied-geology investigations;

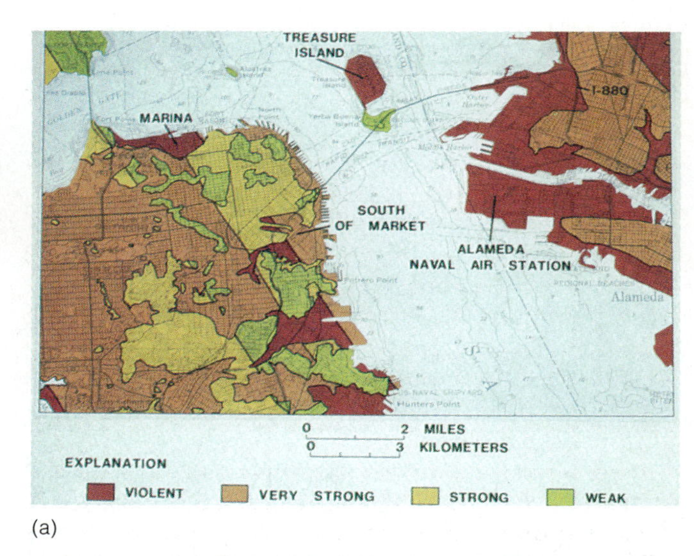

(a)

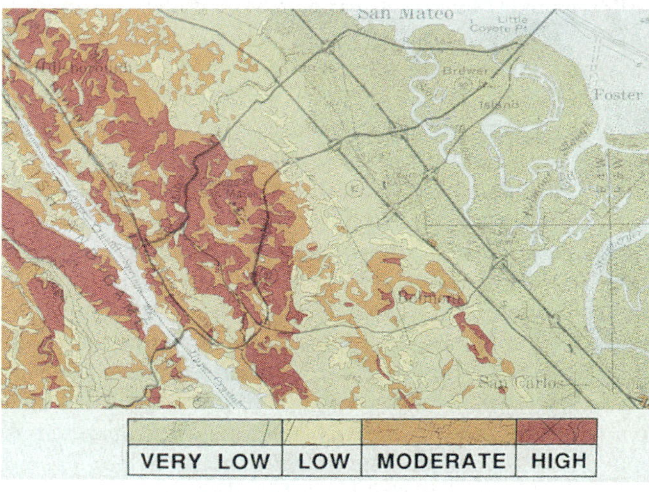

(b)

➢ **FIGURE 1.4** **(*a*) Maximum predicted ground shaking in the San Francisco Bay region due to strong earthquake motion. Predicted ground shaking is a function of the natural geologic foundation at the site, such as soft bay muds in the Marina District (violent shaking) or hard bedrock on Nob Hill (weak shaking). (*b*) Hazards due to earthquake-induced landslides along the San Andreas fault (dotted line) adjacent to San Francisco Bay.**

water-resource planning; and land-use planning for waste dumps, coastal housing development, and transportation." Geologic maps are used extensively in the planning process and beyond. They can be adapted to many special requirements of a given project or long-term plan.

Nature is a good teacher, and many significant advances in environmental geology have resulted from studying environmental geology "case histories"—serious attempts to describe and analyze events that have occurred. Two recent examples—both occurred in 1992—are the underground flood in downtown Chicago (see

(a)

(b)

➢ **FIGURE 1.5** (*a*) **Interstate highway I-880 in Oakland after the 7.1-magnitude Loma Prieta earthquake of 1989.** (*b*) **This high rise building tipped over during the 8.1-magnitude Mexico City earthquake of 1985.**

Chapter 9) and Hurricane Andrew, which caused widespread damage in Florida and Louisiana (see Chapter 11). In the Chicago case, about 250 million gallons of water leaked from the Chicago River through a drill hole for a bridge foundation into tunnels beneath Chicago and flooded the lower levels of major high-rise buildings. This half-billion-dollar "mistake" generated intense reexamination of engineering specifications for tunneling and bridges in the Chicago area. Hurricane Andrew, in August 1992, was the most destructive storm to hit the U.S. mainland to date, with property damage estimated at as much as $20 billion. The heavy damage to homes and other structures led building officials to study the adequacy of the existing building codes and inspections for private dwellings in South Florida. After the earthquakes in Mexico City in 1985 and San Francisco in 1989, the designs of freeways and high-rise buildings were reevaluated (➢ Figure 1.5 and Chapter 4). Many years ago the U.S. Air Force built a base in the Philippines near a supposedly dormant volcano, Mount Pinatubo. Mount Pinatubo's eruption in 1991 forced the facility to be abandoned (➢ Figure 1.6 and Chapter 5).

➢ **FIGURE 1.6 Volcanic ash rises several kilometers above Mount Pinatubo, the Philippine Islands, June 1992.**

Laws now require underground gasoline-storage tanks to be examined for leaks. If any are found, nearby underground water supplies must be examined for possible contamination. New technologies for sanitizing the gasoline-contaminated water and soils discovered by these mandatory inspections have resulted. Disposal of human-generated solid waste is a major problem, which requires identifying geologically acceptable disposal sites—including sites that are safe for sequestering hazardous substances. (The toxic-waste tragedy of Love Canal and its remediation are discussed in Chapter 15.) Extracting water from beneath the land surface for human use may seem innocent enough, but in Venice, Italy, this is causing the land beneath the city to subside. This sinking causes periodic flooding, with serious consequences for the residents and their largest source of revenue, tourism. Similarly, subsidence has altered the flow of surface water and storm drains over large areas of the U.S. Gulf of Mexico and the Great Valley of California due to extraction of underground water.

At the core of these and many other environmental problems is a burgeoning world population of more than 5 billion people. Every one of these persons wishes to share in the good life, but this growing population exists on a planet with limited air, land, water, and mineral resources.

■ Too Many People

The concept of **carrying capacity**—the idea that a given amount of land or other critical resource can support a limited number of people or animals—has intrigued scientists, economists, and philosophers for centuries. Thomas Malthus postulated in 1812 that populations grow geometrically, whereas food supplies increase only linearly, which leads to a gap between demand (population) and supply (food) that eventually results in famine. Humans are inextricably linked to resource utilization, environmental quality, and economic well-being. Thus, a good starting point for any environmental study is to learn how populations grow, a study referred to as **population dynamics.** World population has increased dramatically this century. In 1900, the population of the earth was estimated at 1.6 billion people. By 1990, this figure had more than tripled, to 5.2 billion. Pogo, a popular comic strip character of the 1960s, once said, "Yep, son, we have met the enemy, and he is us," and this applies to the global population dilemma.

Population growth rate is determined by subtracting a population's death rate from its birthrate, with death and birth rates expressed as deaths and live births per 1,000 people per year, respectively. For example, a birthrate of 20 *minus* a death rate of 10 would yield a growth rate of 10 per 1,000 persons, or 1.0 percent. World growth rates were low at the turn of the century (about 0.8%), and they have fluctuated with time, reaching a high of about 2.06 percent during the 1960s; on average they have slowly declined since the 1960s. At first glance you might think that a growth rate of 1 percent would yield a population **doubling time** of 100 years (since 1% per year × 100 years = 100%). This is not the case, however. Populations grow like savings accounts with compounded interest. Just as you can earn interest on interest, the people that are added produce more people, resulting in a shorter doubling time. A very close approximation of doubling time can be obtained by the "rule of 70"; that is,

$$\text{Population doubling time} = 70 \div \text{growth rate (\%)}$$

Thus, the doubling time for a 1 percent growth rate would be 70 years; that for a 2 percent growth rate (close to the present world growth rate) would be 35 years, and so forth. Table 1.1 provides some examples. The global population growth and the average annual growth rate for 1950 to 2000 are shown in Table 1.2. Although some progress has been made in slowing the growth rate, the number of people added each year continues to grow larger. The world population is projected to grow by 959 million in the decade 1990–2000, the largest increment ever for a single decade. In the United States the growth rate was 0.7 percent in 1985, about half of what it had been in 1960. This is laudable progress toward **zero population growth,** which exists when, on average, as many people die as are born each year. Even at this 0.7 percent growth rate, however, the population would double in 100 years, and this does not account for population growth by immigration—which adds significantly to the U.S. population, particularly in such states as Florida and California. For the earth as a whole, immigration is not a factor in the planet's population dynamics, yet.

TABLE 1.1 How Populations Grow

GROWTH RATE (%)	DOUBLING TIME, YEARS*
1.0	70.0
2.0	35.0
3.0	23.3
4.0	17.5
5.0	14.0
6.0	11.7
7.0	10.0

*Calculated by using the formula 70 ÷ growth rate (%), which yields a close approximation up to a growth rate of 10%.

TABLE 1.2 World Population and Growth Rate, 1950–2000

YEAR	GROWTH RATE, %*	POPULATION, MILLIONS
1950	—	2,516
1955	1.77	2,745
1960	1.95	3,027
1965	1.99	3,344
1970	1.90	3,772
1975	1.84	4,079
1980	1.81	4,448
1985	1.80	4,851
1990	1.76	5,292
1995	1.66	5,733
2000	1.56	6,199
2025	—	8,504

*Average annual growth rate for the previous 5 years.
SOURCE: United Nations 1978 Assessment. Department of International Economic and Social Affairs, Statistical Office, Demographics Yearbook 1985 (1987). Demographics Yearbook 1990 (1992); Department of Commerce, Bureau of the Census, World Population 1950–1983.

In 1990 about 90 million people were added to a world population of 5.2 billion people. If this rate of growth continues (1.76% in 1990, down from 1.81% in 1980), our planet will need to support 10 billion people by the year 2030. Slower growth has been reported in all parts of the world except Africa. Growth-rate reduction measures, largely family planning, are being implemented or encouraged throughout the world. The growth rate of China, whose population has broken the billion mark, has decreased more than 30 percent, but it was still 1.45 percent for 1985–1990.

Due to variations in growth-rate momentum, world population projections for the middle of the next century range from a low of 9 to 10 billion to a high of 15 billion people (see ➢ Figure 1.7). The population growth curve in the figure is the typical **J-curve** for exponential growth of any type. In such curves, growth is very slow at first (the flat part of the curve) and then increases rapidly (the steep part of the curve). World population presently exhibits exponential growth, more than tripling in the past 90 years. As the population has grown and demanded more from the earth, 20 percent of arable lands have suffered serious topsoil erosion, 20 percent of tropical rain forests have been obliterated, and many thousands of plant and animal species have disappeared. As population increases, marginal lands are tilled because of the need for more food, and thus, poor agricultural practices become exaggerated. In addition, atmospheric carbon dioxide (CO_2) levels have increased 13 percent over the past fifty years, which may have implications for global warming, and the protective stratospheric ozone layer has decreased 2 percent worldwide in the same period (and more than that over Antarctica; see Chapter 12). None of these problems is insurmountable, but human ingenuity will certainly be taxed to overcome the challenges they present.

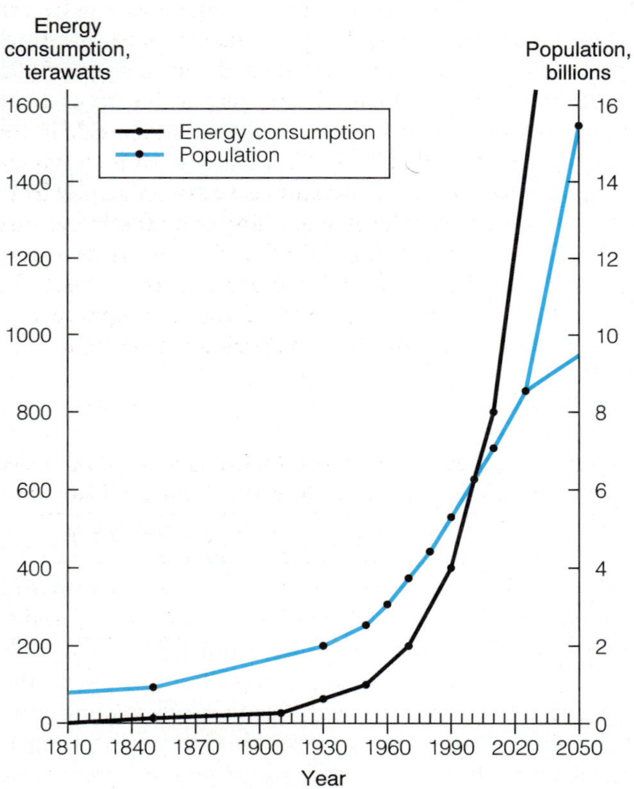

➢ **FIGURE 1.7 Projected population and energy consumption to the year 2050. The two projections for population are based on minimum and maximum birthrate estimates. A terawatt is a thousand billion (trillion) watts, or a billion kilowatts.**

SOURCE: Blue Planet, 1991.

Overpopulation Scenario One: The Sky Is Falling!

We know that catastrophic extinctions have occurred at particular geologic periods in the earth's history. One study estimates that during an interval of catastrophic extinction in the late stages of the Cretaceous Period (Appendix 1), the extinction rate averaged one species per 25 years. In contrast, it is estimated that more than 1,700 species were pushed to extinction in 1991 alone, either from loss of habitat or human predation (that's 4.6 species per day). Granted that our biological inventory of the Cretaceous is fragmental and that modern estimates are biased by our better knowledge of present-day species, but such estimates clearly indicate the malady: too many people and too few resources. We must face the facts that the present world population is stressing the environment and that the carrying capacity of the earth has either been reached or is fast approaching. Furthermore, students of population dynamics, economics, and environmental science generally agree that the earth cannot sustain the 9–10 billion people projected for 2050, assuming the lower future growth rates. Drastic action to stem population growth must be implemented now, for if nothing is done, future population reductions will occur by increased fatalities from natural disasters, mass starvation, and ecological degradation.

The population problem has been attributed to what is loosely called the "north–south disparity," the economic disparity between developed nations, largely in the Northern Hemisphere, and lesser developed or emerging nations, many of which are in the Southern Hemisphere—India being an important exception. Increasingly, the world's people are being divided into two societies, one rich and one poor. Twenty percent of the world's people live in developed countries, and they use 10 times the energy and produce 20 times the waste of the eighty percent who live in the underdeveloped or emerging nations.

Regardless of life-styles, the earth is overpopulated both north and south. The burning of oil, gas, and coal—the so-called **fossil fuels**—doubled every 20 years between 1890 and 1970, mostly to support the enhanced quality of life in the north (➢ Figure 1.8). Burning fossil fuels depletes finite resources and creates problems that adversely affect air and water quality, most importantly by producing the "greenhouse" gas carbon dioxide. If we are in fact at the threshold of human-induced global warming—not all scientists agree this is so—sea level could rise and inundate coastal areas, including all or parts of many large cities. If we desire a healthy and sustainable habitat for future generations, we must stop uncontrolled growth, runaway resource consumption, and degradation of the environment. This will require many changes in our way of life that most of us "northerners" are likely to perceive as sacrifices.

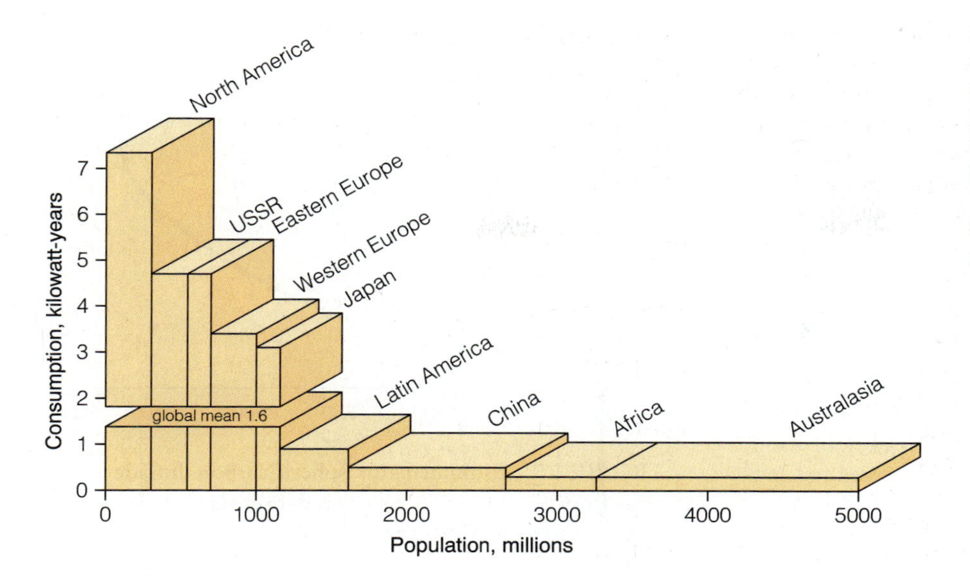

➢ **FIGURE 1.8 Per capita energy consumption in kilowatt-years (kwy) and population by geographic or political region for 1987. Note that the industrialized nations consumed 2–4 times the global per capita mean and 10–20 times that of Third World areas such as Australasia and Africa.**

SOURCE: Blue Planet, 1991.

Overpopulation Scenario Two: The Gaia Hypothesis

British physicist James Lovelock and his work with atmospheric processes and chemistry are respected in the scientific community. He developed an apparatus for measuring trace (parts per trillion) amounts of atmospheric gases and was the first person to establish that chlorofluorocarbons (CFCs) were persistent and long-lived in the atmosphere. This led to speculations on the impact of CFCs on the ozone layer, which has become an area of intense scientific research and debate.

The name of Gaia, the Greek goddess of the earth, has been applied by Lovelock and U.S. biologist Lynn Margulis to their hypothesis that the earth is a superorganism whose environment is controlled by the plants and animals that inhabit it, rather than vice-versa. According to this hypothesis, the planet is "self-adjusting," so to speak. The superorganism regulates its environment with a complex system of mechanisms and buffers, just as an animal adjusts to varying temperatures, chemistry, and other environmental factors. For example, carbon dioxide, a gas that is required for photosynthesis and therefore all of life, also acts as an insulating blanket in that it holds the sun's heat close to the earth. Terrestrial and marine plants act as regulators that "pump out" excess CO_2 from the atmosphere when great amounts are added by volcanic activity, which, if it were to accumulate, would smother humans and other animals. Were it not for the development early on in geologic time of bacteria and algae, which remove carbon dioxide from the air, the earth's atmosphere would have evolved much differently and would be poisonous to life as we know it—much like the atmospheres of Mars and Venus.

According to the Gaia hypothesis, the last ice age ended because of a doubling of atmospheric CO_2 caused by a sudden failure of the "pumps," not by the slow processes of geochemistry as invoked by conventional scientific theory. The amount of carbon dioxide being added to the atmosphere by humans today is comparable to the natural rise that terminated the last ice age and resulted in global warming. (Table 1.3 presents estimates of atmospheric carbon emissions for selected countries in 1989.) Hypothetically, the earth, being responsive to change, should be able to absorb the excess carbon dioxide in the oceans, soils, and plants—if it is left alone. Unfortunately, it may be that the natural system is being prevented from establishing a new tolerable equilibrium, mostly because deforestation is removing the plant life that is necessary to counteract the change. Similar arguments are made for other, less crucial constituents of the atmosphere that have remained constant for long periods of geologic time.

TABLE 1.3 Annual Estimated Atmospheric Carbon Emissions from Industrial Processes; Selected Countries, 1989

	ESTIMATED CARBON EMISSION	
COUNTRY	*Total Amount, Millions of Tons*	*Tons per Capita*
U.S.A.	4,869	19.7
Canada	456	17.3
Australia	257	15.5
Soviet Union	3,804	13.3
Japan	1,040	8.5
China	2,389	2.2
Brazil	207	1.4
India	652	0.8
Zaire	3.8	0.1

SOURCE: World Resources 1992–1993; World Resources Institute and The United Nations Environment Programme, Oxford University Press, N.Y.

The Gaia hypothesis is appealing because it offers a solution for environmental problems. However, even its strongest supporters admit that by sometime between 2050 and 2100, the amount of carbon dioxide in the atmosphere will have doubled. According to Lovelock, "It is a near certainty that the new state will be less favorable for humans than the one we enjoy now."

Toward a Sustainable Society

A **sustainable society** is one that satisfies its needs without jeopardizing the needs of future generations. The current generation *should* strive to pass along its legacy of food, fuel, clean air and water, and mineral resources for material needs to the generations of the twenty-first century (see Case Study 1.1). We know what needs to be done to accomplish this, but knowing what to do and implementing a long-range plan for carrying it out are entirely different. Some optimism is found in the birthrates of the United States and Western Europe; they are approaching **replacement reproduction:** the rate at which just enough babies are born to replace their parents. Nonetheless, these populations are still growing because of the momentum of population growth and—in the United States—because of net immigration. Also encouraging are indications of changing attitudes about fertility among people in less-developed countries such as India, Bangladesh, Indonesia, Egypt, and China. A promising development in industrialized nations is the growing respect for basing an economy on steady-state conditions, rather than on continual growth. The conceptual ideal is sometimes referred to as the *spaceship* economy, since all resources must be conserved, shared, and recycled in space travel. This, together with population control, could lead in time to a sustainable society.

Energy and Global Warming

After population growth, global warming is potentially the most serious long-term environmental problem facing humankind. It is closely associated with the increase in atmospheric carbon dioxide (CO_2) produced by the burning of coal, oil, and natural gas (➢ Figure 1.9). Other greenhouse gases, such as methane, nitrous oxide, ozone, and chlorofluorocarbons (CFCs), are also increasing in the atmosphere, and this, too, is closely associated with population growth. It is estimated that these gases may contribute to about half of the global warming projected for the next century.

Analysis of air bubbles trapped in deep cores of glacial ice shows that CO_2 in the atmosphere has increased 25 percent since the eighteenth century. Calculations suggest that by reducing world CO_2 emissions from the 1992 level of 22 billion tons/year to 8 billion tons/year, the atmosphere (and possibly the climate?) would stabilize at

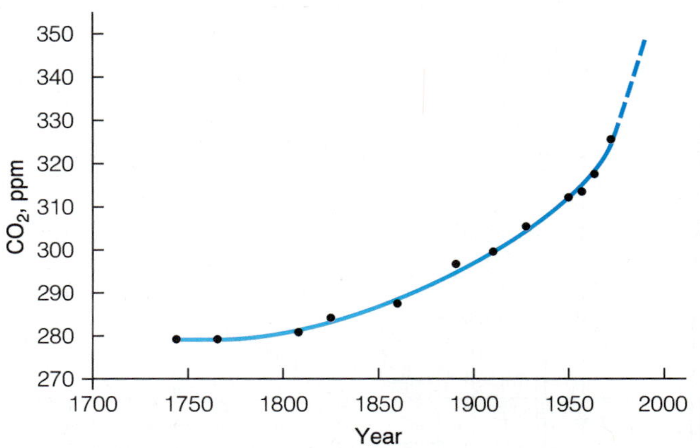

➢ FIGURE 1.9 Estimated atmospheric carbon dioxide changes from 1700 to 1980 based on data from air bubbles trapped in Antarctic ice cores and on atmospheric measurements.

SOURCE: Nat. Oceanographic and Atmospheric Administration (NOAA).

acceptable CO_2 levels. In the developed nations this would require an 85 percent reduction per capita in the consumption of fossil fuels, which translates to the personal sacrifice of conventional fossil-fuel-driven automobiles and other machines. In the early 1990s 500 million motor vehicles were registered in the world—each with an average gasoline consumption of 2 gallons per day—and that number was expected to quadruple by 2025. Thus, alternative fuels *must* be developed. Perhaps these will be liquid fuels derived from coal, biomass (e.g., alcohol), or solar energy.

As developing nations raise their standards of living, their energy consumption will also increase, and competition between the developed and the developing nations for energy supplies will increase. This competition will surely increase as the world's coal and oil reserves decrease. In the decade 1982–1992 energy consumption increased more than 50 percent in Asia, 40 percent in Africa, and 30 percent in Latin America, compared to increases of 8 percent or less in the United States and Europe. Per capita oil consumption actually decreased 16 percent during the 1980s in the United States. The developing nations were using 30 percent of the world's energy in the early 1990s, a usage that was predicted to double by 2015. It is in the best interest of the industrialized nations to undertake research and development of alternative energy sources and to expand their current methods of reducing energy usage.

The only options for large-scale power production sufficient to drive economic growth and human requirements in the near future are solar-energy conversion and nuclear energy, including both fission and fusion. Solar-derived electricity at present costs one-third less than electricity from a nuclear power plant and it lacks the hazards

CASE STUDY 1.1

Summit to Save the Earth?

Twenty years after the first Earth Summit in Stockholm, Sweden, the United Nations Conference on the Environment and Development was held in Rio de Janiero in June 1992. World leaders met to approve a Declaration of Principles, a part of which is quoted here:

> . . . [W]ith the goal of establishing a new and equitable global partnership through the creation of new levels of cooperation among states . . . we proclaim that: [Principle 4] In order to achieve a sustainable development, environmental protection shall constitute an integral part of the development process and cannot be considered in isolation from it and that, [Principle 8] To achieve a sustainable development and higher quality of life for all people, States should reduce and eliminate unsustainable patterns of production and consumption and promote appropriate demographic policies.

Overpopulation is seen as the most formidable problem facing the planet, because population reduction is against human nature. Because the tide of population growth will take decades to stabilize and because environmental pollution continues to increase, the delegates agreed in principle that (1) polluters should bear the immediate cost of cleanup, (2) poverty should be eradicated, and (3) family planning should be promoted. Delegates drafted a 600-page agenda for saving the planet with a $600 billion annual price tag, but they made no plans for raising the money. They agreed to try to hold emissions of greenhouse gases to 1990 levels (a nonbinding agreement) by the end of this century. Toward this goal, the United States committed $75 million to help developing nations control their emissions of greenhouse gases.

The disagreements that surfaced at the conference highlighted economic differences between peoples and nations. Brazil correctly noted that developed nations emit far more greenhouse gases to the atmosphere than do burning forests in developing nations. Poor countries find it difficult to understand why they should stop development for a problem that they did not create. People in developing nations want cars, refrigerators, and television sets just as much as people in the "north" do, and they are willing to exploit their resources to attain these things. China, one of the world's fastest-growing polluters, feels that Western nations should foot at least 20 percent of the bill to clean up the earth's environment. Although China is the fastest-growing polluter among developing nations, its per capita emissions of greenhouse gases are still only 11 percent of per capita U.S. emissions, since transportation in China is mostly by bicycle and bus. Rich nations value and wish to protect their life-styles; thus, the United States is afraid to commit to a definite reduction in CO_2 emissions for fear of industry cutbacks and job losses. The U.S. delegates asked for a ban on all logging in tropical rain forests, and the nations affected retaliated by demanding a ban on logging in the forests of our Northwestern states and Canada.

Although the great hopes for the 1992 Rio conference were not realized, it was clear to all that the world's peoples have common problems that must be addressed by the global community. It may take some kind of eco-disaster to put teeth into the often-voiced calls for population control and environmental protection. To be effective, any corrective actions will require individuals to make sacrifices, sacrifices that will have the greatest impact on citizens of developed nations. Life-styles and attitudes in all affluent countries will have to change.

associated with radioactive substances. On the negative side, with present technology, solar-energy conversion requires vast areas of land and an effective storage system. Alternative and more sophisticated sources of energy will be developed or expanded—sources such as biomass, geothermal, wind, ocean thermal gradient, and perhaps tidal energy—but these are expected to supply only a small percentage of the total need (see Chapter 13).

The Land

The amount of cultivatable soil limits the number of humans earth can support, because soil determines grain production and therefore food supplies. Arable soils worldwide suffered slight to severe degradation between 1945 and 1990 due to poor agricultural practices, natural erosion, and erosion accelerated by deforestation (Table 1.4). The world's farmers lost an estimated 500 million tons of topsoil in the 1970s and 80s, an amount equal to the tillable area of India and France combined. Waterlogging and salt-contamination of soils are reducing the productivity of at least a fourth of the world's cropland, and human-generated smog (ozone) and acid rain also are taking their toll on crops. It is the lack of soil, sometimes combined with drought, that is the cause of famine, not soil infertility. (Soils and degradation of the atmosphere are discussed in Chapters 6 and 12.)

Soil fertility can be improved by applying manufactured chemical and biochemical fertilizers, whereas drought or a lack of soil is difficult to remedy. China, for

TABLE 1.4 Estimated World Soil Degradation, 1945–1990

REGION	DEGRADED LAND AREA AS A PERCENT OF VEGETATED LAND		
	Total	*Light Erosion**	*Moderate to Extreme Erosion***
World	17	7	10
Asia	20	7	13
South America	14	6	8
Europe	23	6	17
Africa	22	8	14
North America (U.S., Canada)	5	1	4
Central America-Mexico	25	1	24

* Light: crop yields reduced less than 10%.
**Moderate: crop yields reduced 10–50%. Severe: crop yields reduced more than 50%. Extreme: no crop growth possible. About 9 million hectares worldwide exhibit extreme erosion, less than 0.5 percent of all degraded lands.
SOURCE: From various sources compiled in World Resources 1992–93, World Resources Institute and the United Nations.

example, has developed and encourages environmentally sound agricultural practices, such as fertilizing rice paddies with ferns that have nitrogen-fixing capability. This practice has spread throughout Southeast Asia.

As populations grow, humans are forced to occupy and to attempt to cultivate less-desirable lands, and lands that may be subject to various life-threatening geologic hazards. Scientific approaches are now available for assessing the risks of flooding, massive landslides, earthquakes, volcanic eruptions, and land subsidence. This book explains these phenomena, as they are presently understood, and the means that have been developed for alleviating the risks that they pose to human life.

Forests

Deforestation is one of the hottest environmental issues of the day. Although most deforestation is occurring in the tropics and remains largely unmeasured, the Food and Agricultural Organization (FAO) of the United Nations estimated in 1991 that 17 million hectares (42.5 million acres) per year of tropical rain forest were being deforested, up 50 percent from the early 1980s rate. Deforestation leads to loss of habitat and decreased biodiversity, contributes to climate change by adding CO_2 to the atmosphere, and often results in soil degradation. ➢ Figure 1.10 shows late 1980s estimates of annual rates of total deforestation in Asia, Latin America, and parts of Africa. Deforestation estimates do not include forests that are logged out or clear-cut and then replanted or simply left to rejuvenate.

One reason for clearing forests is to convert land to agricultural use. In Brazil, an estimated 10-year annual average of 1.7 million hectares (4.3 million acres) are converted, mostly to cattle-grazing pastures, which benefit relatively few Brazilians. At one time the Brazilian government offered tax credits to landowners who cleared their forest, but these credits were stopped in 1987. Since then, the rate has slowed from an estimated peak of 8 million hectares (20 million acres). More than half of the tropical-forest lands destroyed each year are in Brazil. It should be noted, however, that Brazil plans to set aside 1.6 million km^2 (618,000 mi^2), about 20 percent of its land area, as public reserves and parks. The Indonesian government, in a full-page advertisement in a U.S. magazine in 1989, explained that the aspirations of its 170 million people were the same as those of citizens in developed nations, and that in order to better their lives, 20 percent of Indonesian forests must be converted to plantations of teak, rubber, rice, and coffee. This conversion has been accomplished successfully. A poor African farmer, on the other hand, has little choice but to clear land to grow crops just for subsistence. How a country uses its land is a complex political and sociological issue, and there are no simple ways to reconcile the conflicting human needs that the issue involves.

Logging is another cause of deforestation, but it has been demonstrated that selective logging in tropical forests is sustainable if the logged lands are allowed to rejuvenate. In 1988 Thailand was inundated with 40 inches of rain in 5 days. Thousands of logs left on hillsides to dry floated off in mud and water that engulfed entire villages, killing 350 people. The Thai government banned logging, and now it is caught between the companies that want compensation for lost business and the landless poor that were given farms in the previously logged areas.

A third cause of deforestation is the demand for wood fuel and forest products. Many of the forests in India and the Sahel of West Africa have been decimated for use as domestic fuel and in light industry.

Area Cleared Annually, thousands of hectares

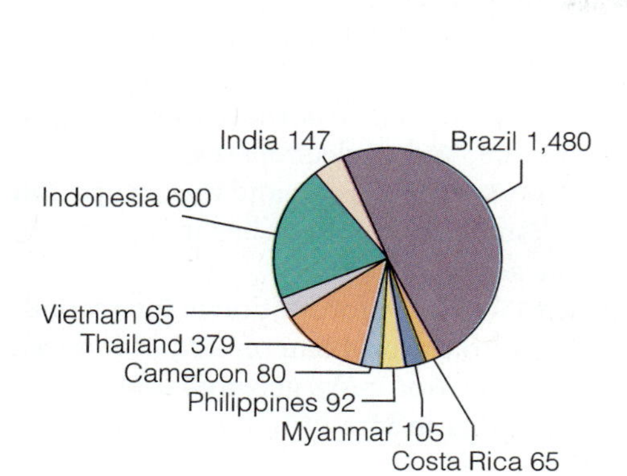

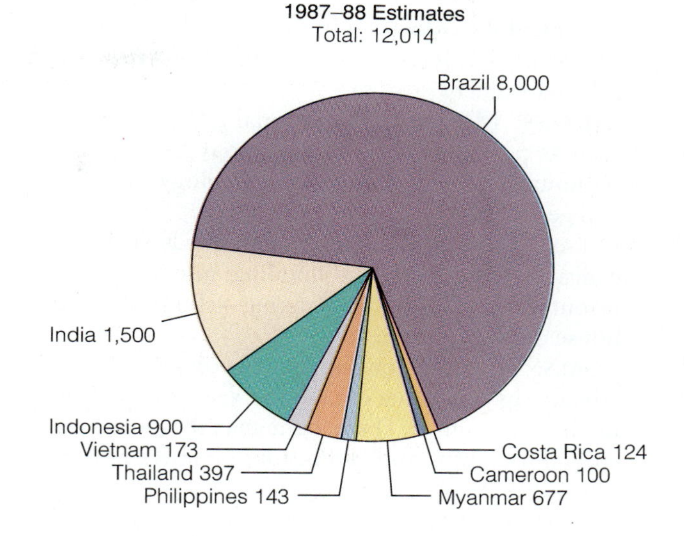

(a)

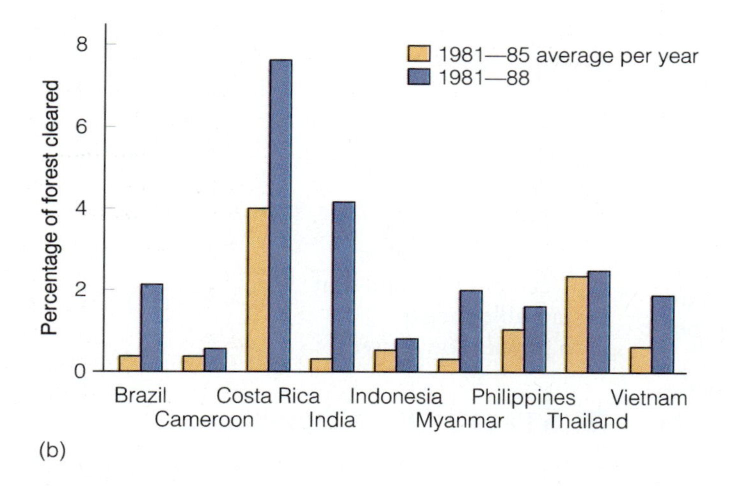

(b)

➢ **FIGURE 1.10** **(*a*) Estimated area lost to deforestation each year in nine tropical countries, 1981–85 versus 1987–88. (*b*) Annual percentage of forest cleared in the nine countries, 1981–85 versus 1987–88. Note that Costa Rica has the largest percentage lost but the second-smallest area lost.**
SOURCE: U.N. Food and Agriculture Organization.

Deforestation is second only to the burning of fossil fuels as a source of atmospheric CO_2. Forests store 450 billion metric tons of carbon. When cleared, they can no longer sequester the carbon, of course; it is released rapidly to the atmosphere when trees are burned, or slowly if they are left to decay. This is truly a global problem and it is discussed in more detail in Chapter 12.

Resources

Our throw-away society, though certainly a convenient one, must become a recycling society as natural resources dwindle and waste-disposal sites become scarce and expensive to operate. Most of what we use is discarded after one use—paper, plastic, glass, aluminum, and even steel. Trash disposal has become so expensive, at least in urban settings where most trash is generated, that recycling is close to being cost-effective. Oil reserves will be exhausted—if not in your lifetime, certainly in that of your children—if alternative energy sources are not utilized to their maximum capacity (Chapter 13). U.S. coal reserves are estimated at a 400-year reserve, but such pollution considerations as smog, acid rain, and global warming will continue to limit its use. World reserves of metallic and nonmetallic resources, many of which are in abundant supply at present, can certainly be extended by recycling. The geological and environmental consequences of energy and mining and the problems associated with waste disposal are considered in Chapter 15.

SUMMARY

Environmental Geology

DEFINED: The interaction of humans and the geological environment.

UTILIZES: Most of the traditional geologic specialties such as petrology, structural geology, sedimentology, economic geology, engineering geology, geophysics, and geochemistry.

AREAS OF INTEREST: Geologic hazards such as volcanism, earthquakes, and flooding; mineral and water resources; and land-use planning—should I build my house on this hillside?

GOALS: To predict or anticipate geologic problems in advance by collecting data in the field and analyzing it in the laboratory. These data may be portrayed on special-use maps and utilized by planners and public officials.

CASE HISTORIES: Nature is a good teacher, and geologists learn much by studying case histories of geologic events.

Population Dynamics

CARRYING CAPACITY: Defined for our purposes as the number of people that the earth can support. This number is uncertain but lies somewhere between the 1990 population of 5.2 billion and the 10–15 billion projected for the year 2050.

POPULATION GROWTH RATE: Defined as live births per 1,000 persons (that is, the birthrate) *minus* the death rate per 1,000. A birthrate of 20 *minus* a death rate of ten *equals* a growth rate of 10 per 1,000 persons, or 1%.

DOUBLING TIME: The number of years in which a population is doubled. It can be projected closely by dividing 70 by the percentage birthrate. The 1990 world birthrate of 1.76% yields a doubling time of 40 years.

Overpopulation Scenarios

UNCONTROLLED GROWTH, WORST-CASE SCENARIO: The earth becomes uninhabitable as resources dwindle, sea-level rises due to global warming and inundates coastal cities, and hazardous ultraviolet radiation strikes the earth's surface due to depletion of the ozone layer.

GAIA HYPOTHESIS: Life is an important regulator of the earth environment. If left alone, natural systems are self-regulating; for example, as carbon dioxide builds up in the atmosphere, plant growth and the oceans will remove it, thereby reducing global warming. Unfortunately, human intervention into natural systems (by deforestation, for example) has rendered a basic premise of this hypothesis untenable.

Sustainable Society

DEFINED: Society's current needs are satisfied without jeopardizing future generations. To accomplish this, population control is needed.

ENERGY IMPLICATIONS: Increasing energy demands dictate that we seek alternatives to burning the earth's reserves of limited fossil fuels. Alternatives include solar, wind, geothermal, and nuclear energy.

LAND IMPLICATIONS: Soil erosion and the occupation of marginal lands subject to geologic hazards are problems that must be addressed.

FORESTS IMPLICATIONS: Deforestation is the second-greatest contributor to the problem of human-induced carbon dioxide production now accumulating in the atmosphere and results in loss of habitat for humans and animals.

RESOURCES IMPLICATIONS: We must invest our efforts in recycling and conserving resources.

KEY TERMS

carrying capacity
doubling time
engineering geology
environmental geology
fossil fuels
J-curve
population dynamics
population growth rate
replacement reproduction
sustainable society
zero population growth

STUDY QUESTIONS

1. Physical geology is the study of earth materials, the processes that act upon them, and the resulting products. How does environmental geology differ from physical geology? In order to answer this question you might compare textbooks in the two subject areas.
2. Which natural resources are most important in determining the carrying capacity of the earth?
3. What are the major problems facing the earth's human population? Distinguish between the problems that are geological in nature and those that are strictly population-driven. Which problems appear to be most readily solvable; most difficult to solve?
4. The United States has a current growth rate of 0.7 percent. At this rate how much time is required for a population to double? Discuss the impact of a doubling of your community's population in that length of time.
5. Distinguish between engineering geology and environmental geology.

6. Of what value are case histories in solving environmental geological problems?
7. What is the impact of overpopulation on energy use and global warming? How does this bode for the well-being of future generations? In what parts of the world will the largest percentage increases in energy consumption probably occur in the next decade, and why?
8. What are some of the adverse consequences of clearing tropical rain forests?

Further Reading

Brown, L., and others. 1990. *The state of the world, 1990.* New York: W. W. Norton and Co.

Blue Planet Group. 1991. *Blue planet: now or never,* case study 9734. Ottawa, Ontario, KIG 5J4, Canada: Blue Planet Group.

Ehrlich, Paul R., and Anne Ehrlich. 1970. *Population, resources, and environment: Issues in human ecology.* New York: W. H. Freeman and Co.

———. 1987. *Earth.* New York: Franklin Watts.

Gore, Al. 1992. *Earth in balance: Ecology and the human spirit.* Boston: Houghton Mifflin Co.

Joseph, Lawrence E. 1990. *Gaia: The growth of an idea.* New York: St. Martin's Press.

Lovelock, James. 1988. *The ages of Gaia: A biography of the living earth.* New York: W. W. Norton and Co.

Silver, Cheryl S., and Ruth S. DeFries. 1990. *One earth one future: Our changing global environment.* Washington, D.C.: National Academy of Sciences, National Academy Press.

Sowers, G. F. 1981. *There were giants on the earth in those days,* 15th Terzhagi lecture. Geotechnical Division of the American Society of Civil Engineers 107, no. 6T4: 383–419.

United Nations. 1980. *World population trends and policies, 1979 monitoring report, Volume 1: Population trends.* New York: United Nations.

World Resources Institute and the United Nations Development and Environment Programs. 1992–93. *Toward sustainable development: A guide to the global environment.* New York: Oxford University Press.

CHAPTER

2

GETTING AROUND IN GEOLOGY

They cannot scare me with their empty spaces
Between stars—on stars where no human
race is.
I have it in me so much nearer home
To scare myself with my own desert places.

ROBERT FROST, POET (1874–1963)

Geologic materials, such as rocks, valuable mineral deposits, and soils are an integral part of the human geologic environment. Rocks form the geologic foundations and provide building material for many of our structures. In addition, rocks are involved in many surface earth processes that are hazardous to humans, such as landslides and rockfalls. In this chapter we will discuss briefly how the planet earth formed and how its rocky crust and atmosphere arrived at their present condition. Environmental geologists are often asked about future geologic events, and in order to better understand the future a geologist must learn as much as possible about the past. An important concept in geology, known as the Law of Uniformity, is that all geologic processes active today have been active in the past, only at differing rates. The concept can be shortened and paraphrased to "the present is the key to the past." For example, by studying the geologically recent past, we have learned that North America was subject to multiple glacial advances over the past 2.3 million years, a period of time known as the Ice Ages. As a result, scientists are now

View down Lone Pine Canyon toward Valhalla, Sequoia National Park, California. Granites and country rocks of the Sierra Nevada have been sculpted into broad valleys and pinnacles by glaciers that existed in the region less than 20,000 years ago. Bare rock and thin soils testify to the erosive power of ice.

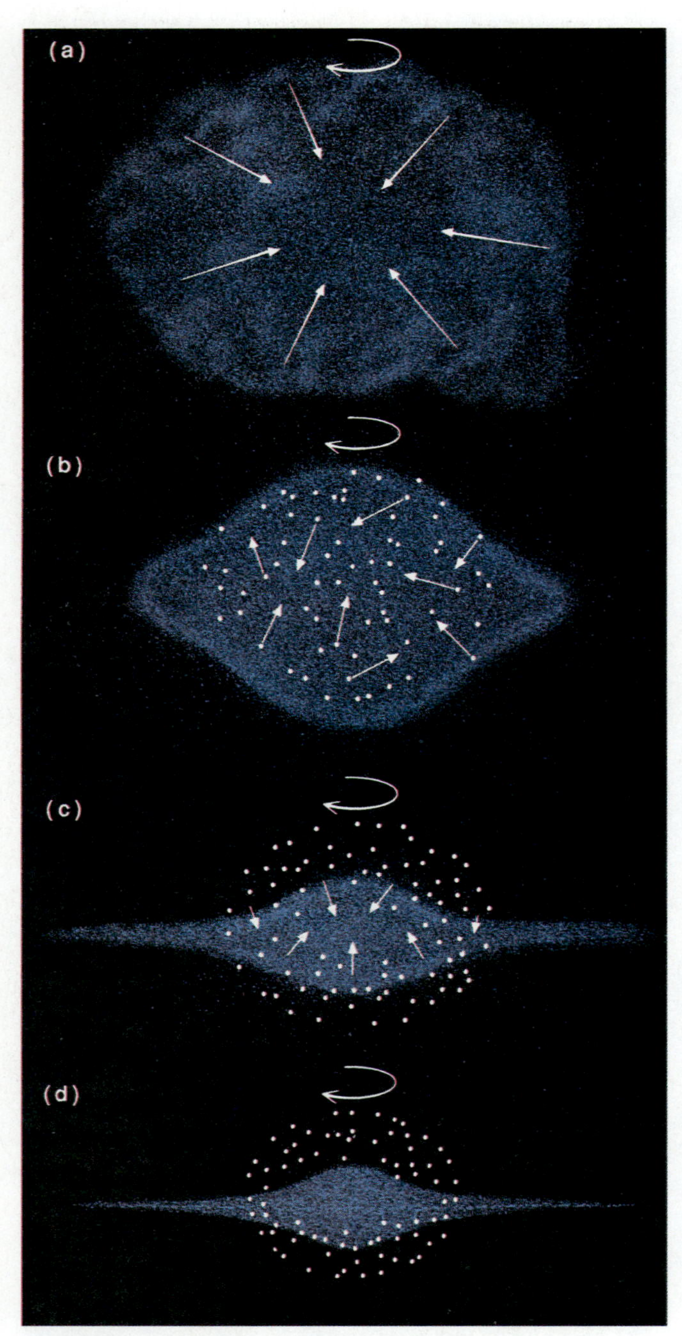

➢ FIGURE 2.1 The formation of the galaxy. (*a*) The pregalactic cloud, rotating slowly, begins to collapse. (*b*) Stars formed before and during collapse have a spherical distribution and noncircular, randomly oriented orbits. (*c*) The collapse leads to a disk with a large central bulge, surrounded by a spherical halo of old stars and clusters. (*d*) The disk flattens further and eventually forms spiral arms. Stars in the disk have relatively high concentrations of heavy elements because they formed from material that had been through stellar nuclear processing.

monitoring and computer modeling weather and climate to anticipate long-term trends and perhaps even the next ice age.

THE BEGINNING

The universe is mostly space, a void, nothingness. Even so, it contains immense quantities of the simplest elements known, hydrogen and helium, which fuel the stars and our sun. This is explained by the widely accepted model that the universe began with a colossal **big bang** when an incredibly dense "cosmic baseball" exploded between 10 and 15 billion years ago, sending matter and energy outward in all directions. Astronomers have inferred the universe's physical conditions from a microsecond after the "bang" until its death, which they calculate should occur about 10^{100} years after its birth. The primordial universe was composed of an expanding cloud of gas containing about 75 percent hydrogen, 23 percent helium, and 2 percent other elements. After about a billion years had passed, the clouds segregated into smaller clouds that eventually formed galaxies and groups of galaxies. Within these protogalactic gas clouds were many far smaller clouds, each held together by gravity. Eventually, as the small clouds or nebulae became more condensed, gravitational attraction caused them to collapse inward, generating tremendous heat in the core of each cloud (about 10 million degrees centigrade) and initiating thermonuclear reactions whereby hydrogen atoms fused to form helium atoms. In this manner the individual stars composing the galaxies were formed (➢ Figure 2.1). It is the tremendous energy released by the fusing of hydrogen atoms to form helium that causes stars to shine and our sun to give off heat and light. There are as many as 50 billion galaxies of tens or even hundreds of billions of stars each, which gives the galaxies a milky appearance in the sky (Greek *galaktos*, "milk"). It has been speculated that there are more stars in the universe than grains of sand on all the beaches on earth! Many of the galaxies are spiral in shape and so large that it takes 100,000 years for light to travel their width (at the speed of light, 186,281 miles per second). Our solar system is in the galaxy we call *the Milky Way*, and our sun is an average-sized star in our galaxy.

Few stars that formed relatively soon after the big bang exist today because they had very high core temperatures and they consumed hydrogen at a prodigious rate. Elements as heavy as iron formed in these large stars, which eventually collapsed into their depleted cores and exploded as **supernovas.** During the explosions, elements heavier than iron formed and were spread about the universe. These were the raw materials for a later generation of stars and for our planetary system.

Our Solar System

Nebulae, or interstellar gas clouds, are usually stable; they do not collapse in on themselves unless they are pushed by an outside event. One such event, a supernova explosion sometimes called the *bing bang,* as opposed to the earlier big bang, may have occurred about five billion years ago. This explosion caused the cloud that was to become our solar system to spin and to start collapsing. As the inner part of the nebula contracted, it was heated by the fusion of hydrogen atoms, and the outer part of the gas cloud was forced into a disc shape by rotation—a condition suggested some 200 years ago by Immanuel Kant and Pierre de Laplace (➢ Figure 2.2). At this stage the hot protosun was surrounded by a cooler, disc-shaped, rotating cloud—the solar nebula. Volatile (low-boiling-point) compounds such as methane and ammonia condensed in the cold outer reaches of the nebula to form the large outer planets, whereas metallic and rocky substances with high melting points condensed closer to the sun and formed the dense inner planets. In time the disc was composed of asteroid-sized lumps of solid matter known as **planetesimals,** which collided and grew into planets by a process called *accretion.* The inner *terrestrial planets* are relatively small, dense, and composed mainly of rock, whereas the outer *Jovian planets* are large and gaseous (➢ Figure 2.3). Pluto, the planet farthest from the sun, is an exception in that it is both small and dense and has an inclined orbit. The planets rotate in the same direction as the overall rotation of the nebular disc with two exceptions. It is postulated that the slow retrograde (opposite) rotation of Venus and the exaggerated tilt of Uranus are accounted for by glancing collisions with large planetesimals. Such an encounter is also thought to have formed the earth's moon. Heat generated by colliding planetesimals and radioactive decay caused the earth to melt and differentiate into layers according to their density. Nickel and elemental iron, being heaviest, sank to form the **core** of the earth, and a lighter, crystalline mush of rock-forming silicates, the **mantle,** formed around the core (➢ Figure 2.4). The earth as a planetary body formed about 4.6 billion years ago, at which time it was too hot to hold an atmosphere. As it cooled, however, gases leaked from the interior—mainly hydrogen, ammonia, methane, and water vapor—which eventually evolved into our oxygen- and nitrogen-rich atmosphere. Thus, the earth's early atmosphere is the result of "outgassing" from the interior. Finally, lighter minerals formed the outer shell of the earth, the **crust**—which includes rocks of both the continents and the ocean basins—about 4.0–4.3 billion years ago. This is the age of the oldest known minerals in continental rocks. (Table 2.5, later in the chapter, summarizes age-dating methods.)

➢ FIGURE 2.2 Formation of the solar system. An interstellar cloud, initially very extended and rotating very slowly, collapses under its own gravitation. This happens most quickly at the center. As the collapse occurs, the internal temperature rises and the rotation rate increases. Eventually the central condensation becomes hot and dense enough to be a star, with nuclear reactions at its core. The hot protosun is surrounded by a cooler, disk-shaped rotating cloud (the solar nebula) from which the planets eventually are formed.

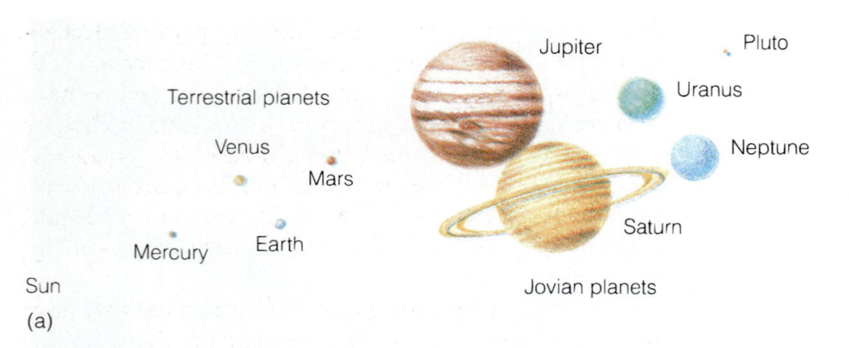

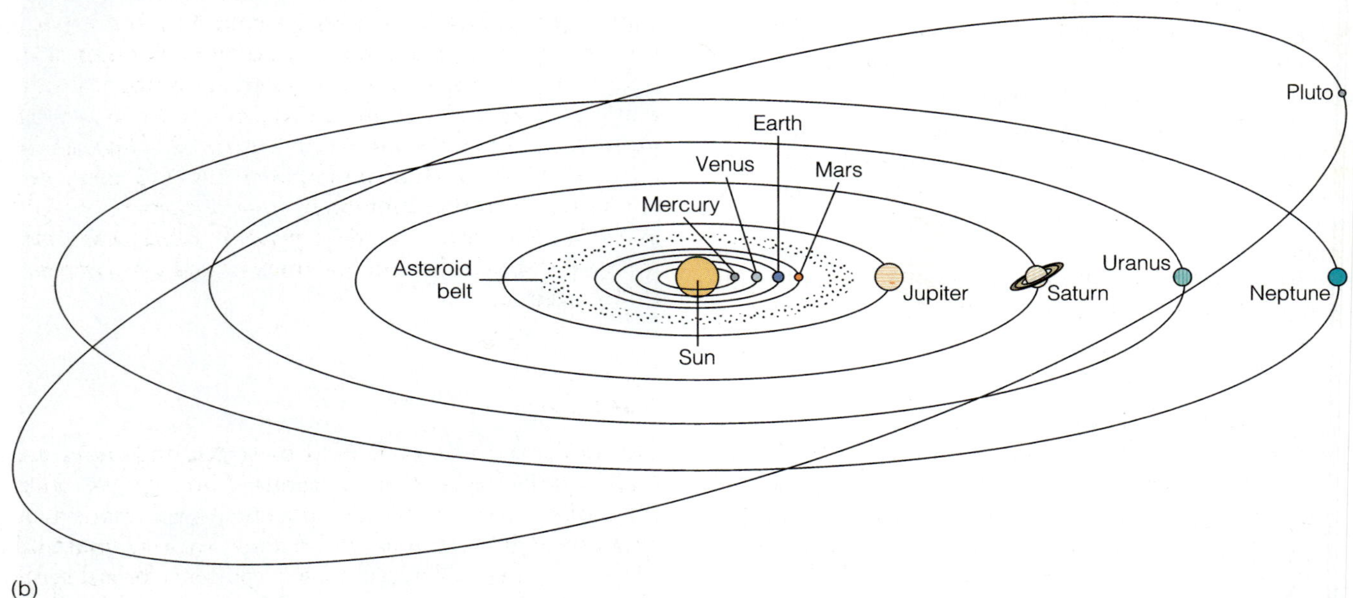

➢ **FIGURE 2.3** (*a*) **Relative sizes of the planets and** (*b*) **their orbital positions in our solar system.**

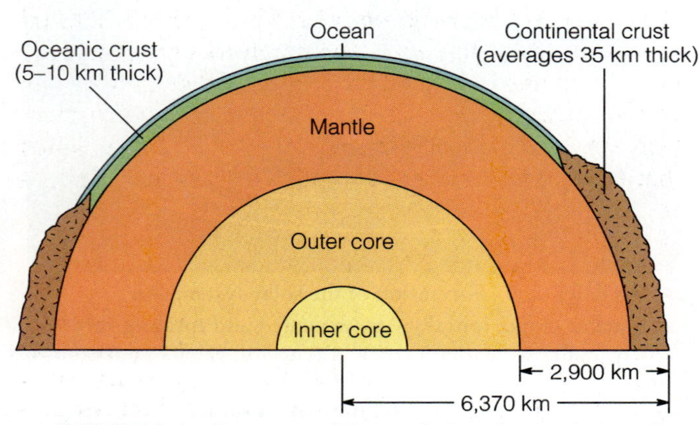

➢ **FIGURE 2.4 Structure of the earth showing oceanic and continental crust, mantle, and outer and inner cores.**

EARTH MATERIALS

Elements, Atoms, and Atomic Structure

Elements are substances that cannot be changed into other substances by normal chemical methods. They are composed of infinitesimal particles called **atoms.** In 1870, Lothar Meyer published a table of the 57 then-known elements arranged in the order of their atomic weights. Meyer left blank spaces in the table wherever elements of particular weights were not known. By about the same time Dmitry Mendeleyev had developed a similar table, but his table was based upon the similar chemical properties of the particular known elements. Where an element's placement based on weight did not group it with elements of similar properties, Mendeleyev did not hesitate to suggest that its weight had been measured incorrectly. He also predicted the properties of the "missing" elements, and very shortly his predictions proved correct with the discovery of the elements gallium, scandium, and germanium.

The weight of an atom of an element is contained almost entirely in its **nucleus,** which contains protons

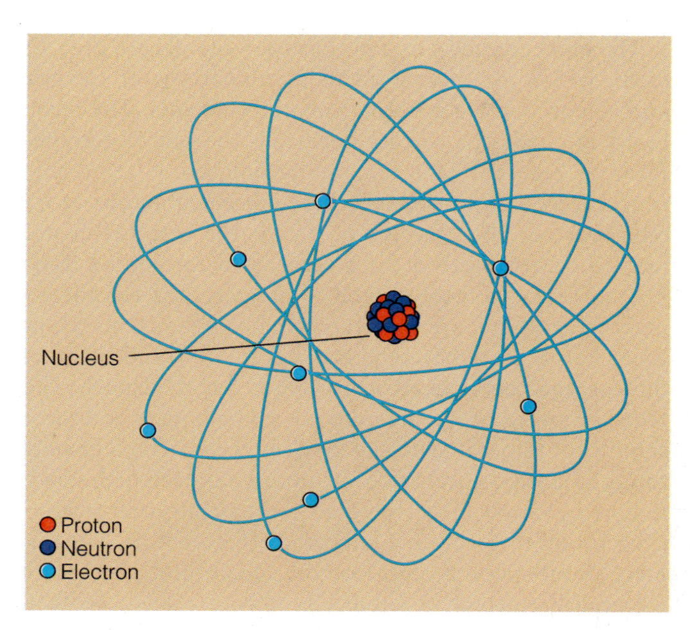

➢ **FIGURE 2.5 The structure of an atom showing the nucleus and its surrounding electron cloud.**

(atomic weight = 1, electrical charge = 1^+) and neutrons (atomic weight = 1, electrical charge = 0). An element's **atomic number** is the number of protons in its nucleus, and this number is unique to that element. The weight of an atom—its **atomic mass**—is the sum of its nuclear protons and neutrons. For example, $_2He^4$ denotes the element helium, which has 2 protons (subscript, left) and two neutrons, for an atomic mass of 4 (superscript, right). Electrical neutrality of the atom is provided by balancing the positive proton charges by an equal number of negatively charged electrons orbiting in shells around the nucleus (➢ Figure 2.5). **Ions** are atoms that are positively or negatively charged owing to a loss or gain of electrons (e^-) in the outer electron shell. Thus sodium (Na) may lose an electron in its outer shell to become sodium ion (Na^+, a *cation)*, and chlorine (Cl) may gain an electron to become chloride ion (Cl^-, an *anion)*. *Valence* refers to ionic charge, and sodium and chlorine are said to have valences of 1^+ and 1^-, respectively. Elemental sodium is an unstable metallic solid, and chlorine is a poisonous gas. When ionized, they may combine in an orderly fashion to form the mineral halite (NaCl), common table salt, a substance necessary for human existence (➢ Figure 2.6).

Isotopes are forms (or species) of an element that have different atomic masses. The element uranium, for instance, always has 92 protons in its nucleus, but varying numbers of neutrons define its isotopes. For example, U^{238} is the common naturally occurring isotope of uranium. It weighs 238 atomic mass units and contains 92 protons and 146 (238 minus 92) neutrons. U^{235} is the rare isotope (0.7% of all uranium) that has 143 neutrons. Similarly, C^{12} is the common isotope of carbon, but C^{13} and C^{14} also exist. A modern periodic table of the elements that shows their atomic number and mass appears in Appendix 2.

Minerals

The earth's crust is composed of rocks, and rocks are made up of one or more minerals. Although we often think of those minerals that have nutritional importance or those with great beauty and value such as gold and silver, the mineral kingdom encompasses a broad spectrum of chemical compositions. **Minerals** are defined as *naturally occurring, inorganic, crystalline* substances, each with a narrow range of *chemical compositions* and characteristic *physical properties*. Thus, neither artificial gems such as zirconia and some sapphires nor organic deposits such as coal and oil are minerals by definition. The crystal form of a mineral reflects an orderly atomic structure, which in turn determines the crystal shape. Solids that have random or noncrystalline atomic structures—

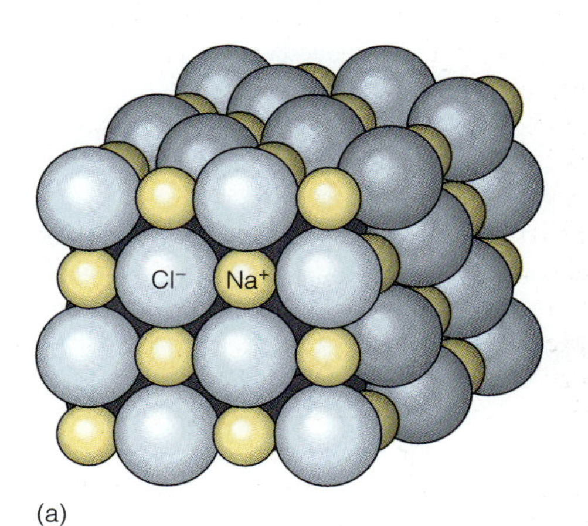

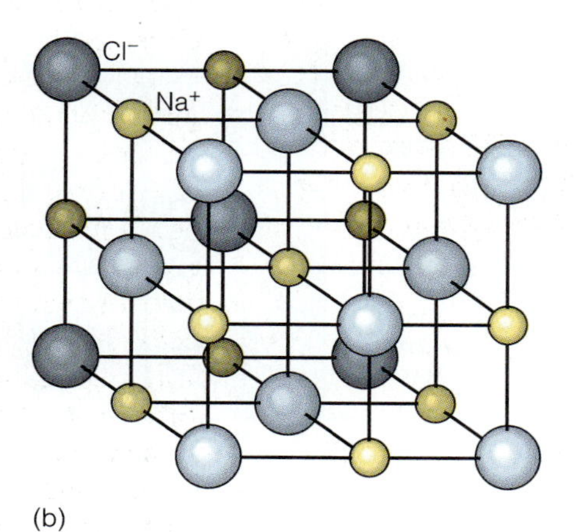

➢ **FIGURE 2.6 Two models of the atomic structure of the mineral halite (NaCl). (*a*) A packing model that shows the location and relative sizes of the sodium and chloride ions. (*b*) A ball-and-stick model that shows the cubic crystal structure of the mineral.**

glass and opal, for example—are described as **amorphous.** Minerals composed of a single element, such as copper or carbon (as in graphite and diamonds), are known as *native elements.* Most minerals, however, are composed of combinations of two or more elements. More than 3,000 minerals have been named and described, but only about 20 make up the bulk of the earth's crustal rocks. The chemistry of minerals is the basis of their classification, and the most common minerals are primarily composed of the elements oxygen (O), silicon (Si), aluminum (Al), and iron (Fe). This should not be surprising, because these are the four most abundant elements on earth.

Geologists classify minerals into groups that share similar negatively charged ions (anions) or ion groups (radicals). Oxygen (O^{-2}) may combine with iron to form hematite (Fe_2O_3), a member of the *oxide group* of minerals. The *sulfide minerals* typically are combinations of a metal with sulfur. Examples include galena (lead sulfide, PbS), chalcopyrite (copper–iron sulfide, $CuFeS_2$), and pyrite (iron sulfide, FeS_2), the "fool's gold" of inexperienced prospectors. *Carbonate minerals* contain the negatively charged $(CO_3)^{-2}$ ion. Calcite ($CaCO_3$), the principal mineral of limestone and marble, is an example.

The most common and important minerals are the *silicates,* which are composed of combinations of oxygen and silicon with or without metallic elements. The basic building block of a silicate mineral is the silicate tetrahedron (➢ Figure 2.7). It consists of one silicon atom surrounded by 4 oxygen atoms at the corners of a four-faced tetrahedron. The manner in which the tetrahedra are packed or arranged in the mineral structure is the basis for classifying silicates. Some of the ways the tetrahedra may be arranged are in layers or sheets (as in mica), in long chains (as in pyroxene and amphibole), and in three-dimensional networks (as in quartz and feldspar). The *feldspar group* of silicate minerals is the largest and most significant mineral group in igneous rocks. The feldspars range in composition from the potassium-rich orthoclase that is common in granites, to the calcium-rich plagioclase found in basalt and gabbro (➢ Figure 2.8). The relative abundances of rock-forming silicate and nonsilicate minerals are shown in Table 2.1.

Serious students of mineralogy are able to identify several hundred minerals without destructive testing. They are able to do this because minerals have distinctive physical properties, most of which are easily determined and associated with particular mineral species. With practice,

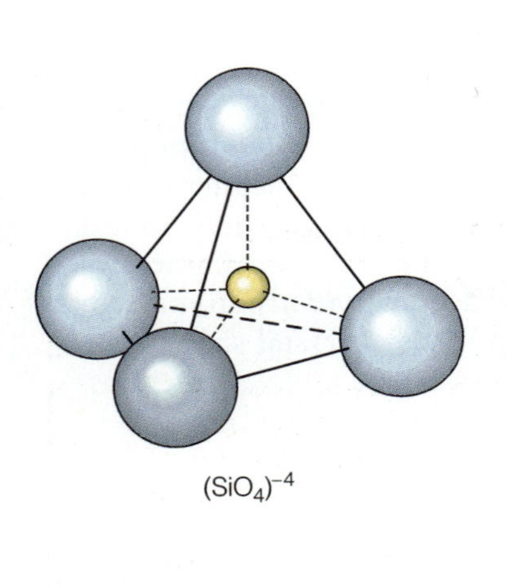

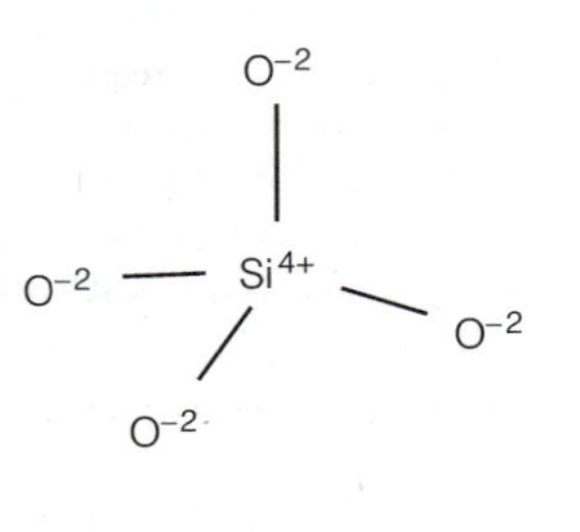

(a)

		Formula of negatively charged ion group	Example
Isolated tetrahedra		$(SiO_4)^{-4}$	Olivine
Continuous chains of tetrahedra		$(SiO_3)^{-2}$	Pyroxene group (augite)
		$(Si_4O_{11})^{-6}$	Amphibole group (hornblende)
Continuous sheets		$(Si_4O_{10})^{-4}$	Micas (muscovite)
Three-dimensional networks	Too complex to be shown by a simple two-dimensional drawing	$(SiO_2)^{o}$ $(Si_3AlO_8)^{-1}$ $(Si_2Al_2O_8)^{-2}$	Quartz Feldspars Feldspars

(b)

➢ FIGURE 2.7 ***(a)*** **The silicate tetrahedron, showing the unsatisfied negative charge at each oxygen that allows it to form *(b)* chains, sheets, and networks.**

(a) (b) (c) (d)

➢ **FIGURE 2.8** (*a*) **Coarse-grained granite, in which feldspar, quartz, and minor amounts of mica and hornblende are visible.** (*b*) **Photomicrograph of a thinly sliced section of granite reveals large interlocking grains.** (*c*) **Minerals in this hand sample of fine-grained basalt are not identifiable.** (*d*) **Photomicrograph of basalt that reveals an interlocking fine-grained texture.**

one can build up a mental catalog of minerals, just like building a vocabulary in a foreign language. The most useful physical properties are hardness, cleavage, crystal form, and to a lesser degree, color (which is variable) and luster. Mineralogists use the **hardness scale** developed by Friedrich Mohs. Mohs assigned a hardness value (H) of 10 to diamond and of 1 to talc, a very soft mineral (Table 2.2). Diamond can scratch almost everything; quartz ($H = 7$) and feldspar ($H = 6$) can scratch glass ($H = 5\frac{1}{2}$–6) and calcite ($H = 3$), and your fingernail can scratch gypsum ($H = 2$). **Cleavage** refers to the characteristic way particular minerals split along definite planes as determined by their crystal structure. Mica has perfect cleavage in 1 direction; this is called *basal cleavage,* because it is parallel to the base of the crystal structure. Feldspars split in 2 directions; halite, which has *cubic* structure, in 3 directions; and so on, as shown in ➢ Figure 2.9. Minerals that have perfect cleavage can be split readily by a tap with a rock hammer or even peeled apart, as in the case of mica. Some minerals do not cleave but have distinctive **fracture** patterns that help one to identify them. The common crystal forms shown in ➢ Figure 2.10

TABLE 2.1 Common Rock-Forming Minerals

MINERAL	ABUNDANCE IN CRUST, %	ROCK IN WHICH FOUND
Plagioclase*	39	Igneous rocks mostly
Quartz	12	Detrital sedimentary rocks, granites
Orthoclase*	12	Granites, detrital sedimentary rocks
Pyroxenes	11	Dark-colored igneous rocks
Micas	5	All rock types as accessory minerals
Amphiboles	5	Granites and other igneous rocks
Clay minerals	5	Shales, slates, decomposed granites
Olivine	3	Iron-rich igneous rocks, basalt
Others	11	Rock salt, gypsum, limestone, etc.

*Feldspar group of minerals

TABLE 2.2 Mohs Hardness Scale

HARDNESS	MINERAL	COMMON EXAMPLE
1	Talc	Pencil lead 1–2
2	Gypsum	
		Fingernail 2½
3	Calcite	Copper penny 3.0
		Brass
4	Fluorite	Iron
5	Apatite	Tooth enamel
		Knife blade
		Glass 5½–6
6	Orthoclase	
		Steel file 6½
7	Quartz	scratches glass
8	Topaz	
9	Corundum	Sapphire, ruby
10	Diamond	Synthetic diamond

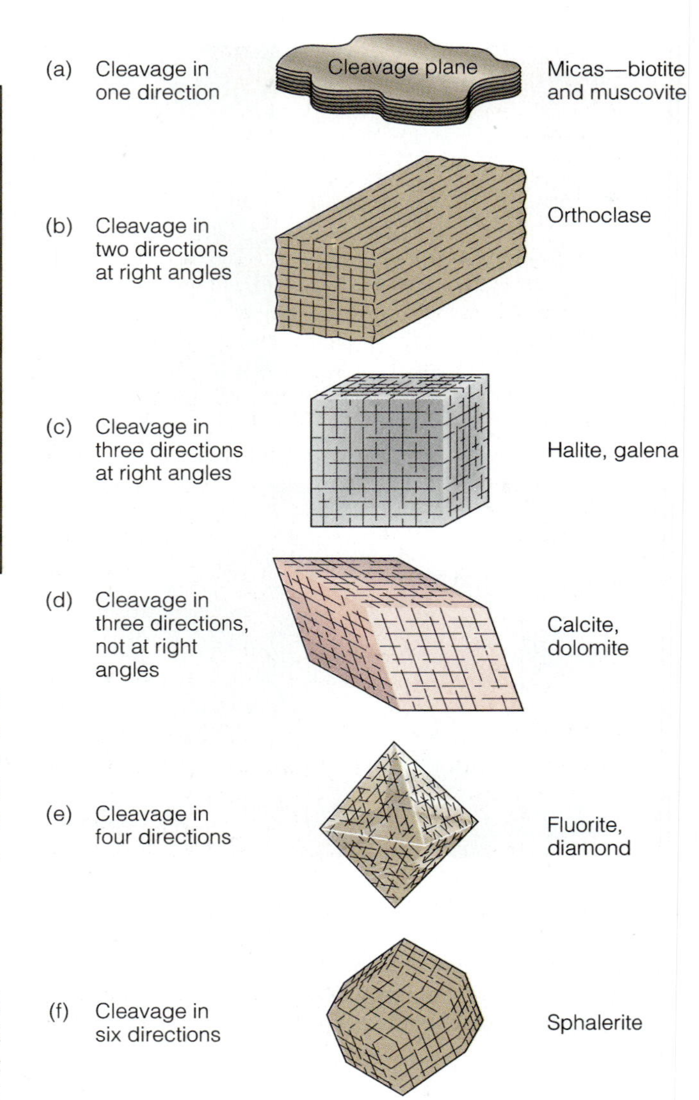

➢ **FIGURE 2.9 Types of cleavage and typical minerals in which they occur.**

can be useful in identifying a particular mineral. Color tends to vary within a given mineral species, so it is not a reliable identifying property. **Luster,** how a mineral reflects light, is useful, however. We recognize metallic and nonmetallic lusters, the latter being divided into types such as glassy, oily, greasy, and earthy. (Case Study 2.1 discusses some gem minerals.)

Rocks

Rocks are defined as consolidated or poorly consolidated aggregates of one or more minerals, glass, or solidified organic matter (such as coal) that cover a significant part of the earth's crust. There are three classes of rocks, based upon their origin: igneous, sedimentary, and metamorphic. **Igneous** (Latin: *ignis*, "fire") rock is crystallized from molten or partly molten material. **Sedimentary** rocks include both *lithified* (that is, turned to stone) fragments of preexisting rock and rocks that were formed from chemical or biological action. **Metamorphic** rocks

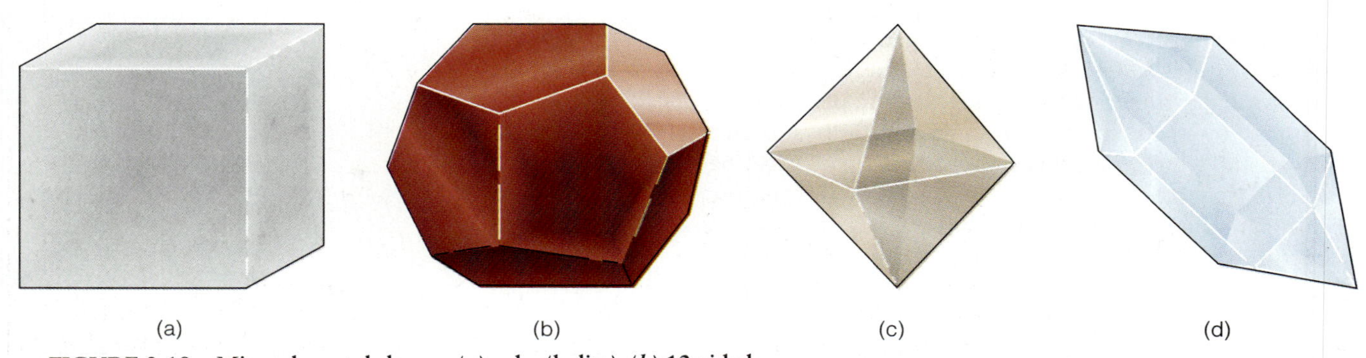

➢ **FIGURE 2.10 Mineral crystal shapes: (*a*) cube (halite), (*b*) 12-sided dodecahedron (garnet), (*c*) 8-sided octahedron (diamond, fluorite), and (*d*) doubly terminated hexagonal prism (quartz).**

CASE STUDY 2.1

Gems—Nature's Artwork

Rarity and beauty are what make precious and semiprecious gems so desirable. Those shown in ➢ Figure 1 are but a sample of several hundred valued gem minerals and their varieties. The largest uncut diamond in the world, the Cullinan diamond (Figure 1a), was recovered in 1905 from the Premier Mine in Kimberley, South Africa; this huge, colorless stone weighed 3,106 carats (1.3 pounds). The smooth face shown in the photograph is a natural cleavage surface, which indicates that this is not the entire stone. Unfortunately, the remainder of it lies undiscovered in a deep volcanic pipe—which is where all diamonds originate. The Transvaal government bought the diamond for 1.6 million dollars and gave it uncut to King Edward VII of England. After months of study, an Amsterdam diamond-cutter's decisive blow was struck, creating 9 large stones and about 100 smaller ones, all of them flawless. All are now in the British crown jewels collection.

Diamonds are the high-pressure form of carbon and are the hardest natural substance in the world. Curiously, graphite, the high-temperature, low-pressure variety of carbon, is among the softest minerals known. Diamonds are also found in Australia, India, South America, and near Murfreesboro, Arkansas, where the initial discovery was made by a farmer in 1906. Thousands of diamonds have been recovered near Murfreesboro since then, one weighing 40 carats, but the deposit has not been found to be commercially viable. For the price of admission, one may now dig for diamonds at Murfreesboro in the "blue ground" of the eroded diamond-bearing volcanic pipe. One lucky hunter uncovered a 15-carat stone. Little known to the general public are the diamond finds in California gold gravels and in the soils and glacial deposits of the Midwest and the East (see ➢ Figure 2). The sources of these diamonds are unknown, but the ones found in ice-age deposits originated somewhere near the boundary between the United States and Canada.

Precious opal (Sanskrit *upala,* "precious stone"; Figure 1b) occurs in shallow surface diggings, and because it lacks crystalline form, it actually is not a mineral. It is amorphous hydrated silica and gets its "fire" from an orderly arrangement of silica spheres interspersed with water molecules. Black and fire opals are the most desirable colors, and each stone is priced individually.

Emerald, the most valuable mineral carat-for-carat, is the precious variety of the silicate mineral, beryl (Figure 1c). It is found in coarse-grained granites called **pegmatites** (see Chapter 14) and in river gravels, mostly in Colombia.

Quartz, plain old SiO_2, although one of the most common minerals, has a host of semiprecious varieties such as purple amethyst (Figure 1d), banded agate, yellow citrine, and smoky and rose quartz. Searching for polishing-quality quartz varieties is the passion of many "rock hounds," and good crystal specimens also are cherished.

(a) (b) (c) (d)

➢ FIGURE 1 (*a*) An exact replica of the Cullinan diamond. The two cut stones are the Star of Africa I and II, which now reside in the British crown jewels collection. (*b*) Precious opal, (*c*) emerald, and (*d*) quartz.

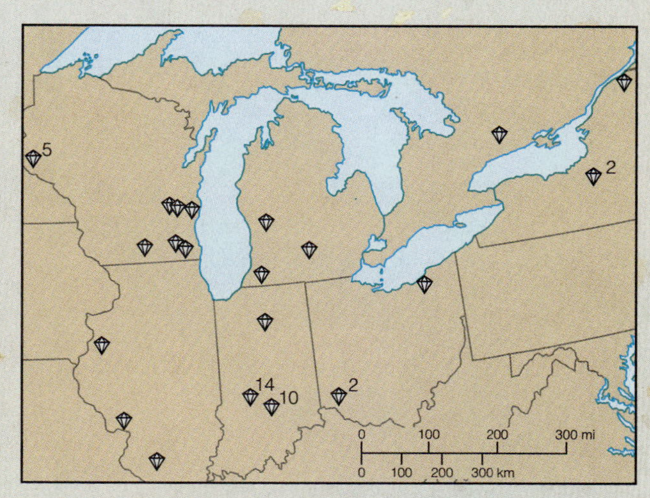

➢ FIGURE 2 Reported diamond finds in glacial drift in the Great Lakes region. The number of diamonds found at a given location is also indicated; unnumbered locations represent single finds.

are those that have been changed, essentially in the solid state, by heat, fluids, and/or pressure within the earth.

The **rock cycle** is one of many natural cycles on earth. The illustration of it in ➢ Figure 2.11 shows the interactions of energy, earth materials, and geologic processes that form and destroy rocks and minerals. The rock cycle is essentially a *closed system;* "what goes around comes around," so to speak. In a simple cycle there might be a sequence of formation, destruction, and alteration of rocks by earth processes. For example, an igneous rock may be eroded to form sediment that subsequently becomes a sedimentary rock, which may then be metamorphosed by heat and pressure to become an igneous rock once more.

Throughout this book we emphasize those rocks which, because of their composition or structure, are involved in geologic events that endanger human life or well-being or that are important as resources. Before going on to take a closer look at the three main types of rocks, you might like to take the side trip provided by Case Study 2.2.

IGNEOUS ROCKS. Igneous rocks are classified according to their texture and their mineral composition. A rock's texture is a function of the size and shape of its mineral grains, and for igneous rocks is determined by how fast or slowly a melted mass cools (see Figure 2.8). If **magma**—molten rock within the earth—cools slowly, large crystals develop and a rock with **phaneritic texture,** such as granite is formed (see Figure 2.8). The resulting large rock mass formed within the earth is known as a **batholith.** Greater than 100 km^2 (39 mi^2) in area by definition, batholiths are mostly granitic in composition and show evidence of having invaded and pushed aside the *country rock* into which they intruded. Also, many batholiths have formed in the cores of mountain ranges during mountain-building episodes. Batholiths are a type of igneous mass referred to as *plutons,* and rocks of batholiths

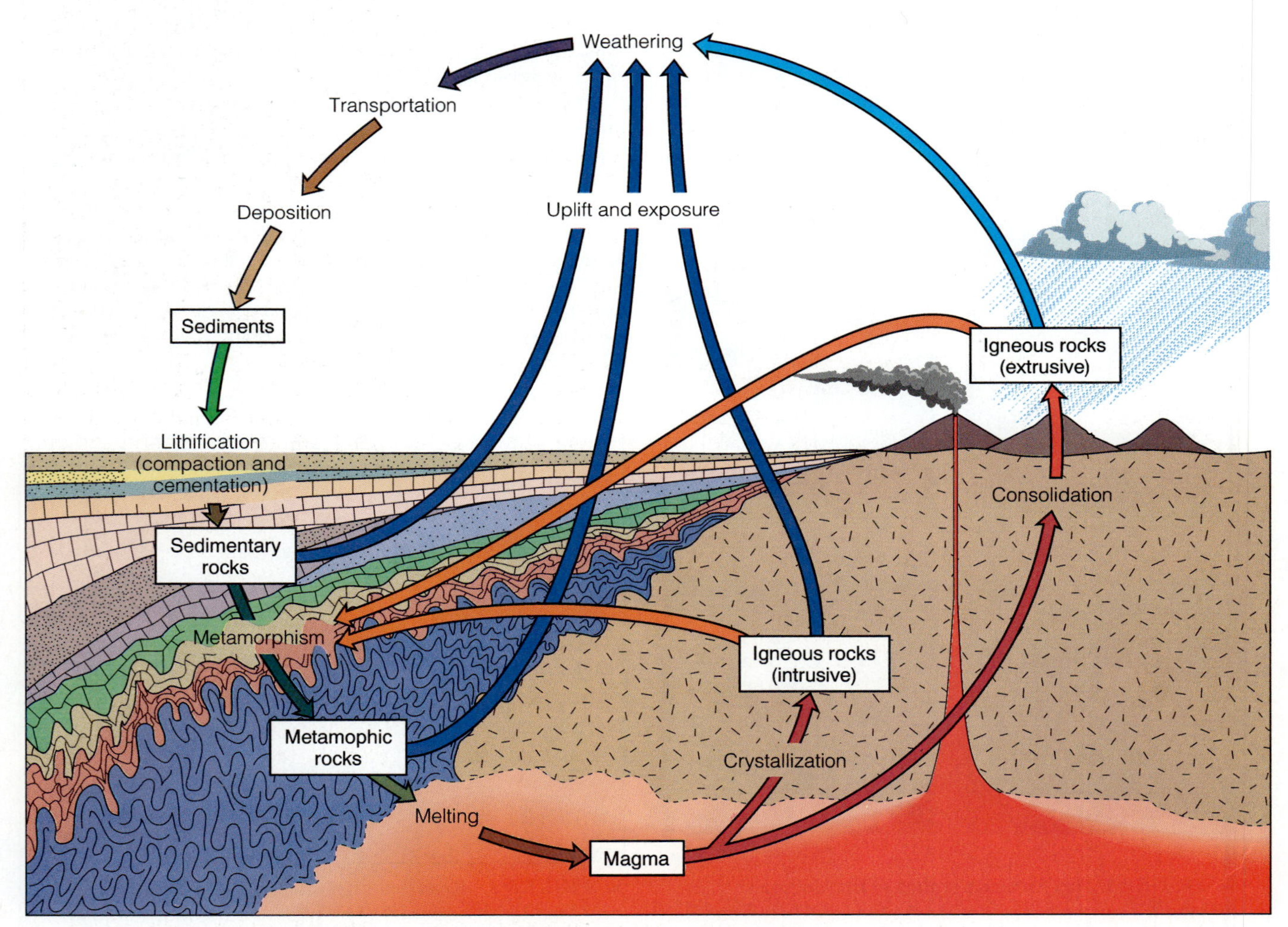

➢ **FIGURE 2.11 The rock cycle shows the relationship between earth processes and earth materials. The three rock types are interrelated by internal and external processes involving the atmosphere, ocean, biosphere, crust, and upper mantle.**

CASE STUDY 2.2

An Urban Field Trip

Many of the buildings on the portion of Chicago's North Michigan Avenue that is called the "Magnificent Mile" are decorated with natural stone. Beautiful examples of the common rock types that you will discuss in class or in the laboratory can be seen there. The Indiana limestone that is sometimes referred to as *Bedford* limestone is widely used in Chicago. The Tribune Building on Michigan Avenue is almost entirely faced with this rock. It is perhaps better known as the stone used for Washington's National Cathedral and many other monuments and buildings in the nation's capital. Rocks from historic sites in the United States are embedded in the Tribune Building's limestone. ➢ Figure 1 shows a piece of sandstone from the Little Bighorn River area in Montana, the site of George Armstrong Custer's infamous "last stand," on the building's exterior.

The orbicular granite on the American Express Building (➢ Figure 2) consists of flesh-colored, rounded crystals of orthoclase (potassium) feldspar mantled by feldspar with more sodium. This texture was originally described from Finland and is known as **Rapakivi** granite. Bloomingdale's is faced with "flamed" granite from Texas (➢ Figure 3). Flaming imparts a rough texture (relief) to the stone surface. The Marshall Field's Building displays beautiful Georgia White Cherokee Marble that is patterned with a darker mineral (➢ Figure 4). Some of the marble slabs are obviously bowed, probably due to the inexorable pull of gravity.

➢ **FIGURE 1 Sandstone from Montana embedded in Indiana limestone; Chicago Tribune Building.**

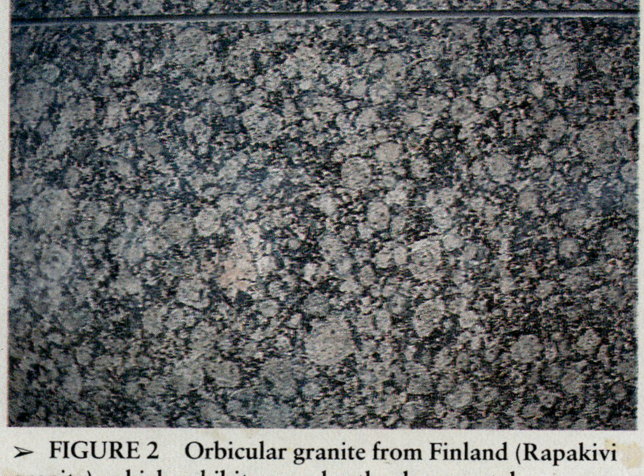

➢ **FIGURE 2 Orbicular granite from Finland (Rapakivi granite), which exhibits round orthoclase crystals; American Express Building.**

➢ **FIGURE 3 Coarse-grained "flamed" granite; Bloomingdale's department store.**

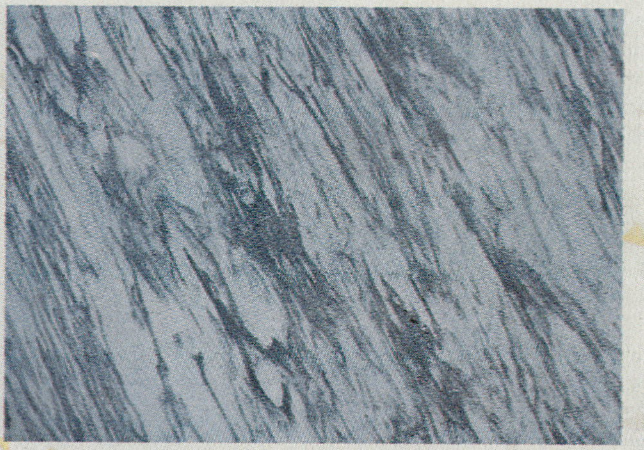

➢ **FIGURE 4 Georgia white "Cherokee" marble, which exhibits uniform crystallinity and streaks of dark-colored minerals; Marshall Field's Building.**

➢ **FIGURE 2.12 The Sierra Nevada batholith and Mount Whitney, looking west from Owens Valley; east-central California.**

are described as **plutonic** (after Pluto, the Greek God of the Underworld), because they were formed at great depth. The Sierra Nevada of California is an example of an uplifted and eroded mountain range with an exposed core composed of many plutons (➢ Figure 2.12).

Lava is molten material at the earth's surface produced by volcanic activity. Lava cools rapidly, resulting in restricted crystal growth and a fine-grained, or **aphanitic,** texture. An example of a rock with aphanitic texture is rhyolite, which has about the same composition as granite but a much finer texture. The grain size of an igneous rock can tell us if it is an **intrusive** (cooled within the earth) or an **extrusive** (cooled at the surface of the earth) rock. The classification by texture and composition of igneous rocks is shown in ➢ Figure 2.13. Note that for each composition there are pairs of rocks that are distinguished by their texture. Granite and rhyolite, diorite and andesite, and gabbro and basalt are the most common pairs found in nature. It should be noted that mineral composition ranges over a continuum from light-colored granites to dark gabbros, with many identified intermediate rock types between them. The end member shown in Figure 2.13, peridotite, is rarely found because it originates deep within the earth. Peridotite is called an *ultramafic* rock because it contains mostly *ferromagnesian* minerals—silicates that are rich in iron (Fe) and magnesium (Mg), such as certain pyroxenes and olivine—that give it a dark or greenish color. Three important volcanic igneous rocks not included in Figure 2.13 are obsidian and pumice, which have glassy textures, and tuff, which

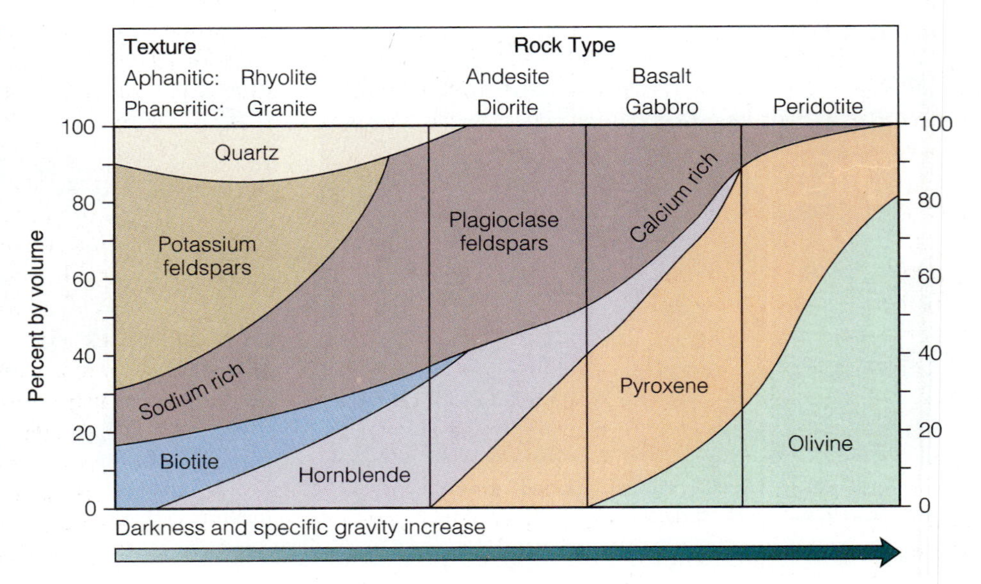

➢ **FIGURE 2.13 Classification of common igneous rocks and their mineralogy. Relative proportions of mineral content are shown for each rock type.**

(a)

(b)

(c)

➢ **FIGURE 2.14 Igneous rocks: (*a*) glassy obsidian; (*b*) pumice, which is formed from gas-charged magmas; and (*c*) tuff. The lavas that form obsidian and pumice cool so rapidly that crystals do not grow to any size. Scale is 5 cm long.**

is consolidated volcanic ash or cinders (see ➢ Figure 2.14). Tuff's fragmental texture is the result of explosive volcanic activity. Geologists describe this texture as **pyroclastic** (literally "fire-broken").

SEDIMENTARY ROCKS. Sediment is particulate matter derived from the physical or chemical weathering of materials of the earth's crust and by diverse organic processes. (Chapter 6 discusses this.) It may be transported and redeposited by streams, glaciers, wind, or waves. Sedimentary rock is sediment that has become **lithified**—turned to stone—by pressure from deep burial, by cementation, or by both of these processes. Detrital sedimentary rocks are those that are composed of **detritus,** fragments of preexisting rocks and minerals. Their texture, which makes this composition quite obvious, is described as **clastic.** Sedimentary rocks are classified according to their grain size. Thus sand-size sediment lithifies to sandstone, clay to shale, and gravel to conglomerate (➢ Figures 2.15 and 2.16a, b, and c). Some sediments are the result of chemical or biological activity. Chemical sedimentary rocks, which may be clastic or nonclastic, include chemically precipitated limestone, rock salt, and gypsum. Biogenic sedimentary rocks are produced directly by biological activity, such as coal (lithified plant debris), some limestone and chalk ($CaCO_3$ shell material), and chert (siliceous shells) (➢ Figure 2.16d and e). Table 2.3 illustrates the classification of sedimentary rocks.

A distinguishing characteristic of most sedimentary rock is bedding, or **stratification** into layers. Because shale is very thinly stratified, or *laminated,* it splits into thin sheets. Some sedimentary rocks, such as sandstone and limestone, may occur in beds that are several feet thick. Other structures in these rocks give hints as to how the rock formed. **Cross-bedding**—stratification that is inclined at an angle to the main stratification—indicates the influence of wind or water currents. Thick cross-beds and frosted quartz grains in sandstone indicate an ancient desert sand-dune environment (➢ Figure 2.17). Some shale and claystone show polygonal *mud cracks* on bedding planes, similar to those found on the surface of modern dry lakes; these indicate desiccation in a subaerial environment (➢ Figure 2.18). *Ripple marks,* low, parallel

➢ FIGURE 2.15 Manner in which sediments are transformed into clastic sedimentary rocks. The sequence of the lithification process is deposition to compaction to cementation to hard rock.

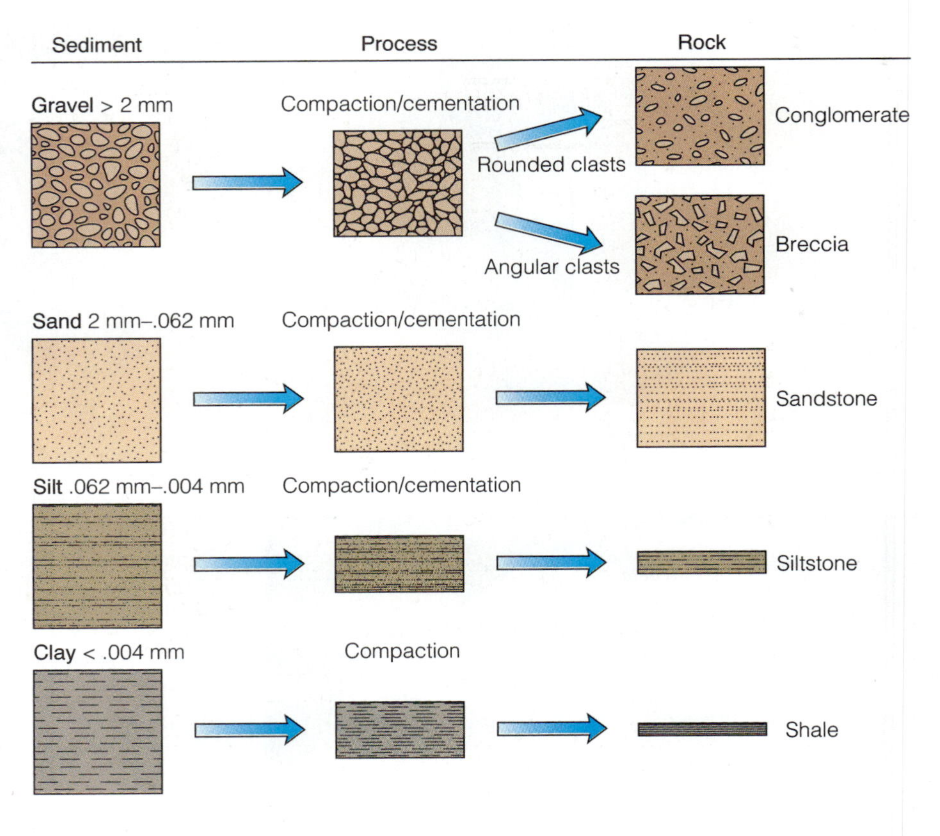

➢ FIGURE 2.16 Common clastic sedimentary rocks: (*a*) shale, (*b*) sandstone, (*c*) conglomerate. Biogenic rocks: (*d*) limestone composed entirely of shell materials (called *coquina*), (*e*) coal.

TABLE 2.3 Classification of Sedimentary Rocks

CLASTIC SEDIMENTARY ROCKS		
Sediment	*Description*	*Rock Name*
Gravel (>2.0 mm)	Rounded rock fragments	Conglomerate
	Angular rock fragments	Breccia
Sand (0.062–2.0 mm)	Quartz predominant	Quartz sandstone
	>25% feldspars	Arkose
Muds		
Silt (0.004–0.062 mm)	Quartz predominant, gritty feel	Siltstone
Clay, mud (<0.004 mm)	Laminated, splits into thin sheets	Shale
	Thick beds, blocky	Mudstone
CHEMICAL SEDIMENTARY ROCKS		
Texture	*Composition*	*Rock Name*
Clastic	Calcite ($CaCO_3$)	Limestone
	Dolomite [$CaMg(CO_3)_2$]	Dolostone
Crystalline	Halite ($NaCl$)	Rock salt
	Gypsum ($CaSO_4 \cdot 2H_2O$)	Rock gypsum
BIOGENIC SEDIMENTARY ROCKS		
Texture	*Composition*	*Rock Name*
Clastic	Shell calcite, skeletons, broken shells	Limestone & Coquina
	Microscopic shells ($CaCO_3$)	Chalk
Altered	Microscopic shells (SiO_2), recrystallized silica	Chert
———	Consolidated plant remains (largely carbon)	Coal

➢ **FIGURE 2.17 Wind-blown sandstone exhibiting large cross-beds; Zion National Park.**

➢ **FIGURE 2.18 Mud cracks in an old clay mine; Ione, California.**

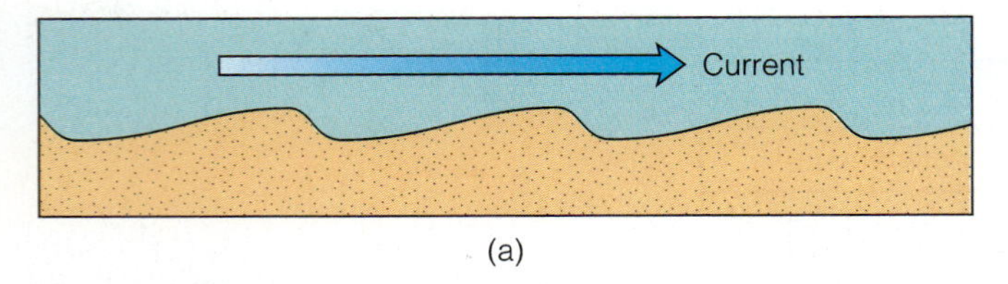

(a)

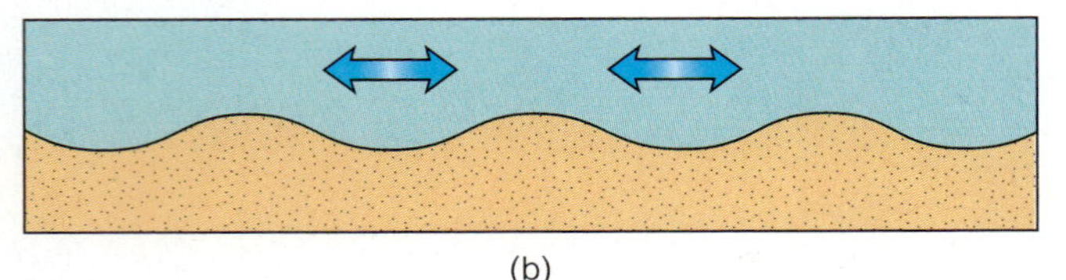

(b)

➢ FIGURE 2.19 (*a*) Asymmetrical ripple marks are common in streambeds and are also found on bedding planes in sedimentary rocks. (*b*) Symmetrical ripple marks are due to oscillating water motion; that is, wave action. Both types of ripple marks are found in sedimentary rocks.

ridges in deposits of fine sand and silt, may be asymmetrical or symmetrical (➢ Figure 2.19). Asymmetrical ripples indicate a unidirectional current; symmetrical ones, the back-and-forth motion produced by waves in shallow water. Only in sedimentary rock do we find abundant fossils, the remains or traces of life.

METAMORPHIC ROCKS. Rocks that have been changed from preexisting rocks by heat, pressure, or chemical processes are classified as metamorphic rocks. The process of metamorphism results in new structures, textures, and minerals. **Foliation** (Latin *folium,* "leaf") is the flattening and layering of minerals by nonuniform stresses. Foliated metamorphic rocks are classified by the development of this structure. Slate, schist, and gneiss are foliated, for example, and they are identified by the thickness or crudeness of their foliation (➢ Figure 2.20). Metamorphic rocks may also form by **recrystallization.** This occurs when a rock is heated and strained by uniform stresses so that larger, more perfect grains result, or new minerals

(a)

(b)

(c)

➢ FIGURE 2.20 Metamorphic rocks: (*a*) coarsely foliated gneiss, (*b*) mica schist showing wavy or crinkly foliation, (*c*) finely foliated slate showing slaty cleavage.

(a)

(b)

➢ **FIGURE 2.21** (*a*) **Limestone recrystallizes to form a white marble. (Streaks are due to disseminated minerals, mostly the green mineral epidote.)** (*b*) **Quartzite, the hardest and most durable common rock, is metamorphosed sandstone.**

form. In this manner a limestone may recrystallize to marble or a quartz sandstone to quartzite, one of the most resistant of all rocks (➢ Figure 2.21). Table 2.4 summarizes the characteristics of common metamorphic rocks.

Rock Defects

Some rocks have structures that geologists view as "defects"; that is, surfaces along which landslides or rockfalls may occur. Almost any planar structure, such as a bedding plane in sedimentary rock or a foliation plane in metamorphic rock, holds potential for rock slides or falls. The orientation of a plane in space, such as a stratification plane, a fault, or a joint, may be defined by its **dip** and **strike.** (Case Study 2.3 explains dip and strike.) **Joints** are rock fractures without displacement and they occur in all rock types. They are commonly found in parallel sets spaced several feet apart (➢ Figure 2.22). **Faults** are also fractures in crustal rocks, but they differ from joints in that some movement or displacement has occurred along the fault surface. Faults also are found in all rock types and they are potential surfaces of "failure." We will examine the relationships of rock defects to geologic hazards when we discuss landslides, subsidence, and earthquakes.

Geologic Time

Interest in extremely long periods of time sets geology and astronomy apart from other sciences. Geologists think in terms of billions of years for the age of the earth and its oldest rocks, numbers which, like the national debt, are

TABLE 2.4 **Common Metamorphic Rocks**

Foliated or Layered		
ROCK	PARENT ROCK	CHARACTERISTICS
Slate	Shale and mudstone	Splits into thin sheets
Schist	Fine-grained rocks	Mica minerals usually crinkled
Gneiss	Coarse-grained rocks	Dark and light layers of aligned minerals
Nonfoliated or Recrystallized		
Marble	Limestone	Interlocking crystals
Quartzite	Sandstone	Interlocking, almost fused quartz grains

➢ **FIGURE 2.22** **Joints in basalt flows, near Chico, California.**

CASE STUDY 2.3

Planes in Space—Dip and Strike

Geologists describe the orientation of a plane surface by the plane's "strike" and "dip." Geologists measure such tangible planes as those made by sedimentary strata, foliation, and faults, because it allows them to anticipate hazards during excavations for such things as tunnels, dam foundations, and roadcuts. Accurate surface measurements of planes also enable geologists to predict the underground locations of valuable resources such as oil reservoirs, ore deposits, and water-bearing strata.

Strike is the direction of the trace of an inclined plane on the horizontal plane, and the direction is referred to north. In the example illustrated in ➢ Figure 1, the strike is 30° west of north. In geologic notation, this is written *N30°W.*

Dip is the vertical angle that the same inclined plane makes with the horizontal plane. Dip is assessed perpendicular to the strike, because any other angle will give an "apparent dip" that is less than the true dip. Because a plane that strikes northwest may be inclined to either the southwest or the northeast, the general direction of dip must be indicated. In our example, the dip is 60° to the northeast, which is notated *60°NE.* Thus, a geologist's notes for these bedding planes would describe their orientation as "strike N30°W, dip 60°NE." All other geologists would understand that the plane has a strike that is 30° west of north and a dip that is 60° to the northeast.

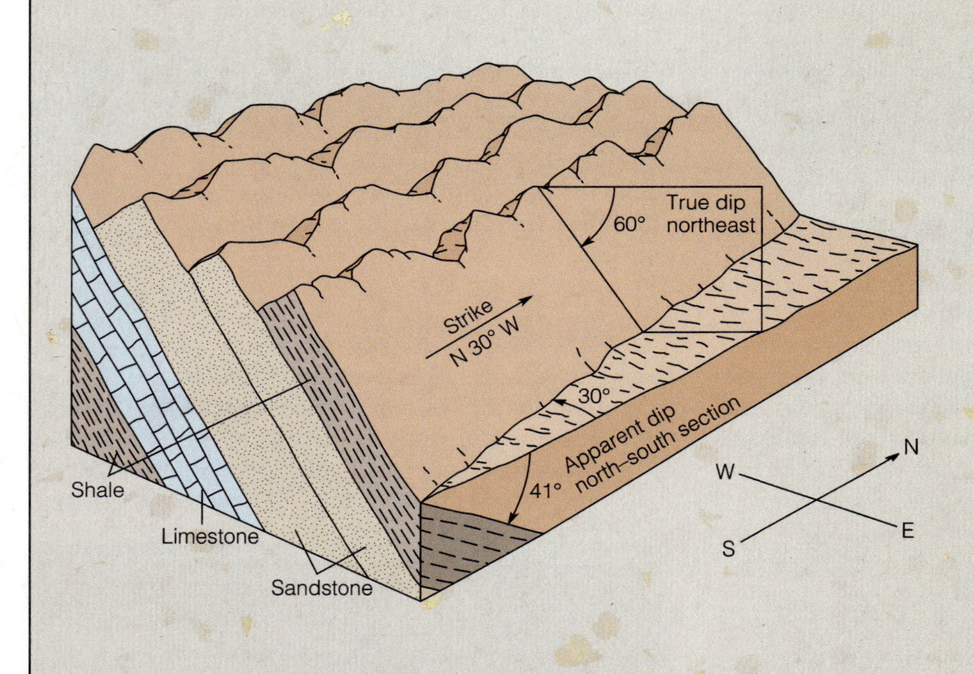

➢ **FIGURE 1 Block diagram showing sedimentary rock that is striking N30°W and dipping 60° northeast. The apparent dip is shown on the north–south cross section at the lower front of the diagram.**

not easily comprehended. Time is such an important element in earth science that it has its own designation: **geologic time.** Why is the time of ancient events relevant to environmental geology? Ongoing research is being directed toward determining recurrence intervals for large earthquakes and volcanic eruptions. It is postulated that if we know the average time interval between catastrophic events over the past several thousand years, we can predict more reliably when the next such events will occur.

Geologists use two different means of dating rocks and geologic events. In the field we normally use **relative age dating,** which establishes the sequence of geologic events. For example, we can establish that "these" rock strata are oldest, that "this" igneous rock is younger than that rock, and so forth. In the laboratory, **absolute age dating** can be carried out. This yields an age in years before the present for an event, rock, or artifact. Relative age is determined by applying geologic laws based on the same reasoning. For instance, the Law of Superposition tells us that in a stack of undeformed sedimentary rocks, the stratum (layer) at the top is the youngest. Using the same reasoning, the Law of Cross-cutting Relationships tells us that a fault is younger than the youngest rocks it displaces or intrudes. Similarly we know that a pluton is younger than the rocks it intrudes (➢ Figure 2.23).

Using these laws, geologists have assembled a great thickness of sedimentary rocks and their contained fossils representing an immense span of geologic time. The geologic age of a particular sequence of rock was then determined by applying the Law of Fossil Succession, the observed chronologic sequence of life forms through geologic time. This allows fossiliferous rocks from two widely separated areas to be correlated by matching key fossils or groups of fossils found in the rocks of the two areas (Figure 2.23). Using such indicator fossils and radioactive dating methods, geologists have developed the geologic time scale to chronicle the documented events of earth history. (Refer to the geologic time scale in Appendix 1.) Note that the scale is divided into units of time during which rocks were deposited, life evolved, and significant geologic events such as mountain building occurred. Eons are the longest time intervals, followed, respectively, by eras, periods, and epochs. The Phanerozoic ("revealed life") Eon began 570 million years ago with the Cambrian Period, the rocks of which contain the first extensive fossils of organisms with hard skeletons. Because of the significance of the Cambrian Period, the informal term *Precambrian* is widely used to denote the time before it, which extends back to the formation of the earth, 4.6 billion years ago. Note that the Precambrian is divided into the Archean and Proterozoic Eons, with the Archean Eon extending back to the oldest known in-place rocks, about 3.9 billion years ago. Precambrian time accounts for 88 percent of geologic time, and the Phanerozoic for a mere 12 percent. The eras of geologic time correspond to the relative complexity of life forms: Paleozoic (oldest life), Mesozoic (middle life), and Cenozoic (most recent life). Environmental geologists are most interested in the events of the past few million years, a mere heartbeat in the history of the earth.

Absolute age-dating requires some kind of natural clock. The ticks of the clock may be the annual growth rings of trees or established rates of disintegration of radioactive elements to form other elements. At the turn of the twentieth century, American chemist and physicist Bertram Borden Boltwood (1870–1927) discovered that the ratio of lead to uranium in uranium-bearing rocks increases as the ages of the rocks increase. He developed a process for determining the age of ancient geologic events that is unaffected by heat or pressure—radiometric dating. The "ticks" of the radioactive clocks are radioactive decay processes—spontaneous disintegrations of the nuclei of heavier elements such as uranium and thorium to lead. A radioactive element may decay spontaneously to another element, or to an isotope of the same element. This decay occurs at a precise rate that can be determined experimentally. The most common emissions are **alpha**

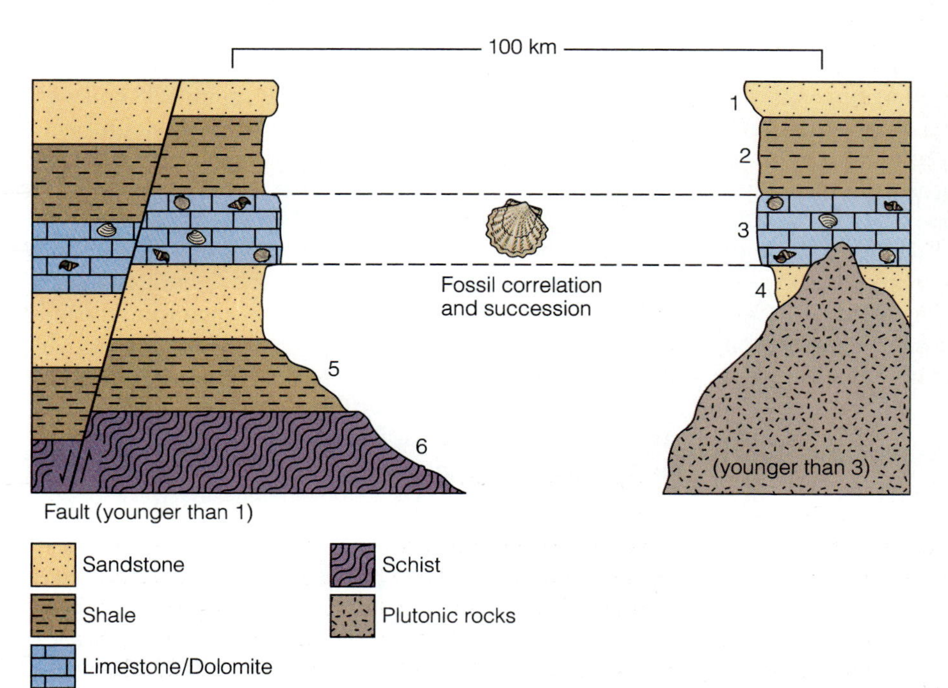

➢ FIGURE 2.23 Geologic cross sections illustrating the Laws of Superposition, Cross-cutting Relationships (intrusion and faulting), and Fossil Succession. The limestone beds are correlative because they contain the same fossils. Numbers indicate relative ages, with 1 being the youngest.

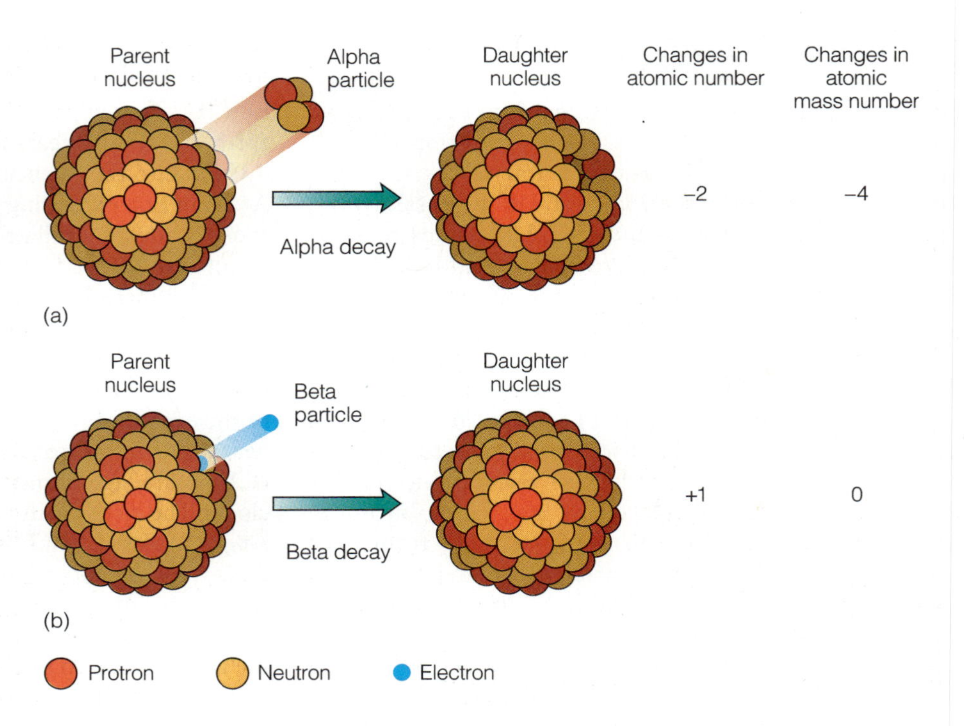

➢ FIGURE 2.24 (*a*) Alpha emission, whereby a heavy nucleus spontaneously emits a helium atom and is reduced 4 atomic mass units and 2 atomic numbers. (*b*) Emission of a nuclear electron (β^- particle), which changes a neutron to a proton and thereby forms a new element without a change of mass.

($_2\alpha^4$) particles, which are helium atoms, and **beta** (β^-) particles, which are nuclear electrons. New radiogenic "daughter" elements are the result of both these decay schemes (➢ Figure 2.24).

For example, although there are three isotopes of carbon, C^{12}, C^{13}, and C^{14}, only C^{14} is radioactive, and this radioactivity can be used to date events between a few hundred and a few tens-of-thousands of years old. Carbon-14 is formed continually in the upper atmosphere by neutron bombardment of nitrogen, and it exists in a fixed ratio to the common isotope, C^{12}. All plants and animals contain radioactive C^{14} in equilibrium with the atmospheric abundance until they die, after which time C^{14} begins to decrease in abundance, and along with it, the object's radioactivity. Thus by measuring the radioactivity of an ancient parchment, log, or charcoal and comparing the measurement with the activity of a modern standard, the age of archaeological materials and geological events can be determined (➢ Figure 2.25). Carbon-14 is formed by the action of cosmic neutrons on N^{14}, and decays back to N^{14} by emitting a nuclear electron (β^-).

$$_7N^{14} + \textit{neutron} \longrightarrow {}_6C^{14} + \textit{proton}$$
$$_6C^{14} \longrightarrow {}_7N^{14} + \beta^-$$

Both carbon—the radioactive "parent" element—and nitrogen—the radiogenic "daughter"—have 14 atomic mass units. However, one of C^{14}'s neutrons is converted to a proton by the emission of a beta particle, and the carbon changes (or *transmutes*) to nitrogen. This process proceeds at a set rate that can be expressed as a **half-life,** the time required for half of a population of radioactive atoms to decay. For C^{14} this is about 5,730 years.

Radioactive elements decay exponentially; that is, in two half-lives one fourth of the original number of atoms remain, in three half-lives one eighth remain, and so on (➢ Figure 2.26). So few parent atoms remain after seven or eight half-lives (less than 1%) that experimental uncertainty creates a limit for radiometric methods.

Whereas the practical age limit for dating carbon-bearing materials such as wood, paper, and cloth is about 40,000 to 50,000 years, U^{238} disintegrates to Pb^{206} and has a half-life of 4.5×10^9 years. Uranium-238 emits alpha particles (helium atoms). Since both alpha and beta disintegrations can be measured with a Geiger counter, we have a means of determining the age of a geological or archaeological sample. Table 2.5 shows radioactive parents, daughters, and half-lives commonly used in age-dating. Note that methods for radioactive age-dating provide half-lives in the thousands (10^3) of years and the millions (10^6) and billions (10^9) of years, but not in the hundreds of thousands (10^5) of years. Thus we might say that our radioactive age-dating clock has an hour hand and a split-second hand, but no minute hand. There are, however, nonradiometric methods for determining ages on the order of 10^5 years.

■ AGE OF THE EARTH

Before the advent of radioactive dating, determining the age of the earth was a source of controversy between established religious interpretations and early scientists.

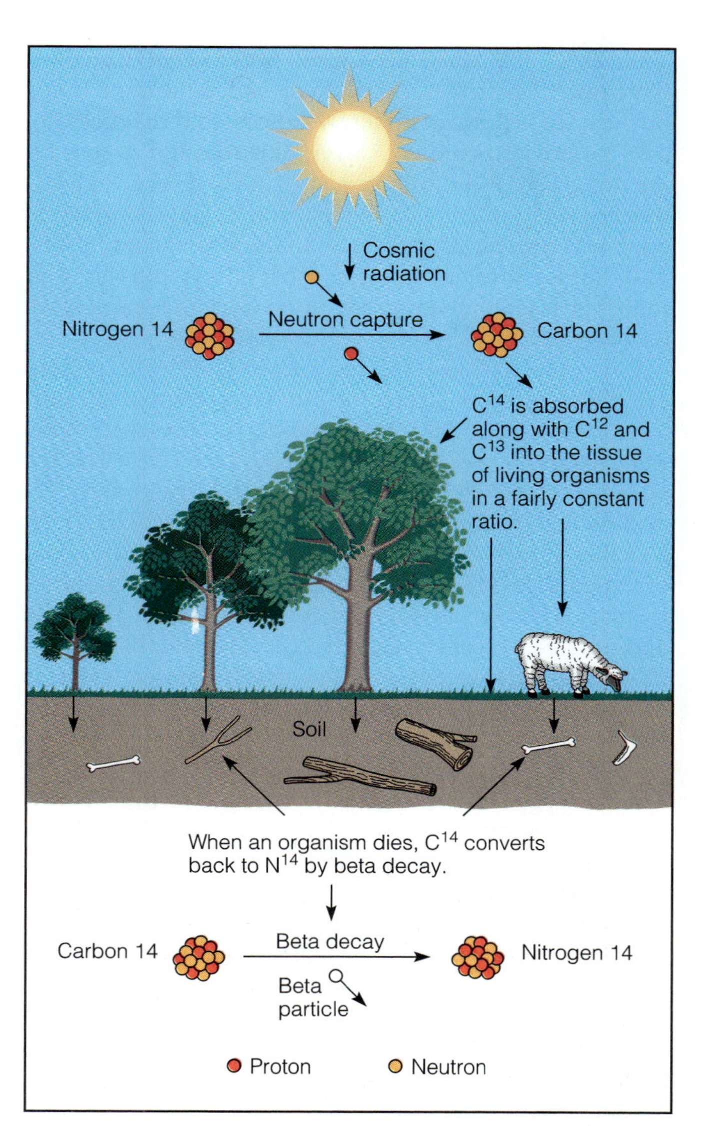

➢ **FIGURE 2.25 The formation of carbon-14 from nitrogen-14 by neutron capture and subsequent proton emission, and the decay of carbon-14 back to nitrogen-14 by emission of a nuclear electron (β^-).**

➢ **FIGURE 2.26 Decrease (decay) of a radioactive parent element with time. Each time unit is one half-life. Note that after two half-lives one fourth of the parent element remains, and that after three half-lives one eighth remains.**

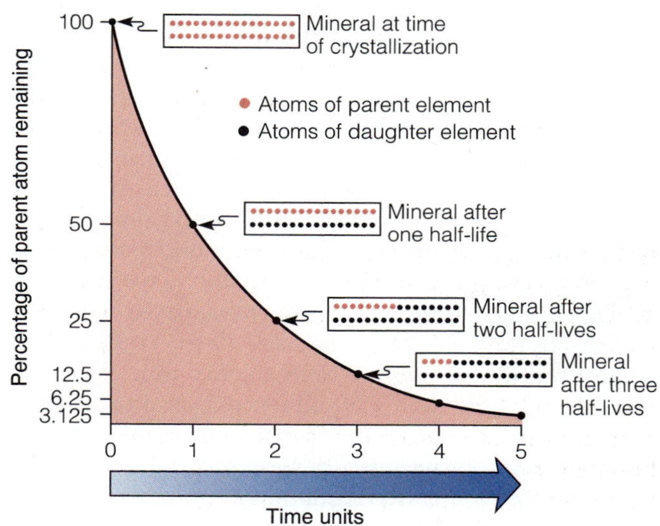

TABLE 2.5 Age-Dating Methods and Some Important Ages

METHOD				
Parent Isotope		*Daughter Isotope*	*Half-life, years*	*Range, years*
$_{92}U^{238}$	→	$_{82}Pb^{206} + 8_2\alpha^4 + 6\beta^- + Q$*	4.5×10^9	10^7–4.5×10^9
$_{90}Th^{232}$	→	$_{82}Pb^{208} + 6_2\alpha^4 + 4\beta^- + Q$	1.5×10^{10}	10^7–4.5×10^9
$_{92}U^{235}$	→	$_{82}Pb^{207} + 7_2\alpha^4 + 4\beta^- + Q$	7.1×10^8	10^7–4.5×10^9
$_{37}Rb^{87}$	→	$_{38}Sr^{87} + \beta^- + Q$	5.0×10^{10}	10^7–4.5×10^9
$_{19}K^{40}$	→	$_{18}Ar^{40} + \beta^- + Q$	1.3×10^9	10^6–4.5×10^9
$_6C^{14}$	→	$_7N^{14} + \beta^- + Q$	5,730	100s to 50,000

Oldest known rocks: Acasta gneisses, Canada; zircon minerals in the gneisses have been dated as 3.96 billion years old.

Oldest known crust: zircon minerals in rocks of western Australia derived from older crust have been dated as 4.0–4.3 billion years old.

Oldest known fossils: algae and bacteria have been dated as 3.5 billion years old.

* Q = heat

Archbishop Ussher (1585–1656), the Archbishop of Armagh and a professor at Trinity College in Dublin, declared that the earth was formed in the year 4004 B.C. Not only did Ussher provide the year but also the day, October 23, and the time, 9:00 A.M. Although Ussher has many detractors, Stephen J. Gould of Harvard University, though not proposing acceptance of such a date, is not one of them. Gould points out that Ussher's work was good scholarship for its time because other religious scholars had extrapolated from Greek and Hebrew scriptures that the earth was formed in 5500 B.C. and 3761 B.C., respectively. Ussher based his age determination on the verse in the Bible that says "one day is with the Lord as a thousand years" (2 Peter 3:8). Because the Bible also says that God created heaven and earth in 6 days, Ussher arrived at 4004 years B.C.—the extra four years because he believed Christ's birth year was wrong by this amount of time. Ussher's age for the earth was accepted by many as "gospel" for almost 200 years.

By the late 1800s geologists believed that the earth was on the order of 100 million years old. Geologists reached their estimates by dividing the total thickness of sedimentary rocks (tens of kilometers) by an assumed annual rate of deposition (mm/year). Evolutionists such as Charles Darwin thought that geologic time must be almost limitless in order that minute changes in organisms could eventually produce the present diversity of species. Both geologists and evolutionists were embarrassed when the British physicist William Thomson (later Lord Kelvin) demonstrated with elegant mathematics how the earth could be no older than 400 million years, and maybe as young as 20 million years. Thomson based this on the rate of cooling of an initially molten earth and the assumption that the material composing the earth was incapable of creating new heat through time. He did not know about radioactivity, which adds heat to rocks in the crust and mantle.

Bertram Boltwood postulated that older uranium-bearing minerals should carry a higher proportion of lead than younger samples. He analyzed a number of specimens of known relative age, and the absolute ages he came up with ranged from 410 million to 2.2 billion years old. These ages put Lord Kelvin's dates based on cooling rates to rest and ushered in the new **radiometric dating** technique. By extrapolating backward to the time when no radiogenic lead had been produced on earth, we arrive at an age of 4.6 billion years for the earth. This corresponds to the dates obtained from meteorites and lunar rocks, which are part of our solar system. The oldest known intact terrestrial rocks are found in the Acasta gneisses of the Slave geological province of Canada's Northwest Territories. Analyses of lead to uranium ratios on the gneisses' zircon minerals indicate that the rocks are 3.96 billion years old. However, older detrital zircons on the order of 4.0–4.3 billion years old have been found in western Australia, indicating that some stable continental crust was present as early as 4.3 billion years ago (Table 2.5). Suffice it to say that the earth is very old and that there has been abundant time to produce the features we see today.

Environmental geology deals mostly with the present and the recent past—the time interval known as the *Holocene Epoch* (10,000 years ago to the present) of the *Quaternary Period*. However, because the Great Ice Age advances of the Pleistocene Epoch of the Quaternary Period (which preceded our own Holocene Epoch) had such a tremendous impact on the landscape of today, we will investigate the probable causes of the "ice ages" in order to speculate a bit about what the future might bring. In addition, Pleistocene glacial deposits are valuable sources of underground water, and they have economic value as sources of building materials. We will see that "ice ages" have occurred several times throughout geologic history (see Chapter 12).

What impresses one most about geologic time is how short the period of human life on earth has been. If we could compress the 4.6 billion years since the earth formed into one calendar year, *Homo sapiens* would appear about 30 minutes before midnight on December 31st (➢ Figure 2.27). The last ice-age glaciers would begin wasting away a bit more than 2 minutes before midnight, and written history would exist for only the last 30 seconds of the year. Perhaps we should keep this calendar in mind when we hear that the dinosaurs were an unsuccessful group of reptiles. After all, they endured 50 times longer than hominids have existed on the planet to date. At the present rate of population growth, there is some doubt that humankind as we know it will be able to survive anywhere near that long.

The earth we see today is the product of earth processes acting on earth materials over the immensity of geologic time. The response of the materials to the processes has been the formation of mountain ranges, plateaus, and the multitude of other physical features found on the planet. Perhaps the greatest of these processes has been the formation and movement of platelike segments of the earth's surface, which influence the distribution of earthquakes and volcanic eruptions, global geography, mineral deposits, mountain ranges, and even climate. These *plate tectonic* processes are the subject of the next chapter.

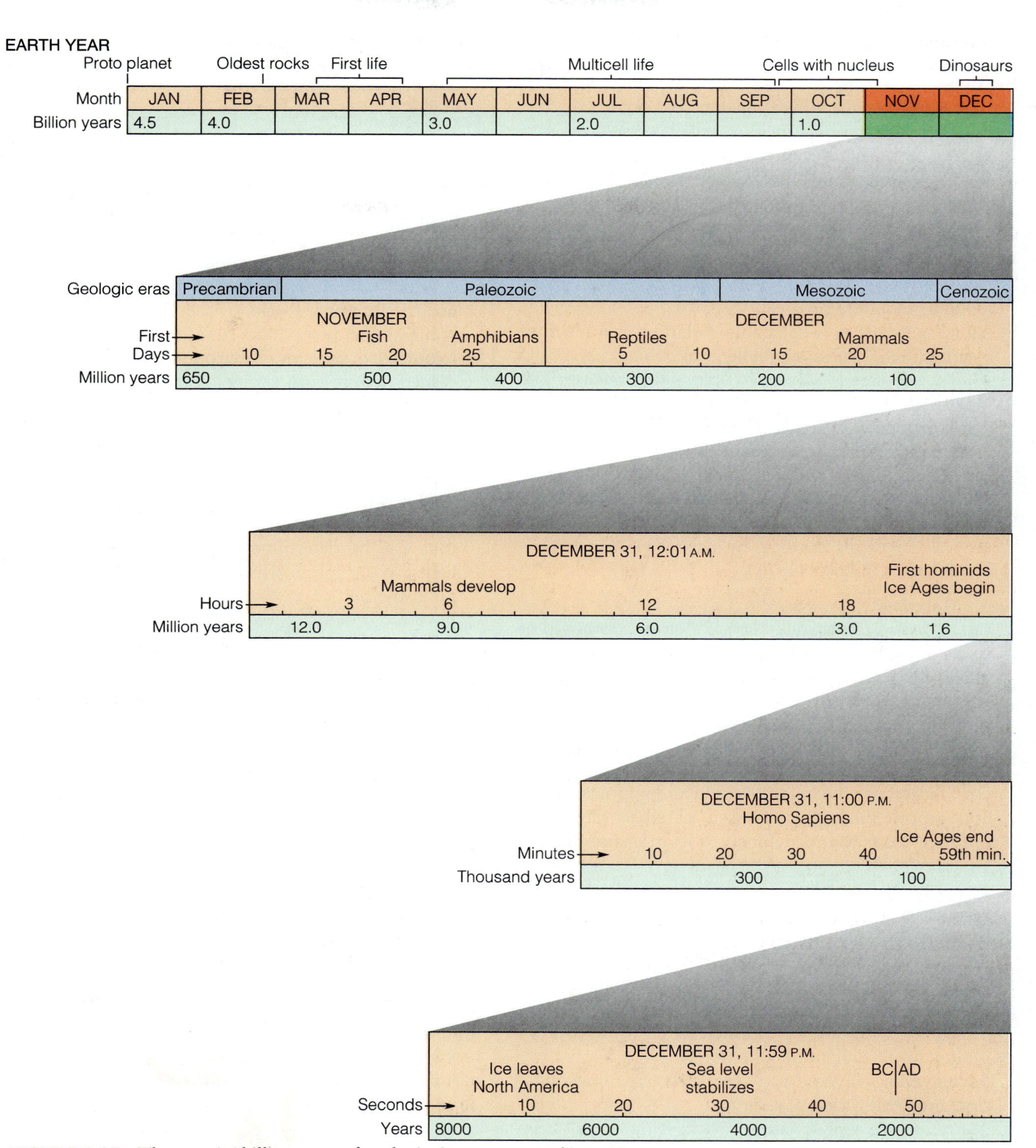

➢ **FIGURE 2.27 The past 4.6 billion years of geologic time compressed into 12 months. Note that the first hard-shelled marine organism does not appear until about November 1, and that humans as we know them have been around only since the last hour before the New Year.**

SUMMARY

Big Bang

DESCRIPTION: Rapid expansion of dense matter, mostly hydrogen and helium, which segregated into smaller "clouds" of matter to form galaxies and stars 10–15 billion years ago. Gravitational collapse of clouds raised temperatures so that thermonuclear reactions occurred, forming heavier elements.

Solar System

DESCRIPTION: Formed 5 billion years ago from an intersteller cloud (nebula).

PROBABLE CAUSE: The nebula collapsed as a result of a supernova explosion (called the "bing bang" by some). The nebula started to spin and condense. The inner part of the cloud became the hot protosun and it was surrounded by a cooler, disc-shaped cloud (the solar nebula). Volatile substances in the outer reaches of the nebula formed the large outer planets; rocky substances closest to the sun formed the dense inner planets.

EARTH: Probably formed by accretion of asteroid-sized masses (planetesimals).

Earth Materials

CRUST: Outermost rocky layer of earth composed of rocks that are aggregates of minerals.

MINERALS: Naturally occurring inorganic substances with a definite set of physical properties and a narrow range of chemical compositions.

CLASSIFICATION BY: Chemistry of anions (negative ions or radicals), such as oxides, sulfides, carbonates, etc. Silicates are the most common and important mineral group and are composed of silica tetrahedra $(SiO_4)^{-4}$ units that are joined to form chains, sheets, and networks.

IDENTIFICATION: Physical properties such as hardness, cleavage, crystal structure, fracture pattern, and luster.

ROCKS: Consolidated or poorly consolidated aggregates of one or more minerals or organic matter.

CLASSIFICATION BY: Texture and composition.

CLASSES: *Igneous*—rock formed by crystallization of molten or partially molten material.

Sedimentary—layered rock resulting from consolidation and lithification of sediment.

Metamorphic—preexisting rock that has been changed by heat, pressure, or chemically active fluids.

STRUCTURES: Many planar rock structures may be viewed as defects along which landslides, rock falls, or other potentially hazardous events may occur. Stratification in sedimentary rocks and foliation in metamorphic rocks are common planar features. Faults and joints occur in all classes of rock.

ROCK CYCLE: A sequence of events by which rocks are formed, altered, destroyed, and reformed as a result of internal and external earth processes.

Geologic Time

RELATIVE TIME: The sequential order of geologic events established by using basic geologic principles or laws.

GEOLOGIC TIME SCALE: The division of geologic history into eons, eras, periods, and epochs. The earliest 88 percent of geologic time is known informally as *Precambrian* time. Precambrian rocks are generally not fossiliferous. Rocks of the Cambrian Period and younger contain good fossil records of shelled and skeletonized organisms. Environmental geology is concerned mainly with geologic events of the present epoch, the Holocene, and the one that preceded it, the Pleistocene.

ABSOLUTE AGE: Absolute dating methods use some natural "clock." For long periods of time, such clocks are known rates of radioactive decay. A parent radioactive isotope decays to a daughter isotope at a rate that can be determined experimentally. This established rate, expressed as a half-life, enables us to determine the age of a sample of material. We currently have radiometric methods for determining ages of up to 50,000 years (C^{14}) and in the millions and billions of years (U/Pb), but not for determining ages of 100,000s of years old.

AGE OF THE EARTH: Comparing present Pb/Pb ratios with those in iron meteorites we can extrapolate backward and obtain an earth age of 4.6 billion years old.

ROCKS AND CRUST: The oldest known in-place rocks are 3.96 billion years old. Dating of minerals derived from older crust yields an age of 4.0–4.3 billion years for stable continental crust.

KEY TERMS

absolute age dating
alpha decay
amorphous
aphanitic
atom
atomic mass
atomic number
batholith
beta decay
big bang
clastic
cleavage
core
cross-bedding
crust
detritus
dip
element
extrusive
fault
foliation
fracture
half-life
hardness scale
igneous rocks
intrusive
ion
isotope
joint
lava

lithified
luster
magma
mantle
metamorphic rocks
mineral
nucleus
pegmatite
phaneritic texture
planetesimal
plutonic
protostar
pyroclastic
radiometric dating
Rapakivi
recrystallization
relative age dating
rock
rock cycle
sedimentary rocks
stratification
strike
supernova

Study Questions

1. Explain to a friend (or to yourself) the "big bang" theory of how the universe formed. Then explain how our planetary system may have come into existence.
2. What physical and chemical factors are the bases of the rock-classification system? What are the three classes of rocks, and how does each form? Identify one characteristic of each class that usually makes it readily distinguishable from the other rock classes.
3. How and where do batholiths form? What type of rock most commonly forms in batholiths?
4. What are some of the planar surfaces in rocks that may be weak and thus lead to various types of slope failure?
5. How may structures in sedimentary rocks be used to reconstruct past environments?
6. How may absolute age-dating techniques be used to the betterment of human existence?
7. The earth is 4.6 billion years old, and humans have been on earth for only a few hundred thousand years—but a blink of an eye in the great length of geologic time. What changes of a global nature have humans invoked on the earth's natural systems (water, air, ice, the solid earth, and biology) in this short length of time? Which impacts are reversible, and which ones cannot be reversed or mitigated?
8. What is the most common mineral species? Name several rocks in which it is a prominent constituent.
9. What is cleavage? How can it serve as an aid in identifying minerals?
10. The most common intrusive igneous rock is composed mostly of (1) the most common mineral and (2) a mineral of the most common mineral group. Name the rock and its constituent minerals.
11. What mineral is found in both limestone and marble?

Further Reading

Albritton, C. C. 1984. Geologic time. *Journal of Geological Education* 32, no. 1: 29–47.

Bowring, S. A.; I. S. Williams; and W. Compston. 1989. 3.96 Ga gneisses from the Slave province, Northwest Territories, Canada. *Geology* 17: 971–975.

Brown, V. M., and J. A. Harrell. 1991. Megascopic classification of rocks. *Journal of Geological Education* 39: 379.

Cameron, A. G. W. 1975. The origin and evolution of the solar system. *Scientific American* 233, no. 3: 32.

Dietrich, R. V., and Brian Skinner. 1979. *Rocks and rock minerals.* New York: John Wiley and Sons.

Eicher, D. L. 1976. *Geologic time,* 2d ed. Englewood Cliffs, N.J.: Prentice-Hall, Inc.

Hurley, Patrick. 1959. *How old is the earth?* New York: Anchor Books.

Libby, W. F. 1955. *Radiocarbon dating.* Chicago: University of Chicago Press.

National Research Council, National Academy of Sciences. 1993. *Solid-earth sciences and society.* Washington, DC: National Research Council Commission on Geosciences, Environment, and Resources, National Academy of Sciences Press.

Sagan, Carl. 1975. The solar system. *Scientific American* 233, no. 4: 98.

Silk, Joseph. 1989. *The big bang,* 2d ed. New York: W. H. Freeman and Co.

Snow, T. P. 1993, *Essentials of the dynamic universe.* St. Paul, Minn.: West Publishing Company.

Wicander, Reed, and James Monroe. 1989. *Historical geology.* St. Paul, Minn.: West Publishing Company.

Winkler, Erhard M. 1991. Building stone and decay in downtown Chicago. In Killey, Myrna, ed., *Field trip guidebook.* Sudbury, Mass.: Association of Engineering Geologists.

CHAPTER

3

PLATE TECTONICS

Happy the man whose lot it is
to know the secrets of the earth.

EURIPIDES, DRAMATIST (*ca* 484–406 B.C.)

During the 1960s our concepts of the architecture of the earth's crust and how the earth "works" were completely revolutionized by the theory of **plate tectonics** (Greek *tektonikos,* "builder"). The theory maintains that the lithosphere, the outer rocky shell of the earth, consists of a number of rigid plates, 70–150 km (40–90 mi) thick, that slide under, past, and away from one another. Continents, although parts of the moving plates, are but passive passengers. Where plates interact with each other, earthquakes and volcanic activity occur, mountain ranges may form, and internal heat is released to the earth's surface. Plate tectonics integrates two older explanations for the distribution of the earth's physical and geophysical features—the theories of continental drift and sea-floor spreading.

CONTINENTAL DRIFT

What ultimately was to become the theory of plate tectonics began in 1910 when Alfred Wegener (1880–1930), a German meteorologist, presented a paper to the Frankfurt Geological Society in which he expressed his theory that the continents have moved great horizontal distances in the past 100 million years. He called his hypothesis *die Verschiebung der Kontinente,* meaning "continental displacement," which English-speaking scientists translated

View down the axis of a rift valley in Iceland formed by down faulting of the flat valley floor. Iceland is a very large volcano on the mid-Atlantic Ridge, which is part of the longest mountain range on earth, extending through the three major ocean basins. The prominent rift valley along the central axis of Iceland was formed by forces pulling the earth's crust apart (tension) at a divergent boundary between two tectonic plates. As the plates move apart, lavas rise to feed volcanic vents, and faults form along which the rift valley subsides.

to **continental drift,** the name that stuck. Wegener called the late Paleozoic supercontinent—one large, unified landmass consisting of all the continents we know today—**Pangaea** (Greek: "all land"; ➢ Figure 3.1). He theorized that Pangaea split apart during Mesozoic time and that the continents had then moved steadily in all directions to their present positions. Wegener explained that as the continents drifted westward (*Westwanderung)*, they had plowed up mountain ranges such as the Andes and the North American Coast Ranges. Australia became geographically isolated, and the Atlantic Ocean became larger at the expense of a diminishing Pacific Ocean. Strong evidence in favor of the hypothesis, according to Wegener and many Southern Hemisphere geologists, is the distribution of late Paleozoic (Permian Period) glacial deposits and plant fossils known collectively as the *Gondwana Succession.* These distinctive fossil land plants (➢ Figure 3.2) and their associated glacial deposits are found only in South America, Africa, India, Australia, and Antarctica (➢ Figure 3.3). The only explanation for this, it was argued, is that these continents were once connected in the southern part of Pangaea. Some years earlier this hypothetical former landmass with the distinct Permian glaciation pattern had been named **Gondwanaland.** (The northern landmass had been named **Laurasia.**)

Additional paleontological evidence supporting continental drift is supplied by the amphibian and reptile fossils currently found on widely separated Gondwana continents. In particular are the fossil remains of the Permian freshwater reptile *Mesosaurus* found in Brazil and South Africa. It is inconceivable that these freshwater creatures

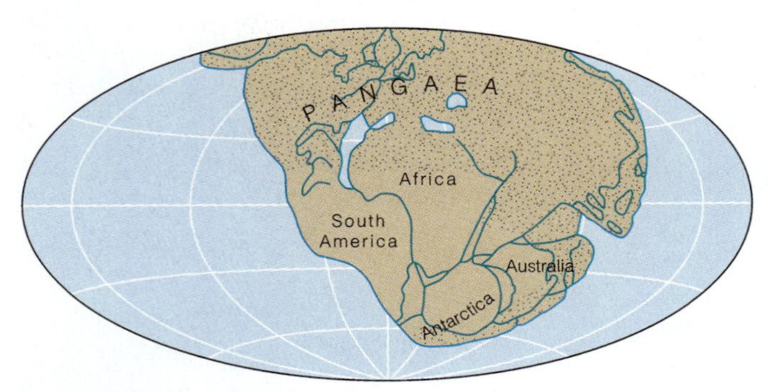

Upper Carboniferous Period

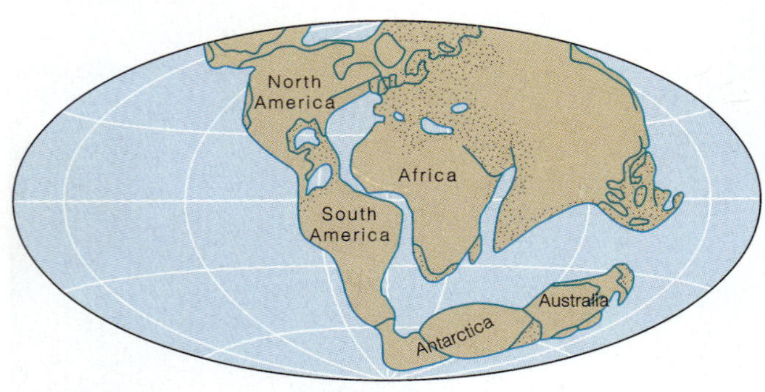

Eocene Period

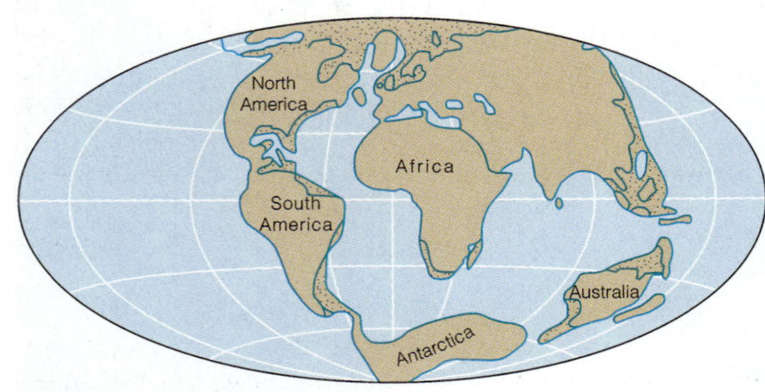

Early Quaternary Period

➢ FIGURE 3.1 Alfred Wegener's interpretation of the breakup of Pangaea and continental drift over the past 270 million years. The stippled areas adjacent to and over the continents are shallow seas receiving sediments, which explains the similarity of geologic formations of the same age on continents now widely separated. Africa is arbitrarily held in its present-day position as a point of reference for the motion of the other continents.

(a)

(b)

➢ FIGURE 3.2 Plant fossils of the Gondwana Succession. These fossils are found only in the southern continents of Gondwana; that is, in present-day South America, Africa, India, Australia, and Antarctica. Shown are *Glossopteris* leaves from Upper Permian (*a*) Dunedoo Formation and (*b*) Illawarra Coal Measures in Australia.

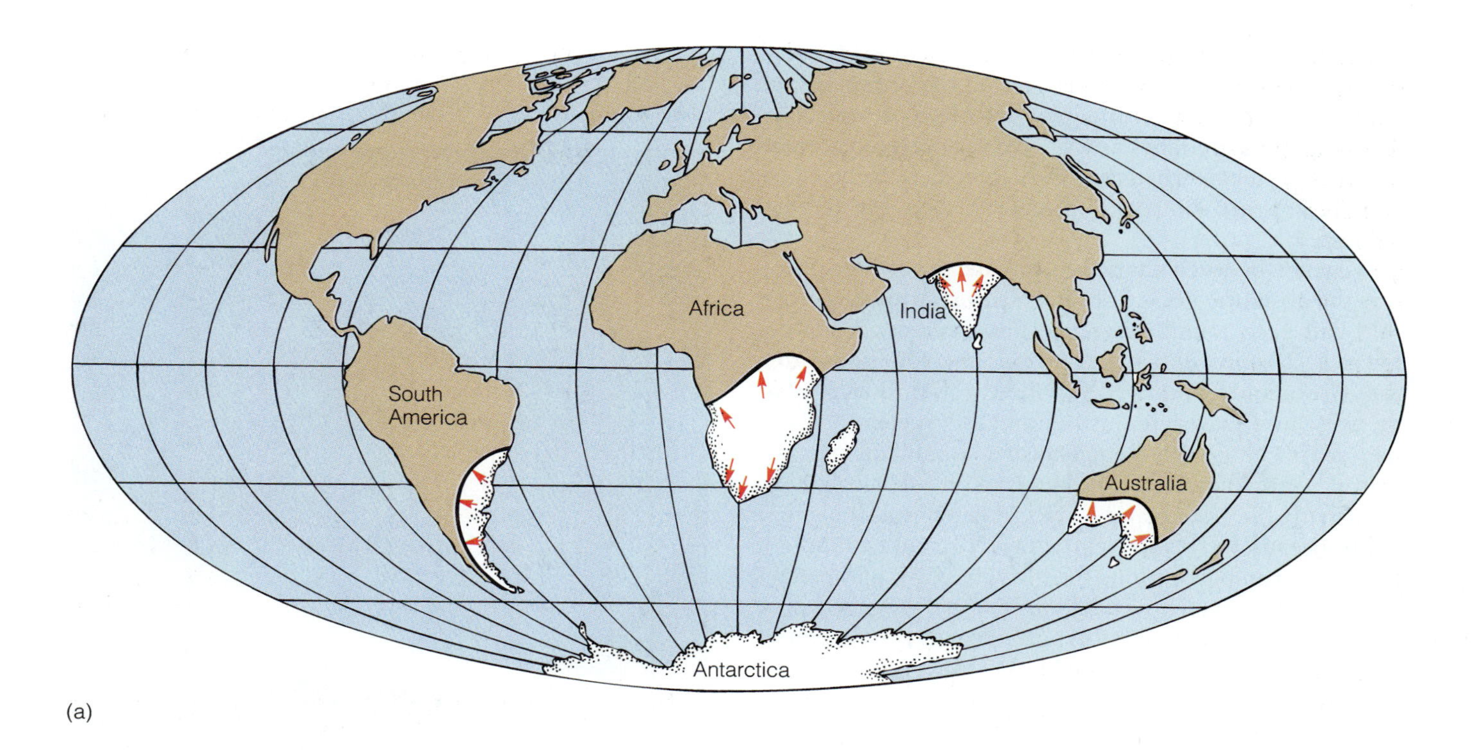

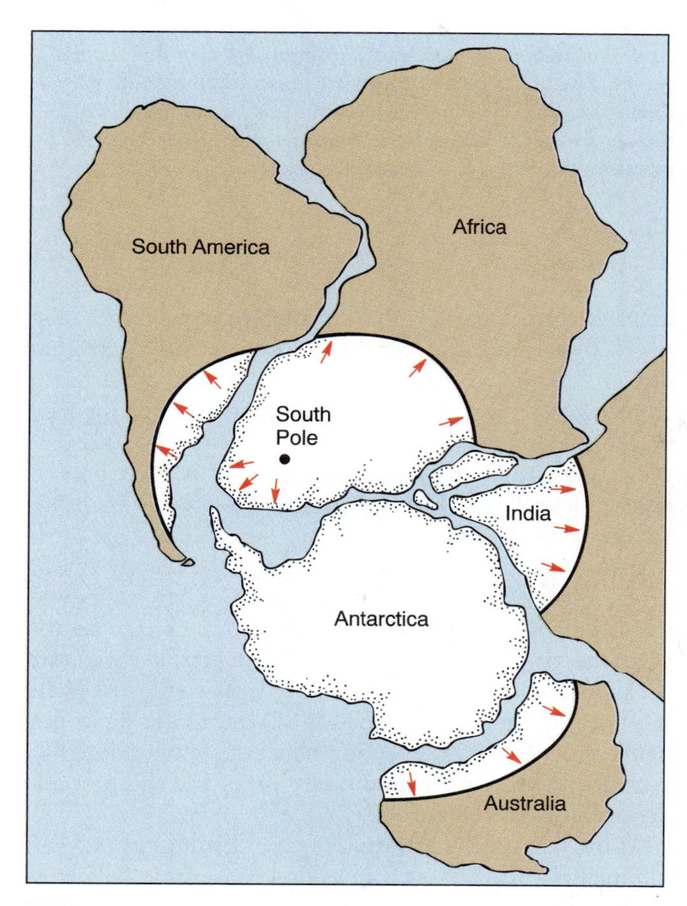

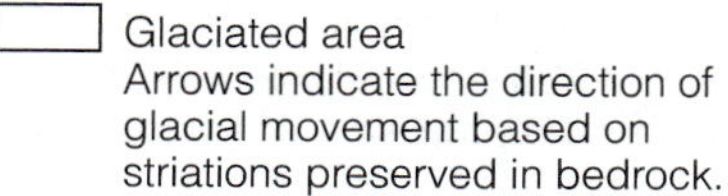

(b)

(c)

➢ FIGURE 3.3 (*a*) The directions of glacial grooves preserved in bedrock on continents as they are now positioned show the direction of glacier movement during Late Paleozoic (Gondwana Succession) time. For a large glacier or a number of glaciers to produce the directions of the observed grooves would otherwise require a source area in the Indian and South Atlantic Oceans. (*b*) With Gondwana continents reunited and an ice sheet over a south pole in South Africa, the directions of glacier motion are resolved. (*c*) Gondwana glacial grooves at Hallet's Cove, Australia. These grooves were formed in Permian time and are more than 200 million years old.

swam or island-hopped across the salty Atlantic Ocean to find an identical freshwater habitat on the opposite side. It seems much more reasonable that *Mesosaurus* inhabited large Permian lakes and/or rivers that spanned parts of both continents when they were joined, and when Pangaea split apart the rocks containing their fossils were separated.

Wegener believed that the ultimate proof of drifting was the absolute motion of Greenland relative to Europe that had been found. He relied upon surveys made between 1823 and 1870—later proved in error—indicating that Greenland had moved 420 meters (one-fourth mi) to the west in 47 years (an astounding 8.9 meters or 29 ft per year). Nonetheless, persuasive arguments against drifting were founded on the lack of a known mechanism for driving the system and on the rather simple logic that if the oceans were sufficiently rigid to cause mountain ranges to crumple up along the forward edges of the moving continents, then how could they *move*. Wegener spent his entire life defending his hypothesis in scientific papers and five editions of his classic book *The Origin of Continents and Oceans*. He died on an expedition to Greenland where, it is said, the theory was born 30 years earlier when he saw a giant iceberg float away from a calving glacier. Unfortunately, he was never able to reconcile the forces necessary to drive continental drift and his hypothesis gained relatively few adherents until thirty years after his death. Nevertheless, there persisted the nagging correlations of fossils, mountain ranges, and ages of rocks and geologic provinces across the Atlantic and Indian Oceans. Wegener deserves our admiration for his dogged determination, scientific insight, and ability to synthesize seemingly disparate data. In order to evaluate Wegener's attractive but controversial concept, one requires some knowledge of the structure and composition of the earth's interior. We examine this subject next.

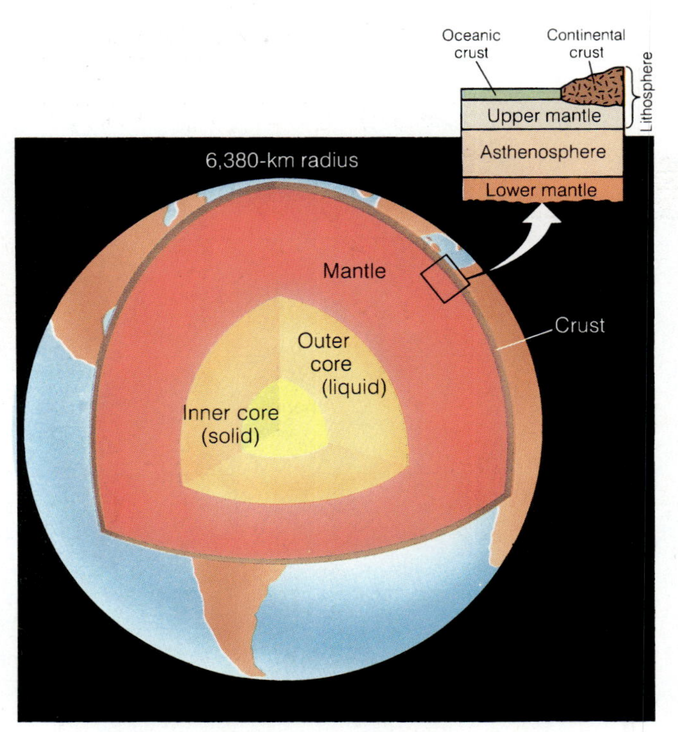

➢ **FIGURE 3.4 The internal structure of the earth. The lower part of the figure shows the layers that are defined by contrasts in density, as deduced from observations of seismic waves. The upper part of the figure illustrates the relative thickness of the crust in continental areas as compared with the sea floor. The lithosphere composes the rigid tectonic plates moving on a plastic asthenosphere.**

Structure of the Earth

The interior of the earth is differentiated into distinct layers separated by boundaries called **discontinuities;** see ➢ Figure 3.4. The model of the earth's interior in the figure is not speculation. Our "X-ray" vision is provided by earthquake (seismic) waves, which give us clues to the physical and chemical nature of the earth's interior. Although seismic waves cannot tell us exactly what composes the earth hundreds of kilometers below its surface, their velocities and travel paths give us information about the physical properties (density, elasticity, etc.) of the materials, from which we can make deductions about their composition. The **lithosphere** is the solid, rigid part of the earth comprising the crust and the upper mantle. It extends 100–150 kilometers (60–90 mi) below the surface of the continents and is 70–80 kilometers (40–50 mi) thick beneath the oceans. What we regard as the plates in plate tectonic theory are composed of lithosphere. The crust, or upper part of the lithosphere, consists of relatively light (granitic) continental crust and somewhat denser (basaltic) oceanic crust. Continental crust has a density of about 2.7 grams per cubic centimeter (gm/cc) and stands topographically above the denser oceanic (basalt) crust with a density of 3.0 gm/cc. This "floating" of lighter (less dense) lithospheric blocks on denser lithosphere is known as **isostasy.** At the base of the crust is the **Mohorovičić discontinuity** (usually called simply the *M* or the *Moho*), which is indicated by a drastic change in the behavior of seismic waves due to a change in the chemical or physical nature of the material (➢ Figure 3.5). The thickness of the crust and depth to the "M" varies, but it is about 35 kilometers (22 mi) under the continents and 10 kilometers (6 mi) beneath the sea floor. The crust is thickest under mountain ranges and thinnest under ocean basins (Figure 3.5).

At the base of the lithosphere, at a depth of about 100 kilometers (60 mi), is a zone of low seismic velocities known as the *plastic layer,* or the **asthenosphere.** It is here that the rigid lithosphere uncouples, or detaches itself, from the underlying mantle and moves about. What drives the plates is poorly understood, but it is most cer-

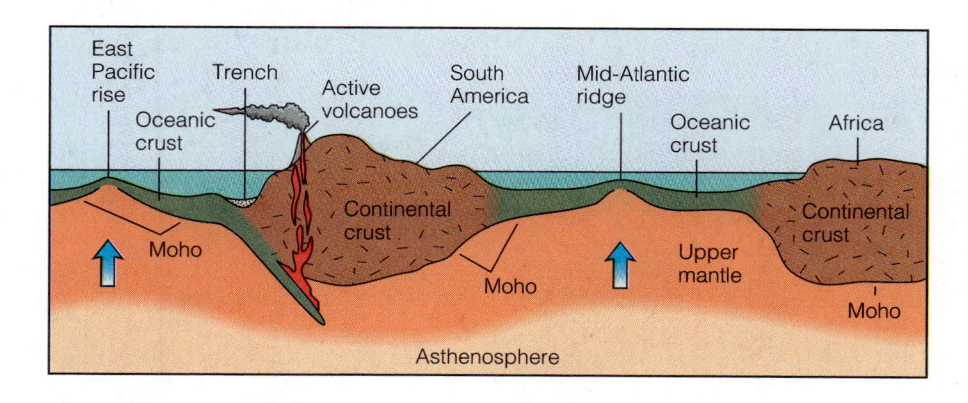

➢ **FIGURE 3.5 Schematic section from Africa through South America to the Pacific Ocean, showing oceanic and continental crust and the "M" discontinuity (not to scale). Mid-ocean ridges and a trench are shown as they are found in this region.**

tainly a combination of gravity acting on the lithosphere and thermal differences within the mantle. Thermal convection—analogous to warm air rising and cool air sinking in closed loops in our atmosphere—has been suggested. (We can produce a similar, observable convective phenomenon by carefully pouring cold milk into hot coffee or by heating tomato soup over the burner on a kitchen stove.) Although thermal convection does not adequately explain all the nuances of plate tectonics, it serves as a first approximation of the driving force. To complete the picture of the earth's interior, we find a very dense molten core and a solid inner core, both believed to be composed of a nickel–iron alloy. This composition is suggested by theory and by analogy to nickel–iron meteorites that strike the earth, which may represent chunks of a disintegrated planetary body.

Sea-Floor Spreading

U.S. geologists Harry Hess and Robert Dietz independently concluded in 1960 that new sea floor was being created at mid-ocean ridges by volcanic activity. Inasmuch as the earth was not growing larger, they said, an equal amount of oceanic crust is probably being lost at trenches (➢ Figure 3.6). They envisioned the driving force to be convection currents in the mantle caused by heat differences within the earth and gravity, an explanation that had been proposed 15 years earlier by British geologist Arthur Holmes. According to this theory, convection carries the rigid new crust away from the mid-ocean ridges like a conveyor belt and drives it into the mantle at trenches. This new theory, called **sea-floor spreading** by Dietz, explained the distribution of earthquakes and volcanoes on earth quite nicely.

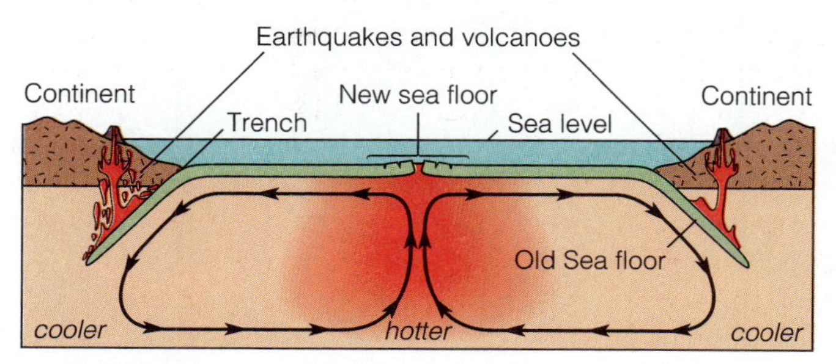

➢ **FIGURE 3.6 Sea-floor spreading driven by convection currents. New crust forms at ridges and is consumed at trenches. Volcanic activity and earthquakes are generated along the sinking oceanic crust.**

Plate Tectonics: The Global Theory

Plate tectonics is a unifying concept in that it provides a coherent model of the outer part of the earth and how it works. According to the theory, the outer part of the earth is composed of seven large and a number of small interacting plates. There are only three ways these plates can interact and, therefore, three types of boundaries: divergent, convergent, and transform. Each of the three boundaries are produced by different stress fields. At **divergent boundaries** there is *tension,* stresses acting in opposite directions away from bodies that tend to pull them apart. At **convergent boundaries** there is *compression,* stresses that converge in the same plane on bodies and tend to shorten or compress them. At transform boundaries there are *shear* forces acting in opposite directions in different planes that cause adjacent parts of a body to slide relative to each other in a direction parallel to a fracture or plane of contact (➢ Figure 3.7).

Where plates move apart, openings are created and molten material fills them, forming new oceanic lithosphere. These boundaries, divergent boundaries, are also known as **spreading centers,** and they occur at mid-ocean ridges. Here we find high heat flow, usually mild volcanic activity, and shallow (< 40 kilometers or 25 mi deep) earthquakes.

Where plates come together or collide, crust is destroyed. These zones are known as convergent boundaries. One form of convergence occurs where an oceanic plate is driven beneath another oceanic plate or a continental plate; this process is called **subduction.** Where

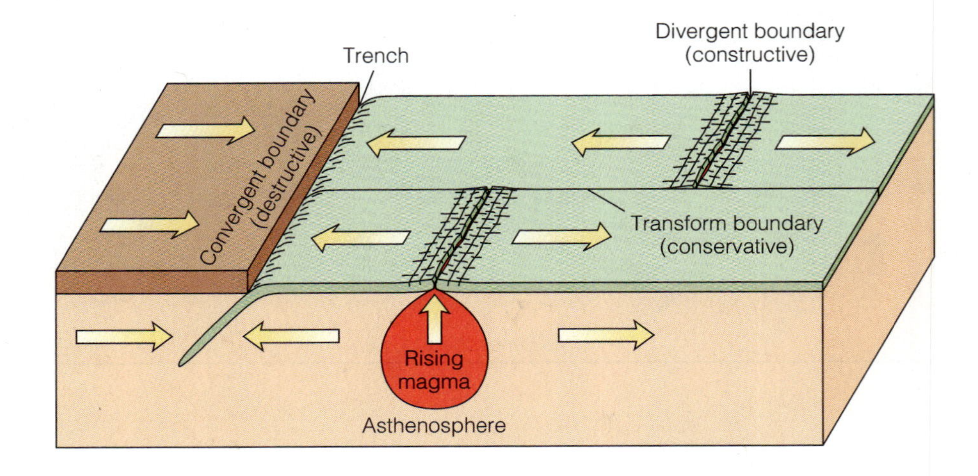

➢ **FIGURE 3.7 Divergent, convergent, and transform plate boundaries. Each type of boundary is characterized by a particular type of plate motion and stress.**

two continental plates collide, mountain ranges may form. The Himalayas formed this way when the sediments forming them were caught in a "vice" between the Indian plate and the Eurasian plate. There we find strong earthquakes but no true subduction, because the lighter continental crust cannot sink into denser lithosphere. Where there *is* subduction, the descending plates may be identified by a zone of earthquake **foci** (points within the earth where the earthquakes originate) inclined at an angle of about 60 degrees to the earth's surface. This zone was discovered by Japanese seismologists in 1927 and is called the **Benioff zone,** in honor of Hugo Benioff who mapped the zones worldwide. Only at subduction zones do we find earthquake foci greater than 400 kilometers (250 mi) deep. This fact is cited as evidence that one rigid plate is sinking beneath another, thereby creating earthquakes at great depth. Subduction is most commonly associated with volcanic island arcs, such as the Japanese and Aleutian Islands, and the most destructive earthquakes and explosive volcanoes occur at these boundaries. (Earthquakes and volcanoes are the subjects of Chapters 4 and 5.)

At **transform boundaries,** plates slide past one another along faults, and crust is neither created nor destroyed. Also known as **transform faults,** we find that portions of mid-ocean ridges and trenches are offset or that the fault connects a ridge and a trench. At the San Andreas fault, an example of a transform boundary, we find only earthquakes and no volcanic activity directly related to the boundary (➢ Figure 3.8). Regardless of the type of boundary, earthquakes and their associated epicenters (the points at the surface of the earth that are directly over earthquake foci) outline the major plates (➢ Figure 3.9).

Plate boundaries may also be described by what happens to the crust at that plate margin rather than the relative direction of plate motion. Where plates diverge at a spreading center, new ocean crust is created, and we have a *constructive* plate boundary. A *destructive* boundary is one where an oceanic plate sinks and disappears at a subduction zone. Finally, at transform boundaries crust is neither created nor destroyed, and we have a *conservative* boundary. Because these designations describe how the margin is modified, they also serve as a memory aid for what processes are taking place there.

Proof of Plate Tectonics

The essential evidence for sea-floor spreading and plate tectonics was to be found by studying a totally unrelated phenomenon, the magnetic forces within the earth. This study determines the direction and strength of the earth's ancient "fossil" magnetic fields—its **paleomagnetism—**by measuring the **polarity** of magnetic minerals contained in ancient rocks, preferably rocks of known age. You see, the earth is essentially a magnetic dipole, like a bar magnet, that has a positive (north) pole and negative (south)

➢ **FIGURE 3.8 The San Andreas fault is a transform fault and is one of the plate boundary types between the North American and Pacific plates.**

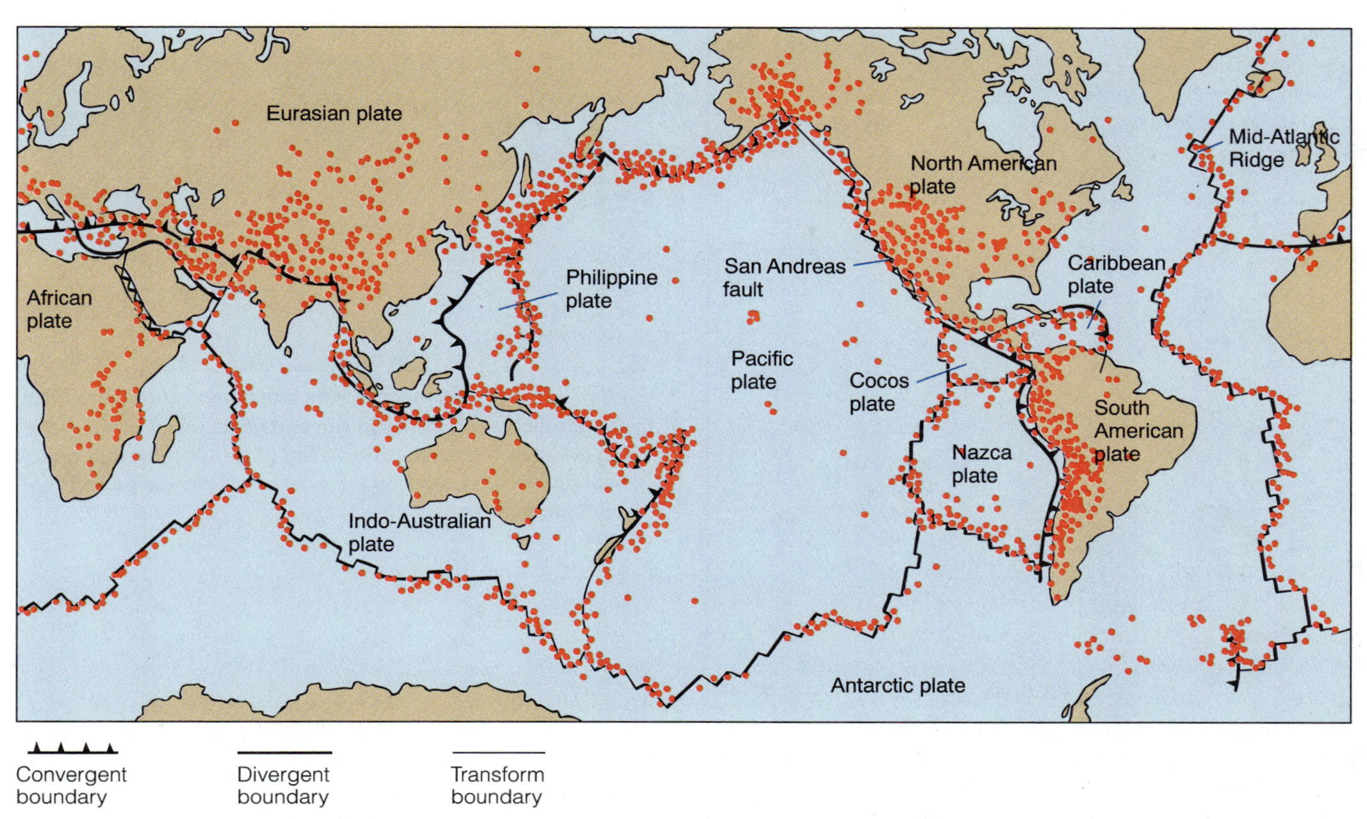

➢ **FIGURE 3.9 The earth's major tectonic plates and several smaller ones. Note that the plates are outlined by earthquake epicenters, each dot on the map representing an epicenter.**

pole. *Polarity* refers to the orientation of a magnet's positive or negative pole that is exhibited when the earth's north magnetic pole attracts the north-seeking end of a compass needle. Studies in the mid-1960s indicated that the earth's magnetic field reversed itself periodically throughout geologic time; that is, the north magnetic pole became the south magnetic pole, and vice versa. Rocks that formed when the earth's magnetic field was the same as today's field yield a strong magnetic signal, or *positive anomaly,* whereas those that formed when the field was the opposite of today's field (reversed) yield a weak signal, a *negative anomaly.* An **anomaly** is a departure from what is expected or from the mean. These anomalous signals can be mapped one sample at a time in a laboratory or on a large scale with an airborne or shipboard instrument called a *magnetometer.* We now know that during the past 5 million years the earth's magnetic poles have flip-flopped four times. The long periods of normal and reversed polarity are called *magnetic epochs,* and within the epochs there were shorter reversals, called *magnetic events* (➢ Figure 3.10).

Oceanographers and geologists had long been puzzled by the striped magnetic-anomaly patterns mapped in sea-floor rocks that parallel mid-ocean ridges (➢ Figure 3.11). Based on U.S. geologists' findings on the timing of magnetic reversals, British researchers reasoned that new sea floor created at mid-ocean ridges should preserve the magnetic field of the time it was formed and that, if sea-floor spreading had occurred, the pattern should change from normal to reversed as one travels away from the ridge. Repeated traverses by ships over several spreading centers were made, and just such a parallel, symmetric pattern of normally and reversely magnetized rocks was confirmed (Figure 3.11). In fact, at least 171 reversals over the past 76 million years are now known. The rate at which the sea floor is moving away from a ridge axis may be calculated by measuring the distance from the ridge to a magnetic anomaly of known age, and then simply dividing the distance by the age of the anomaly. The rate at which the ocean is opening up, becoming larger, is then twice the spreading rate calculated for one side of the ridge, because the spreading movement is presumed to be symmetrical. Plates move at rates on the order of centimeters per year (about the same rate as human fingernail growth), not meters per year as Wegener believed. The East Pacific Rise is spreading at the fastest rate, with a

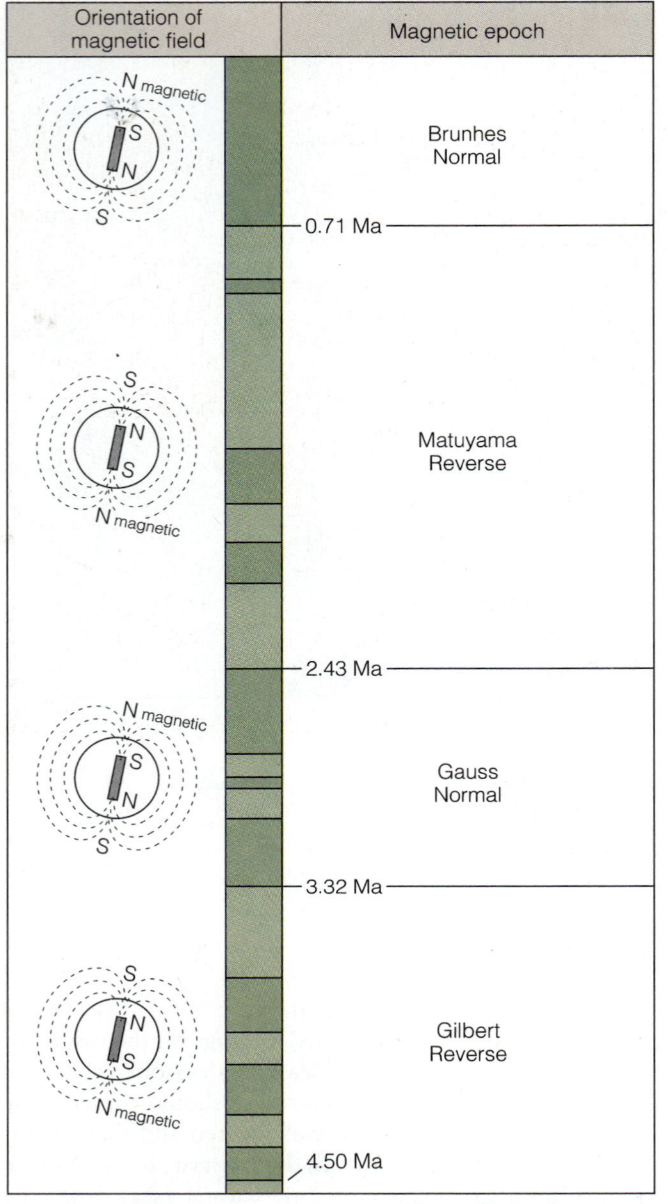

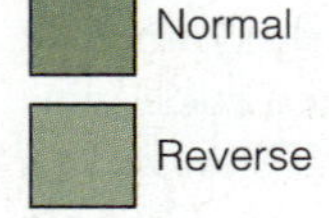

➢ **FIGURE 3.10 The earth's magnetic field is much like the field that would be generated by a bar magnet inclined at 11 degrees from the earth's rotational axis. Reversals of the field have occurred periodically, leaving "fossil" magnetism in rocks that can be dated. Each epoch is named after an important student of earth magnetism.**

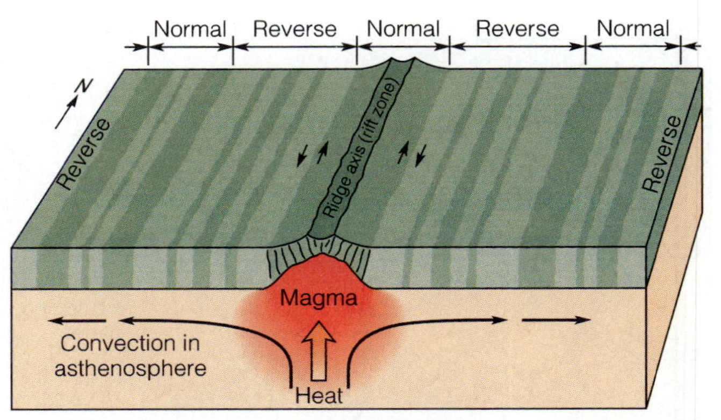

➢ **FIGURE 3.11 Manner in which magnetic anomalies form stripes parallel to a mid-ocean ridge. As lava forms at the ridge axis, it assumes the magnetic polarity of the earth at that time. Normal polarity episodes are dark stripes; reversed polarity episodes are light stripes.**

spreading rate up to 18 centimeters (7 in) per year, whereas the northern Mid-Atlantic Ridge is slowest, at about 2 centimeters (¾ in) per year (➢ Figure 3.12). Recently developed techniques using very-long baseline radio interferometry (VLBI), satellite laser ranging (SLR), and global positioning system (GPS) are providing direct measurements that confirm the theory of plate tectonics. As these techniques' use expands, it will be possible to monitor surface deformation directly, and this will contribute to our understanding of earthquakes.

How Plate Tectonics Works

As can be seen from the model of plate-tectonic movement in ➢ Figure 3.13, plate boundaries determine the distribution and type of volcanic and earthquake activity. This relatively new global theory also explains other facts and features that have puzzled scientists in the past. For instance, at the mid-twentieth century many geologists believed that ocean basins and continents were stable, semipermanent features of the earth's crust. This belief was refuted when ancient sediments could not be found on the sea floor. Only young deposits, no more than 200 million years old (Jurassic age; representing just five percent of geologic time), have been found. Plate tectonics resolves this problem, because it allows that ocean basins have opened and closed many times throughout geologic time as the plates have moved about; therefore, no sediments older than the last opening cycle would be expected. In fact, if we multiply the average spreading rate by the length of the mid-ocean-ridge system, we find that new oceanic crust is formed at a rate of 2.8 km^2 (1.7 sq mi) per year. Because our oceans cover an area of 310

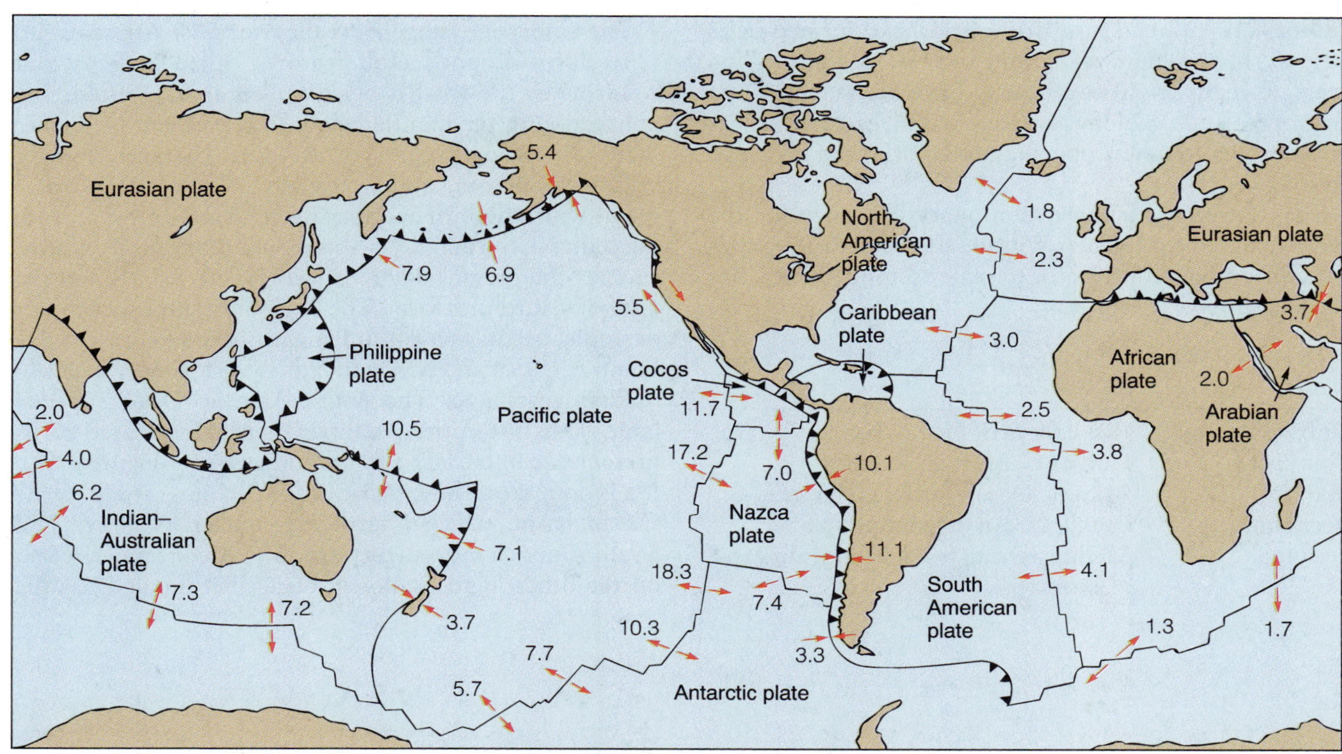

➢ **FIGURE 3.12** **Magnetic anomalies of known age are plotted against the distance from the respective mid-ocean ridge to determine the spreading rate. This map shows the average movement rate (cm/year) and the relative motion of the tectonic plates.**

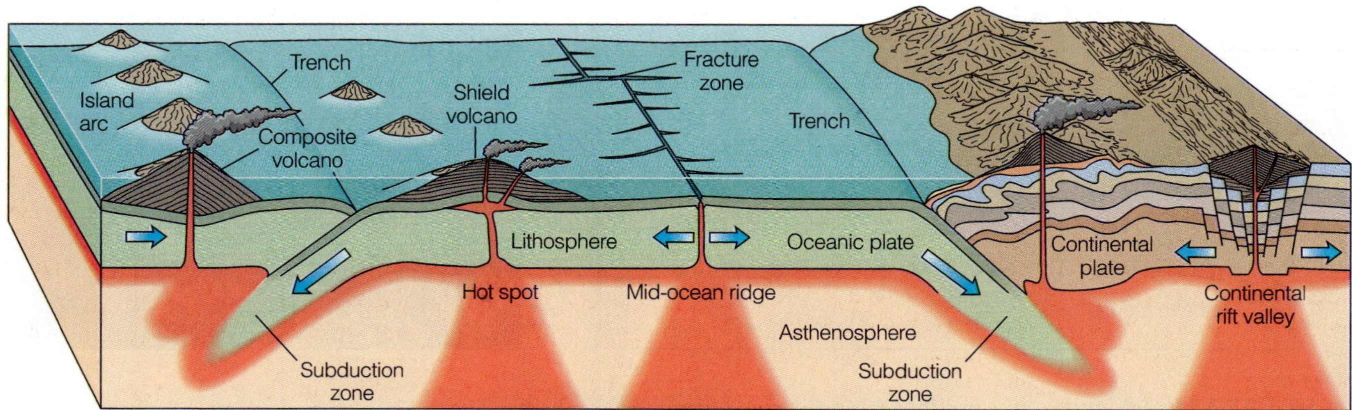

➢ **FIGURE 3.13** **Schematic cross-section illustrating all the elements of the plate tectonic model. Divergent and convergent boundaries, and their relationship to sea-floor and continental volcanoes are shown. Fracture zones on the sea floor are transform boundaries, and the mid-ocean ridge is offset along these features in the diagram.**

million km^2 (120 sq mi), they could have formed in as little as 110 million years, and over the past 2 billion years, as many as 20 oceans may have been created and destroyed! Land geology also supports at least two and as many as five complete openings and closings of all ocean basins.

The *Wilson cycle,* an evolutionary sequence for the oceans named for J. Tuzo Wilson, a Canadian geophysicist and a major contributor to plate tectonic theory, has been developed as follows:

EVOLUTIONARY STAGE	EXAMPLE
Embryonic	Rift valleys of East Africa
Youthful	Gulf of California, Red Sea
Mature	Atlantic Ocean (growing larger)
Declining	Pacific Ocean (becoming smaller)
Terminal	Mediterranean Sea (closing, almost extinct)

The embryonic-stage rift valleys of East Africa exemplify down-dropped fault features called **grabens** that result when the crust is being pulled apart, "rifted," by convection in the mantle below the continent (➢ Figure 3.14). As rifting progresses, the crust becomes thinner, volcanism occurs, and eventually a new ocean basin forms with a floor of oceanic crust and an oceanic spreading center in the middle. In this manner long, narrow seas, such as the present-day Red Sea (Figure 3.14d) and the Gulf of California, form. The Central Atlantic Ocean, for example, began as a rift in Pangaea between present-day South America and Africa in late Triassic time about 180 million years ago. The South Atlantic Ocean opened later. Thus the entire Atlantic basin has matured to its present size in the last 180 million years (➢ Figure 3.15). It has done so at the expense of the declining-stage Pacific Ocean basin, which is growing smaller as North and South America move westward. The Mediterranean Sea, on the other hand, is slowly being "terminated" as the

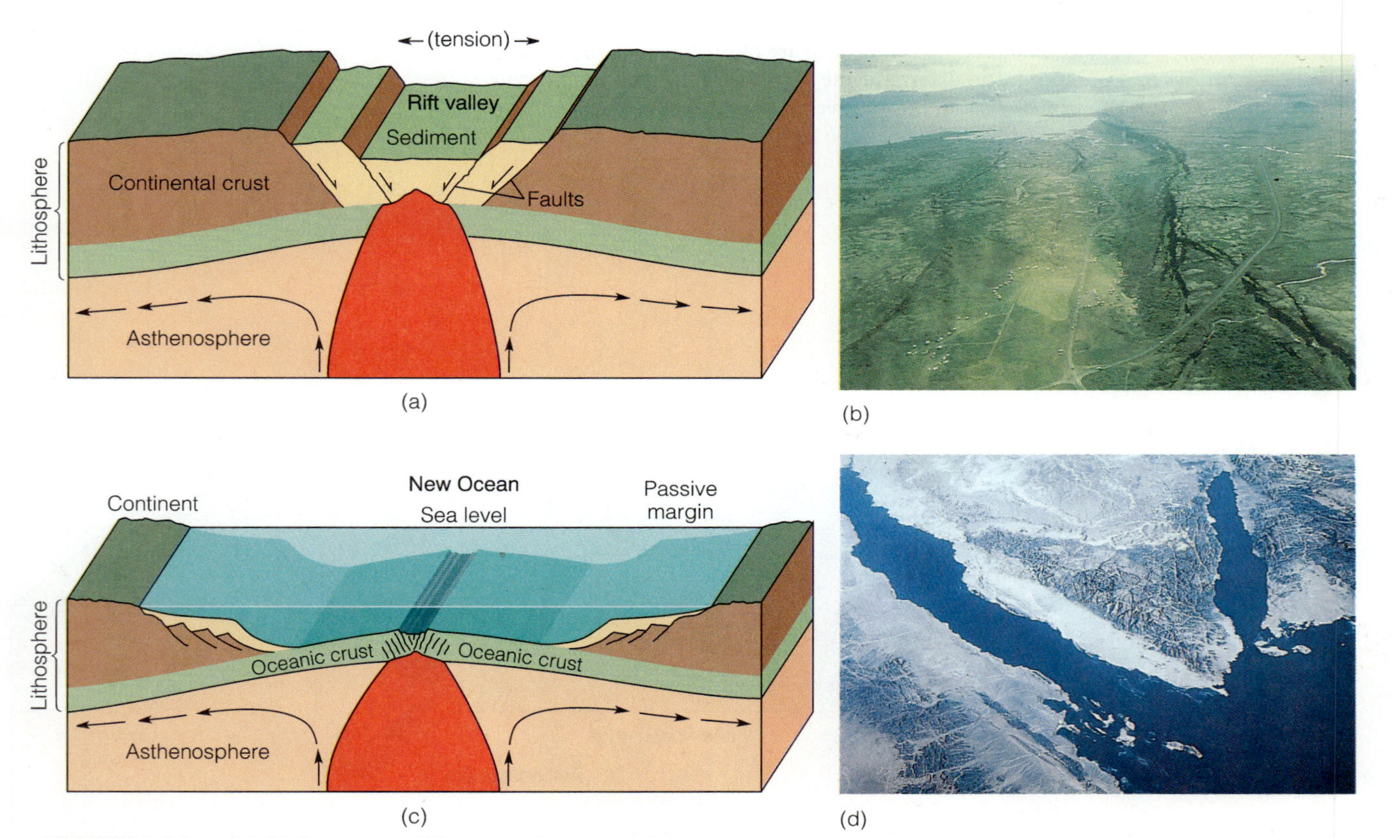

➢ **FIGURE 3.14** **(*a*) Initial stages of rifting (stretching) and formation of a graben like those in East Africa. (*b*) Western boundary of the Thingvellir graben, Iceland, formed by stretching (tension) along the Mid-Atlantic Ridge. This is similar to the mechanism that forms the rift valleys in East Africa. (*c*) An ocean is born as the rift widens and new oceanic crust forms at the spreading center. (*d*) The Red Sea is a spreading center. A youthful ocean basin, it looks today much like the South Atlantic Ocean must have appeared about 170 million years ago.**

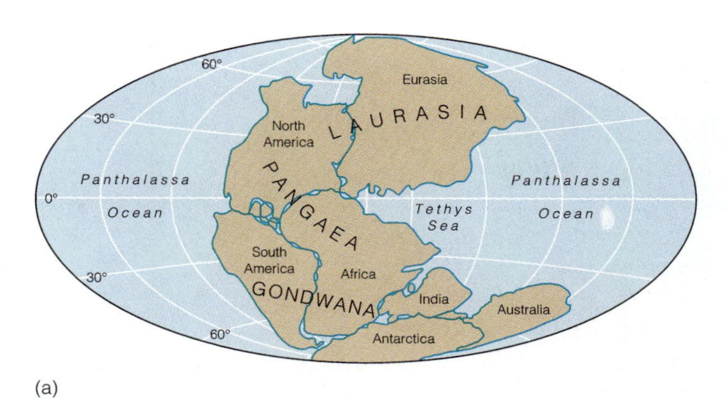

(a)

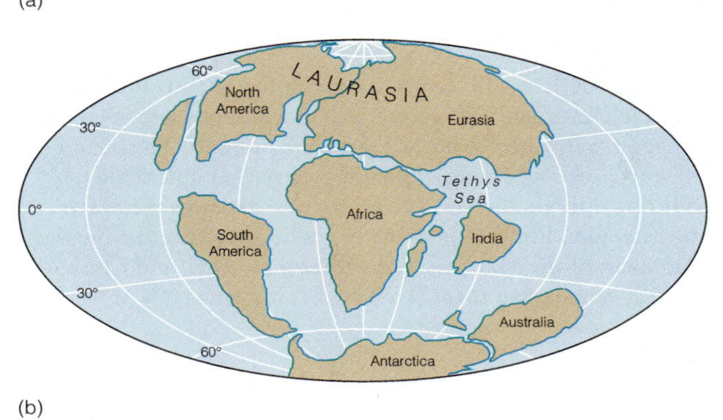

(b)

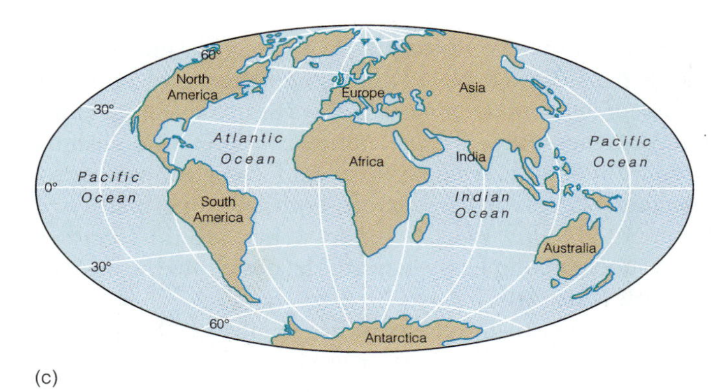

(c)

➢ **FIGURE 3.15** **Paleogeographic reconstruction of continents' movements over the past 180 million years. Panthalassa was the "one" ocean dominating the globe, as Pangaea was the "one" continent. These maps show present-day coastlines, although ancient coastlines did not coincide with these. Compare these maps with those of Figure 3.1 originally published in 1915 and revised by Wegener in 1929 just before his death. (*a*) Early Triassic Period, (*b*) Late Cretaceous Period, and (*c*) as it is today.**

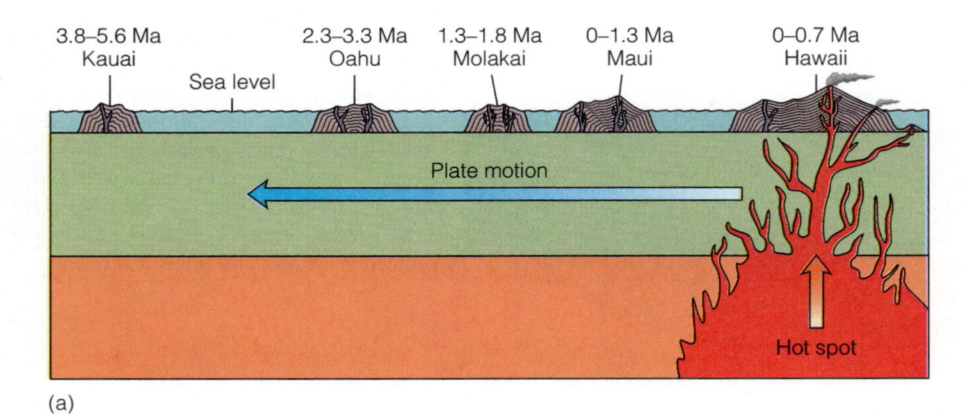

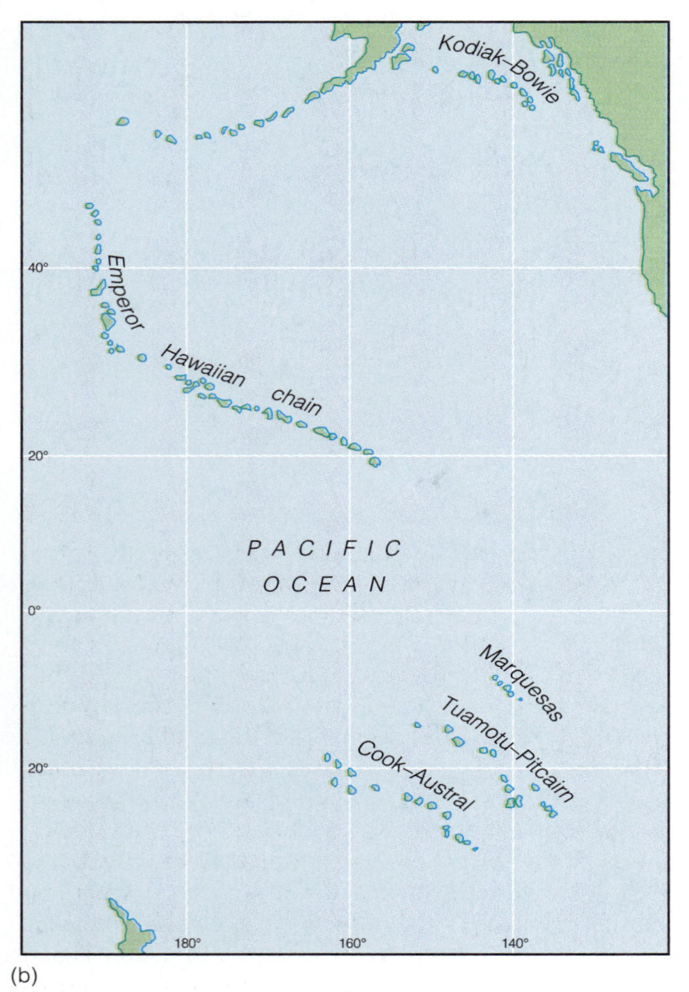

➢ **FIGURE 3.16** (*a*) **Cross section of five islands formed over the Hawaiian "hot spot." The Hawaiian Islands become systematically older in a northwest direction.** (*b*) **The bend in the line at the beginning of the Emperor Seamount chain indicates a more northerly direction of plate movement over the Hawaiian hot spot prior to 45 million years ago. Note the alignment of the Marquesas, Tuamotu–Pitcairn, and Cook–Austral Islands that formed over other hot spots.**

African and Eurasian plates converge on the basin. Evidence for this closure is found around the Mediterranean basin, where we find strong earthquakes, active volcanoes, and the east–west-trending High Atlas and Alpine mountain ranges.

A more dramatic evolution is apparent in Asia. There the Indian plate moved thousands of kilometers northward at about 5 centimeters (2 in) per year and collided with the Eurasian plate, folding, faulting, and uplifting the sediments of the ancient Tethys Sea to form the Himalaya Mountains. Where continent–continent collisions occur, mountain ranges form and catastrophic earthquakes occur. This is because both colliding plates are composed of low-density granitic crust, and thus, neither can subduct completely beneath the other into the heavier mantle (Case Study 3.1).

Within ocean basins there are chains of aligned volcanic islands that become progressively older in one direction. The Hawaiian Islands, Line Islands, and Tuamotus are such chains in the Pacific Ocean. Although the Galápagos Islands off the coast of Ecuador—made famous by the writing of Charles Darwin—are only roughly aligned, they are of similar origin. These lines-of-islands are not associated with plate boundaries, but have formed over rising plumes of lava, or **hot spots,** in the mantle. These rising intraplate plumes penetrate the general convective circulation in the mantle that drives plate motion and are independent of spreading centers. As a tectonic plate moves over one of these plumes, it carries with it the volcano that formed there from the plume, and a new volcano forms on the sea floor in its place. A line of extinct volcanoes forms over time, with their ages increasing with distance from the hot spot. In the case of the Hawaiian chain, each island formed over the stationary hot spot and was then carried northwestward on the moving lithospheric plate (➢ Figure 3.16a and b). The Island of Hawaii is the youngest of the Hawaiian volcanoes and is the only one above a hot spot with active volcanism

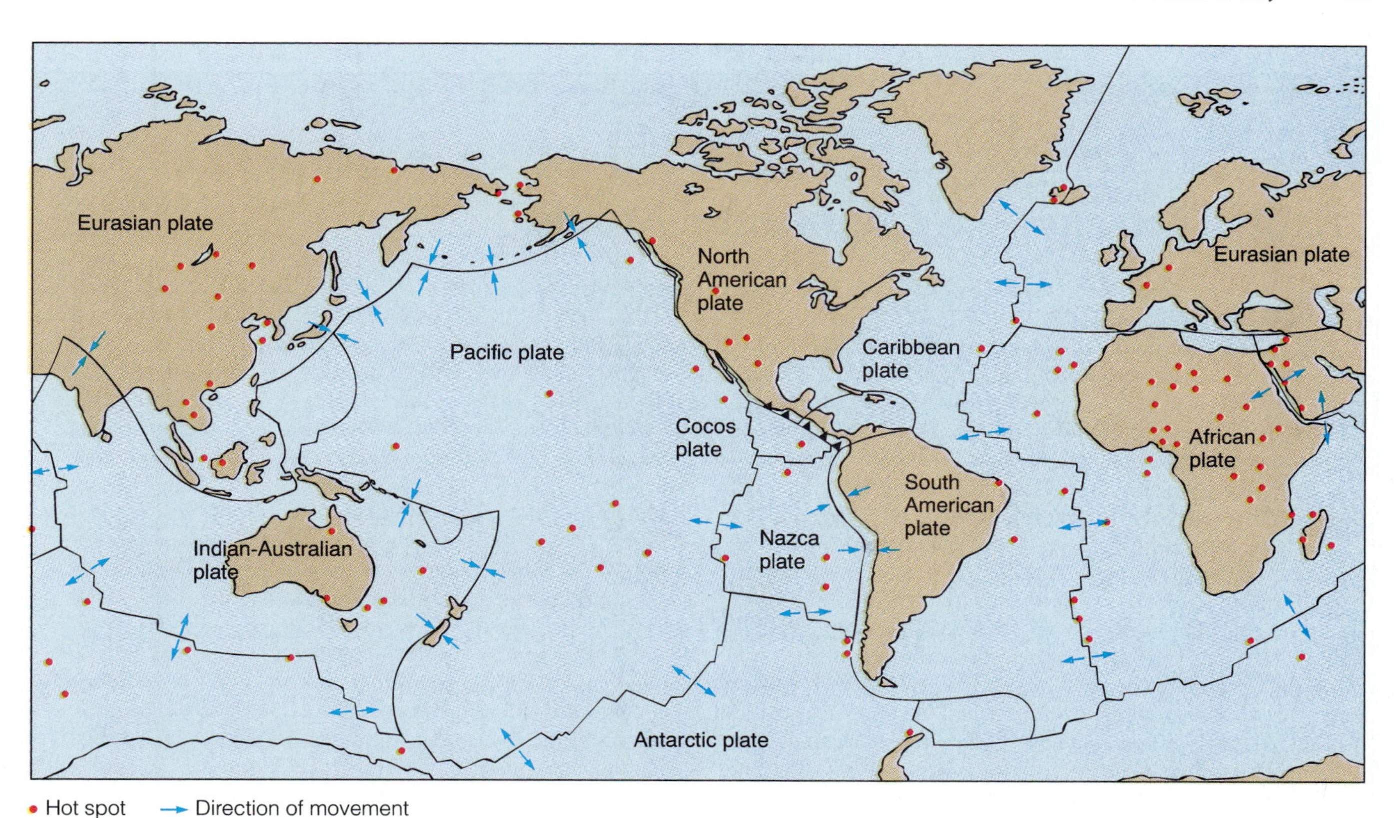

➢ **FIGURE 3.17 The earth's hot spots and their relationship to plates and the plates' direction of movement.**

today. A line drawn from the Island of Hawaii to Midway Island, the oldest of the chain, would represent an arrow marking the direction and rate of movement of the Pacific plate in this region. The track of the Hawaiian Islands forms a "lazy L" with the Emperor Seamount chain, a line of submerged volcanoes that extends northward from Midway and traces a different direction of plate movement prior to 45 million years ago. Of greater environmental concern because of population densities are hot spots that occur beneath the continents. Yellowstone National Park, for example, is situated above a midplate hot spot with a potential for destructive volcanic activity. More than 120 hot spots that have been active within the last 10 million years have been identified, and about a third of them penetrate continental areas (➢ Figure 3.17). Almost 40 volcanic plumes underlie the African plate, suggesting that it is moving very slowly or may even be stationary.

Much of the sea-floor evidence that supports the plate tectonics theory comes from the Deep-Sea Drilling Project that was funded by the National Science Foundation and implemented aboard the research drill ship *Glomar Challenger*. During a series of cruises in the Mediterranean Sea between 1970 and 1975, thick salt deposits were discovered beneath the sea floor that indicated that about 10 million years ago (Miocene time), the basin was isolated from the Atlantic Ocean and that the sea totally evaporated. Access of Atlantic Ocean water to the Mediterranean through the Strait of Gibraltar was cut off by uplift as the African plate moved northward, and also by a global lowering of sea level. Further drilling yielded sediments indicating that the floors of the two Mediterranean basins once resembled the great dry lakes of present-day deserts in the U.S. West. Occasionally they were occupied by water when the connection with the Atlantic was periodically reestablished, but generally they were lifeless salt flats. The Mediterranean must have looked as Death Valley does today from a high vantage point, except that major rivers such as the Rhone and Danube cascaded down the continental slopes to the flats below (➢ Figure 3.18a). About five million years ago the Strait of Gibraltar reopened, and the Atlantic Ocean flowed into the Mediterranean Sea as a waterfall or cascade of great magnitude (➢ Figure 3.18b). This inflow filled the Mediterranean in a few thousand years. The sediments on top of the five-million-year-old salt-flat deposits contain fossils of marine organisms, indicating rapid filling with normal sea water, and no abnormal salinity, which would have occurred if the filling had been slow.

Plate tectonics theory unifies and incorporates the basic concepts of continental drift and sea-floor spreading. In one way or another, plate tectonics explains why nearly

CASE STUDY 3.1

Exotic Terranes—A Continental Mosaic

Because continental plates divide, collide, and slip along faults, it is not surprising that bits and pieces of continental crust existing within the oceans ultimately collide with continents at subduction zones and become stuck there. Because of these pieces' low density, they stand high above oceanic crust (due to isostatic equilibrium), and these microplates or microcontinents become plastered, *accreted*, to larger continental plates when they collide with them. The accreted plates are known as **terranes**, defined as fault-bounded blocks of rock with histories quite different from those of adjacent rocks or terranes.* In size they may be several thousand square kilometers or just a few tens of square kilometers, and they may become part of a continent composed of many terranes. Oceanic crust and the sediment resting on it may be scraped up onto the continents. Terranes may consist of almost any type of rock, but all terranes are fault-bounded, have a paleomagnetic signature that indicates a distant origin, and have little in common with adjacent terranes or with the continental **craton**—the continent's stable core.

The terrane concept developed in Alaska when geologic mapping for land use planning revealed that the usually predictable pattern of rocks and structures was not valid for any distance. In fact, the rocks the geologists found a few kilometers away were almost always of a "wrong" composition and age. Further studies showed that Alaska is a collection of microplates (terranes)—tectonic flotsam and jetsam, if you wish—that have mashed together over the past 160 million years and that are still arriving from the south. One block, the Wrangellia terrane, was an island during Triassic time and it has a paleomagnetic signature of rocks that formed 16° from the equator. It is not known whether Wrangellia formed north or south of the equator, because it is not known whether the magnetic field at the time was normal or reversed. In either case, the terrane traveled a long distance to become part of present-day Alaska. It now appears that about 25 percent of the western edge of North America, from Alaska to Baja California, formed in this way; that is, by bits and pieces being grafted onto the core of the continent and thus enlarging it (➢ Figure 1). In the distant future, part of California may become an exotic terrane of Alaska.

If you've deduced that terrane accretion onto the continents (also known as *docking*) is a characteristic of active continental margins, you are right (➢ Figure 2). The east coast of North America is a rifted (pulled-apart), or *passive*, margin. Material is added to it by river sediment forming flat-lying sedimentary rocks that become part of the continent. Such rocks are not subject to the mountain-building forces of active margins and they accumulate in thick, undisturbed sequences.

**Terrane*, as noted, is a geological term describing the area or surface over which a particular rock or group of rocks is prevalent. *Terrain* is a geographical term referring to the topography or physical features of a tract of land.

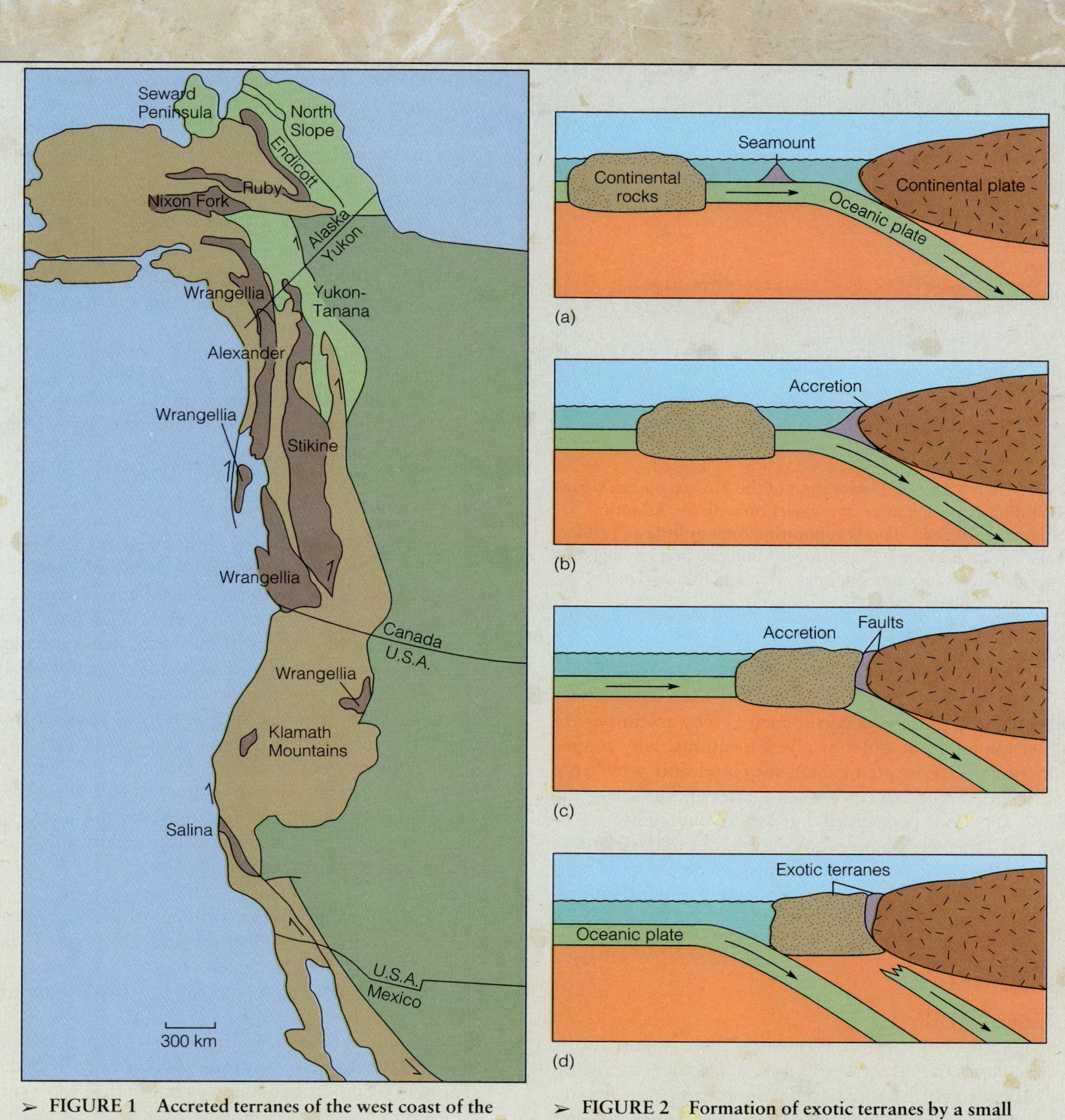

➢ FIGURE 1 Accreted terranes of the west coast of the United States and Canada. The Salina, Wrangellia, Alexander, Stikine, Klamath Mountains, Ruby, and Nixon Fork blocks are accreted terranes that were probably once parts of other continents and have been displaced long distances. The light-green areas are probably displaced parts of North America, and the dark-green area is the stable North American craton. The brown areas represent rocks that have not traveled great distances from their place of origin.

➢ FIGURE 2 Formation of exotic terranes by a small continental mass and a seamount being scraped off the subducting plate and onto the continental mass. Much of the west coasts of the U.S. and Canada have grown in this manner.

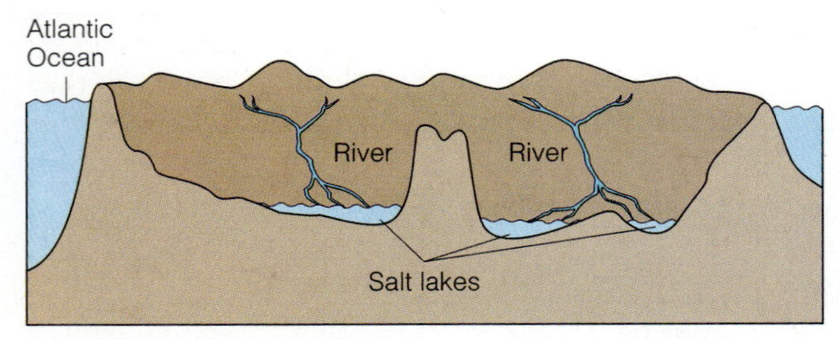

(a) 10 Ma DESICATION

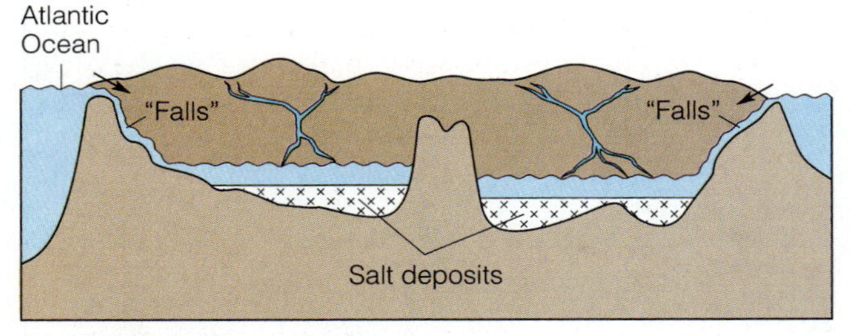

(b) 5 Ma FILLING

➢ **FIGURE 3.18** ***(a)*** **Desiccation of the Mediterranean basin 10 million years ago after its connection with the Atlantic Ocean was blocked. Total evaporation took as little as 1,000 years.** ***(b)*** **Two-km-thick salt deposits indicate that refilling and evaporation occurred many times. The final filling occurred about five million years ago and was rapid.**

all geologic phenomena occur, such as why volcanoes are either explosive or quiet at given locations, why mountain ranges are located where they are, and why large faults such as the San Andreas in California cut across the landscape and periodically generate damaging earthquakes. The theory has revolutionized our understanding of how the earth works and has provided geologists with a useful model in their searches for new mineral and oil deposits.

SUMMARY

Continental Drift

DESCRIPTION: Alfred Wegener's theory that during the Permian Period there was a single supercontinent, Pangaea, which split apart during Jurassic time into individual continents that have since migrated to their present positions.

CAUSES: The lack of an explanation for a driving force was a weakness of the theory.

EVIDENCE:

1. Similar glacial deposits and plant fossils in Permian rocks (the Gondwana Succession) in Africa, South America, India, Australia, and Antarctica.
2. Similarities between fossils, ages of rocks, and geologic provinces on opposite sides of the Atlantic Ocean.
3. The jigsaw fit of the continents if the Atlantic and Indian Oceans were closed.

Sea-floor Spreading

DESCRIPTION: Theory independently proposed by both Harry Hess and Robert Dietz that new oceanic crust is created at mid-ocean ridges and moves laterally across the ocean basins to trenches, where it sinks and returns to the mantle.

CAUSES: The most widely accepted explanation is that convection cells in the mantle generated by thermal and density differences carry the rigid overlying crust (lithosphere) away from the ridges until the crust cools and descends toward the mantle at oceanic trenches.

EVIDENCE:

1. Distribution of deep-focus earthquakes at trenches.
2. Lack of ancient sediments on the sea floor.
3. Increasing age of sea-floor sediments away from mid-ocean ridges.

Plate Tectonics

DESCRIPTION: Theory that unifies and incorporates the earlier theories of continental drift and sea-floor spreading. According to it, the lithosphere is composed of 7 large plates and a number of smaller ones that move about on the surface of the earth. Plates interact in only three ways: they *collide* at convergent boundaries marked by trenches or mountain ranges; they *move apart* and new sea floor is formed at divergent boundaries marked by mid-ocean ridges; or they *slip by one another* along transform faults.

CAUSE: Convection currents in the asthenosphere (the plastic layer of the upper mantle) carry the plates as passive passengers from divergent boundaries to convergent and conservative boundaries.

EVIDENCE:

1. Land geology, ocean drilling, and geophysical data support past openings and closings of the ocean basins (Wilson cycle).
2. Magnetic anomaly patterns ("stripes") on both sides of mid-ocean ridges.
3. Chains of oceanic volcanoes and seamounts that systematically increase in age with distance from the present locations of active volcanism (hot spots) indicate that plates are moving over fixed plumes of lava coming from deep in the mantle.
4. The geological and geophysical evidence for continental drift and sea-floor spreading also support plate tectonics.

EXPLAINS:

1. The thick salt deposits beneath the Mediterranean Sea and its dessication 10 million years ago.
2. The existence of narrow seas and gulfs such as the Gulf of California and the Red Sea as spreading centers.

3. The location and geology of the Himalaya Mountains.
4. The location of violent earthquakes and volcanoes at convergent margins.

Key Terms

anomaly
asthenosphere
Benioff zone
continental drift
convergent boundary
craton
discontinuity
divergent boundary
foci (*focus*, singular)
Gondwanaland
grabens
hot spot (plume)
isostasy
Laurasia
lithosphere
Mohorovičić discontinuity ("M" or Moho)
paleomagnetism
Pangaea
plate tectonics
polarity
sea-floor spreading
spreading center
subduction
terrane
transform boundary
transform fault

Study Questions

1. What geologic and geographic evidence did Alfred Wegener use to demonstrate continental drift?
2. How do we know that the interior of the earth is hot, and why has it not cooled entirely over the last 4.5 billion years?
3. How do density differences control the movement of fluids? In your own words describe heat transfer by convection currents, giving some illustrations from everyday life.
4. What are the major spheres or shells of the earth's interior based upon density? What is meant by the term *lithosphere?* Describe the lithosphere and tell how it is separated from the deep mantle.
5. What is a tectonic "plate"? Give an example (exact geographic location) of each type of plate boundary. How fast do plates move?
6. What geophysical observations and measurement support the model of sea-floor spreading and plate tectonics?
7. How can the change in direction of a line of oceanic islands—such as the turn between the Hawaiian Islands chain and the Emperor Seamount chain—be explained?
8. Draw a cross section of a convergent plate boundary (a subduction zone) and label the following: trench, volcanic areas, Benioff zone, lithosphere, asthenosphere.
9. How might the distribution of the Permian Gondwana Succession be explained without invoking plate tectonics (ancient landmass, land bridges, etc.)? Might an alternative explanation be more reasonable than drifting continents and plate tectonics?
10. How have the narrow bodies of the Red Sea and the Gulf of California formed? Study a map, seeking an obvious relationship between the Red Sea and the Dead Sea.
11. How might knowledge of past and present plate movement be used to the benefit of humankind?

Further Reading

Ballard, Robert. 1983. *Exploring our living planet.* Washington, D.C.: National Geographic Society.
Cox, Allan. 1986. *Plate tectonics: How it works.* Palo Alto, Calif.: Blackwell Scientific Publications.
Hallam, A. 1973. *A revolution in the earth sciences.* Oxford, U.K.: Oxford University Press.
Hsu, Ken. 1983. *The Mediterranean was a desert.* Princeton, N.J.: Princeton University Press.
Moores, E. M., ed. 1990. Shaping the earth: tectonics of continents and oceans. *Scientific American readings.* New York: W. H. Freeman and Co.
Sullivan, Walter. 1974. *Continents in motion: The new earth debate.* New York: McGraw Hill Book Co.
Tarling, D., and Maureen Tarling. 1970. *Continental drift: A study of the earth's moving surface.* New York: Anchor Books.
Uyeda, Seiya. 1977. *The new view of the earth.* New York: W. H. Freeman.
Wegener, Alfred. 1929. *The origin of continents and oceans.* New York: Dover Publications, 1966 translation.
Wilson, J. Tuzo. 1972. Continents adrift. *Scientific American readings.* San Francisco: W. H. Freeman.

Video:

National Geographic Society. 1983. *Born of fire,* with WQED, Pittsburgh, 60 minutes.
National Geographic Society. 1980. *Dive to the edge of creation,* Emmy Award–winner, 60 minutes.

CHAPTER

4

EARTHQUAKES AND HUMAN ACTIVITIES

A bad earthquake at once destroys the oldest associations; the world, the very emblem of all that is solid, had moved beneath our feet like crust over fluid; one second of time has created in the mind a strange idea of insecurity, which hours of reflection would not have produced.

CHARLES DARWIN, 1835
CONCEPCIÓN, CHILE

As millions of Americans watched the opening ceremonies of the 17 October 1989 World Series game at San Francisco's Candlestick Park on television, they witnessed the unfolding of another major earthquake in the populous San Francisco–Oakland Bay area. Although strong ground motion lasted less than a minute, the earthquake caused 67 deaths and 3,500 injuries and it left 12,000 persons homeless (➢ Figure 4.1). At least six million people were directly affected by this strongest event to strike the Bay Area since the great earthquake of 1906, when fewer than a million people lived in the area. This event is known as the Loma Prieta ("dark rolling mountain") earthquake, named after the mountain closest to its epicenter about 100 kilometers (60 mi) southeast of San Francisco.

Bridges damaged during the Loma Prieta earthquake of October 18, 1979. The piers that support the bridges sank into a foundation that was liquified by earthquake shaking, causing the bridge lanes to assume a bathtub shape.

➢ **FIGURE 4.1 Life goes on amid damage due to ground failure in the Marina district of San Francisco, 17 October 1989.**

Let us compare the Loma Prieta earthquake to the one that occurred in Alaska in 1964. The Alaskan earthquake produced ground motion *100 times* stronger than that of the Loma Prieta earthquake and caused 131 deaths, but it resulted in only about one-twelfth the dollar loss. This is because Alaska's population density is only 1 person per square mile, whereas as many as 6,000 people live within a square mile in San Francisco. Thus the magnitude of an earthquake is not the only factor that determines its impact on humans. A small earthquake may do much more damage than a large one if it is centered near a concentration of population. Fortunately, neither New York City (26,000 people/mi^2) nor Hong Kong (150,000 people/mi^2) is in "earthquake country."

We have long known that earthquakes are not randomly distributed over the earth. Plate tectonic theory explains that most earthquakes occur at or near plate boundaries and that the strength of an earthquake is closely related to the type of boundary on which it occurs.

THE NATURE OF EARTHQUAKES

Earthquakes are the result of abrupt movements on **faults**—fractures in the earth's lithosphere. The types of faults and the earth forces that cause them are shown in ➢ Figure 4.2. The movements occur as the earth's crustal plates slip past, under, and away from one another. Because the **stress** (force per unit area) that produces **strain** (deformation) can be transmitted long distances in rocks, faults do not necessarily occur exactly on a plate boundary, but they generally occur in the vicinity of one. The mechanism by which stressed rocks store up strain energy along a fault to produce an earthquake was explained by Harold F. Reid after the great San Francisco earthquake of 1906. Reid proposed a mechanism to explain the shaking that resulted from movement on the San Andreas fault, known as the **elastic rebound theory** (see ➢ Figure 4.3). According to this theory, when sufficient strain energy has accumulated in rocks, they may rupture rapidly—just as a rubber band breaks when it is stretched too far—and the stored strain energy is released as vibrations that radiate outward in all directions (➢ Figure 4.4).

Most earthquakes are generated by movements on faults within the crust and upper mantle that do not produce ruptures at the ground surface. We can thus recognize a point within the earth where the fault rupture starts, the **focus,** and the **epicenter,** the point on the earth surface directly above the focus (Figure 4.4). Elastic waves—vibrations—move out spherically in all directions from the focus and strike the surface of the earth. Damaging earthquake foci are generally within a few kilometers of the earth's surface. Deep-focus earthquakes, on the other hand, those whose foci are 300–700 kilometers (190–440 mi) below the surface, do little or no damage. Earthquake foci are not known below 700 kilometers. This indicates that the mantle at that depth behaves plastically due to high temperatures and confining pressures, deforming continuously as ductile substances do, rather than storing up strain energy.

The vibration produced by an earthquake is complex, but it can be described as three distinctly different types of waves (➢ Figure 4.5). Primary waves, **P-waves,** and secondary waves, **S-waves,** are generated at the focus and travel through the interior of earth; thus, they are known as **body waves.** P- and S-waves are so-named because they are the first (primary) and second (secondary) waves to arrive from distant earthquakes. As these body waves strike the earth's surface, they generate long waves, **L-waves,** which travel only at the surface, being analogous to water ripples on a pond.

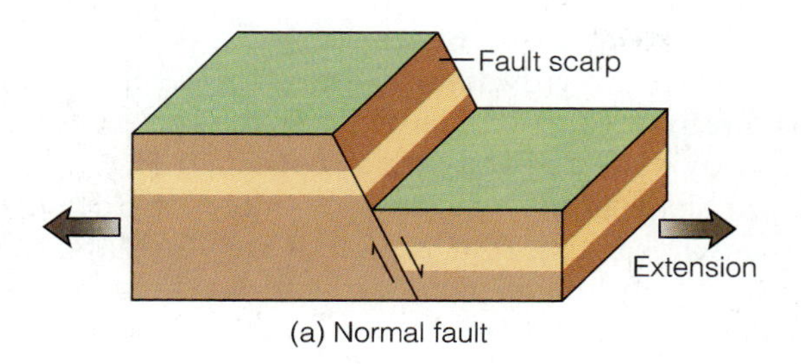

(a) Normal fault

(b)

Compression

(c) Reverse fault

(d)

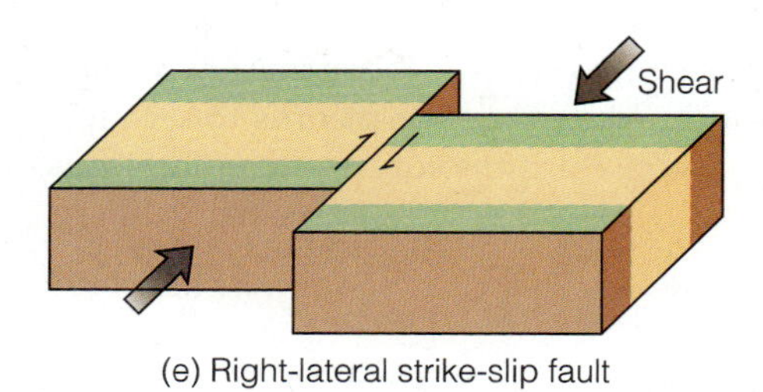

(e) Right-lateral strike-slip fault

(f)

➢ FIGURE 4.2 Three types of faults based on relative displacement and stresses producing faulting. (*a*) Normal fault geometry and (*b*) example in Lower Paleozoic sediments in Big Bend country, Texas. (*c*) Reverse fault geometry and (*d*) example; Death Valley area, California. (*e*) Right-lateral strike-slip fault geometry and (*f*) example; Imperial Valley, California, lettuce field.

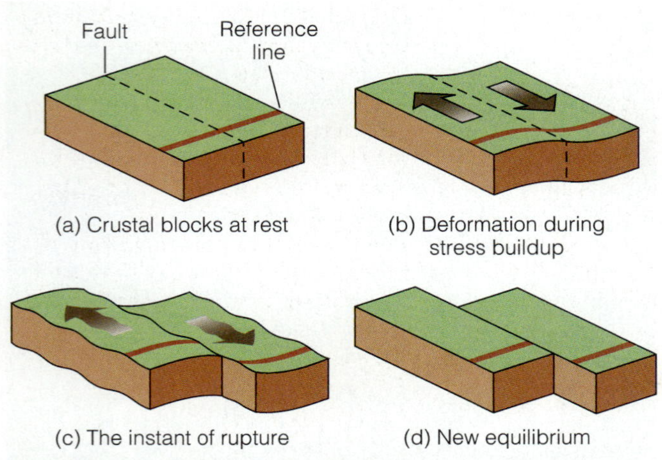

➢ **FIGURE 4.3 The cycle of elastic-strain buildup and release for a right-lateral strike-slip fault according to the elastic rebound theory of earthquakes.**

P-waves (➢ Figure 4.5a) are **longitudinal waves;** the solids, liquids, and gases through which they travel are alternately compressed and expanded in the same direction the waves move. Their velocity depends upon the resistance to changes in volume (compressibility) and shape of the material through which the waves travel. P-waves travel about 300 meters (1000 ft) per second in air, 300–1000 meters (1000–3000 ft) per second in soil, and faster than 5 kilometers (3 mi) per second in solid rock at the surface. P-wave velocity increases with depth in the earth, because the materials composing the mantle and core increase in rigidity with depth. This increasing rigidity causes the waves' travel paths to be bowed downward as they move through the earth. P-waves speed through the earth with a velocity of about 10 kilometers (6 mi) per second. At the ground surface they have very small amplitudes (motion) and cause little property damage. Although they are physically identical to sound waves, they vibrate at frequencies below what the human ear can detect. Earthquake noise has been reported, however, which could be P-waves of a slightly higher frequency or some other vibrations whose frequencies are in the audible range.

S-waves (➢ Figure 4.5b) are **transverse (shear) waves;** they produce ground motion normal (perpendicular) to their direction of travel. This causes the rocks through which they travel to be twisted and sheared. These waves have the shape produced when one end of a garden hose or a loosely hanging rope is given a vigorous flip. They can travel only through material that resists shearing; that is, material that resists two forces acting in opposite directions in different planes. Thus, S-waves travel only through solids, because liquids and gases have no shear strength (try piling water in a mound). S-wave ground motion may be largely in the horizontal plane and can result in considerable property damage. S-wave velocity is approximately 5.2 kilometers (3.2 mi) per second from distant earthquakes; hence they arrive after the P-waves.

Unexpended P- and S-wave energy bouncing off the earth's surface generates the complex surface waves known as L-waves (➢ Figure 4.5c). These waves produce a rolling motion at the ground surface. (In fact, slight motion sickness is a common response to L-waves of long duration.) The waves are the result of a complex interaction of several wave types, the most important of which are *Love waves,* which exhibit horizontal motion normal to the direction of travel, and *Rayleigh waves,* which exhibit retrograde (opposite to the direction of travel) elliptical motion in a plane perpendicular to the ground surface. L-waves may produce large displacements of the ground surface and they are the most destructive of the earthquake waves.

Locating the Epicenter

A **seismograph** is an instrument designed specifically to detect, measure, and record vibrations in the earth's crust (➢ Figure 4.6). It is relatively simple in concept, but some very sophisticated electronic systems have been developed for transmitting earthquake records to seismological stations by radio transmission from remote areas and by

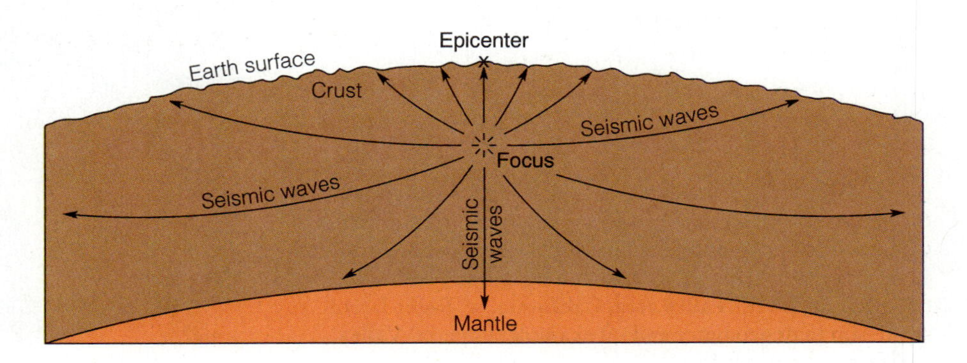

➢ **FIGURE 4.4 Focus, epicenter, and seismic wave paths of a hypothetical earthquake.**

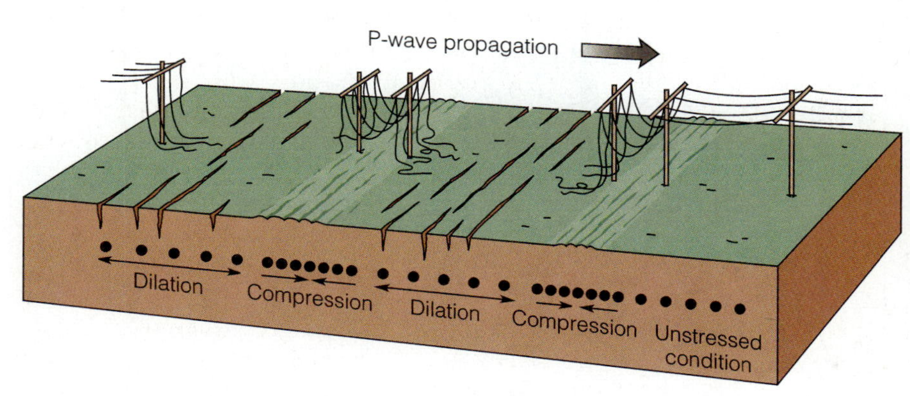

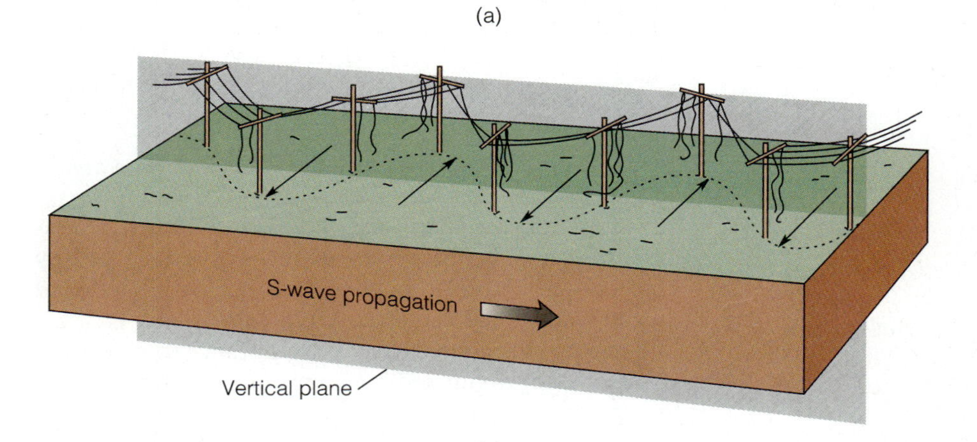

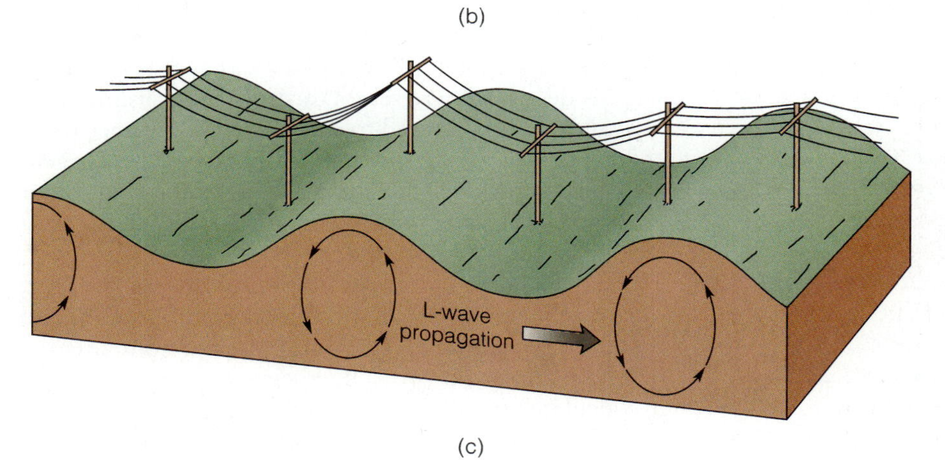

➢ **FIGURE 4.5** **Ground motion during passage of earthquake waves. (*a*) P-waves compress and expand the earth. (*b*) S-waves move in all directions perpendicular to the wave advance, but only the horizontal motion is shown in this diagram. (*c*) L-waves create surface undulations that result from a combination of the retrograde elliptical motions of Rayleigh waves (shown) and Love waves, which move from side to side at right angles to the direction of wave propagation.**

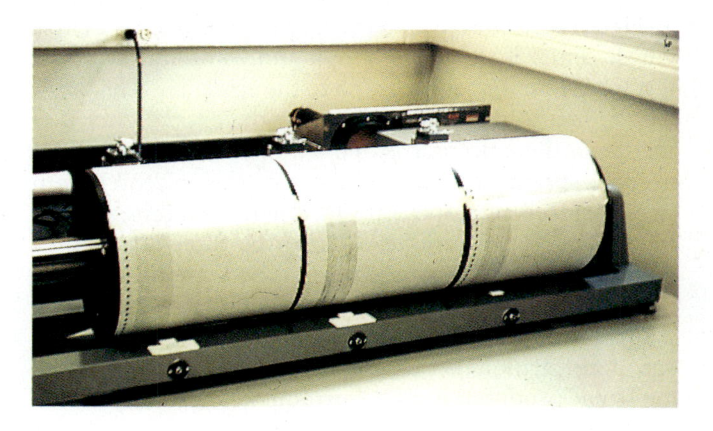

➢ **FIGURE 4.6** **In response to electronic data received by telephone line or radio waves from a seismograph in the field, a recorder traces a seismogram.**

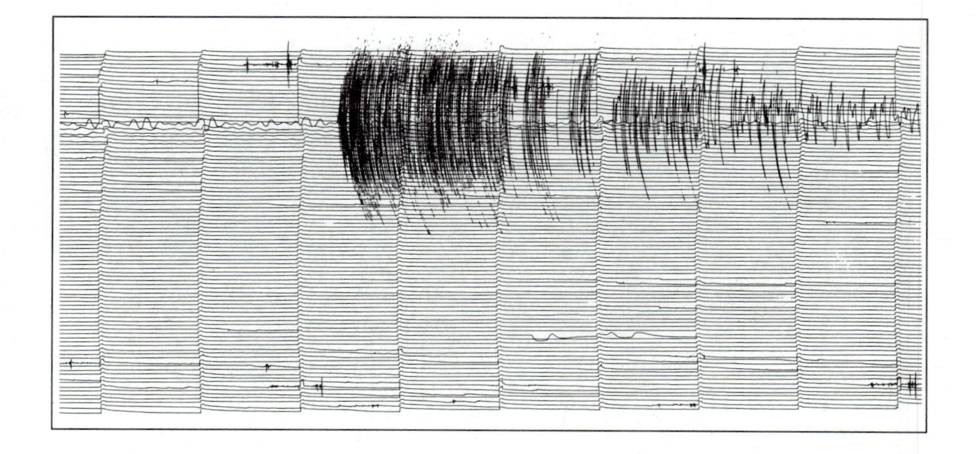

➢ **FIGURE 4.7 Portion of the seismogram of the Loma Prieta earthquake, 17 October 1989 at 5:03 P.M. (PDT), which registered 7.1 on the Richter scale.**

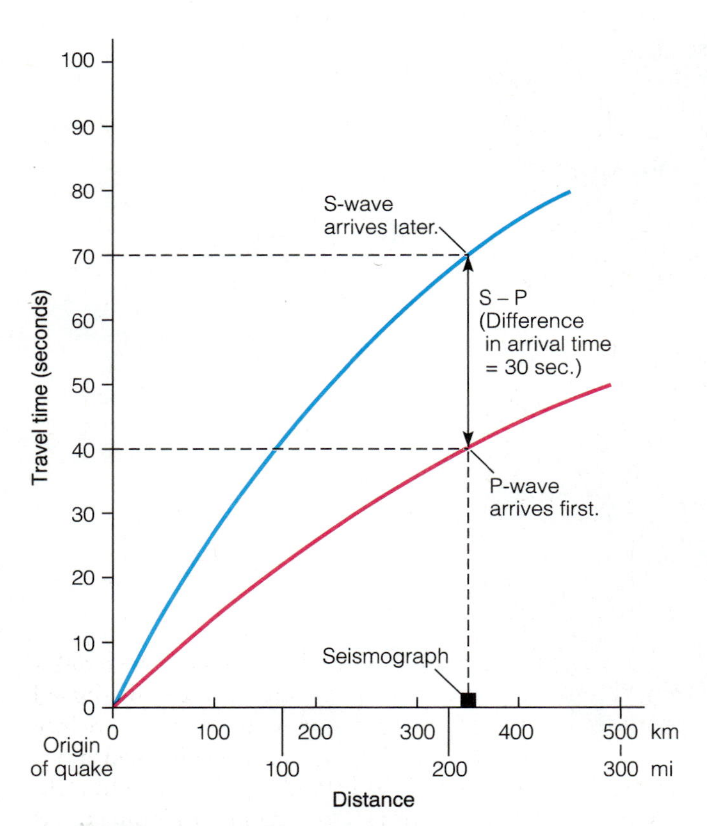

➢ **FIGURE 4.8 Generalized graph of distance versus travel time for P- and S-waves. Note that the P-wave has traveled farther from the origin of the earthquake than the S-wave at any given elapsed time.**

telephone lines from developed areas. Seismic data are recorded onto a **seismogram** (➢ Figure 4.7) and on magnetic tape for later playback and analysis. Because seismographs are extremely sensitive to vibrations of any kind, they are installed in quiet areas such as abandoned oil and water wells, cemeteries, and parks.

When an earthquake occurs, the distance to its epicenter can be approximated by computing the difference in the arrival times of P- and S-waves at various seismograph stations. Although the method actually used by seismologists today is more accurate and determines the depth and location of the quake's focus and its epicenter, what is described here serves to show how seismic-wave arrival times can be used to determine the distance to an epicenter. Because it is known that the two waves are generated simultaneously at the earthquake focus and that P-waves travel faster than S-waves, it is possible to calculate where the waves started.

Imagine that two trains leave a station at the same time, one traveling at 60 km/h and the other at 30 km/h, and that the second train passes your house one hour after the first train goes by. If you know their speeds, you can readily calculate your distance from the station as 60 kilometers. The relationship between distance to an epicenter and the arrival times of P- and S-waves is illustrated in ➢ Figure 4.8. The epicenter will lie somewhere on a circle whose center is the seismograph and whose radius is the distance from the seismograph to the epicenter. The problem is then to determine where on the circumference of the circle the epicenter is. To learn this, more data are needed; specifically, the distances to the epicenter from two other seismograph stations. Then the intersection of the circles drawn around each of the three stations—a method of map location called *triangulation*—specifies the epicenter of the earthquake (➢ Figure 4.9).

In addition to being able to locate the epicenter and focus of an earthquake, we can determine exactly how the fault that produced a given earthquake moved by study-

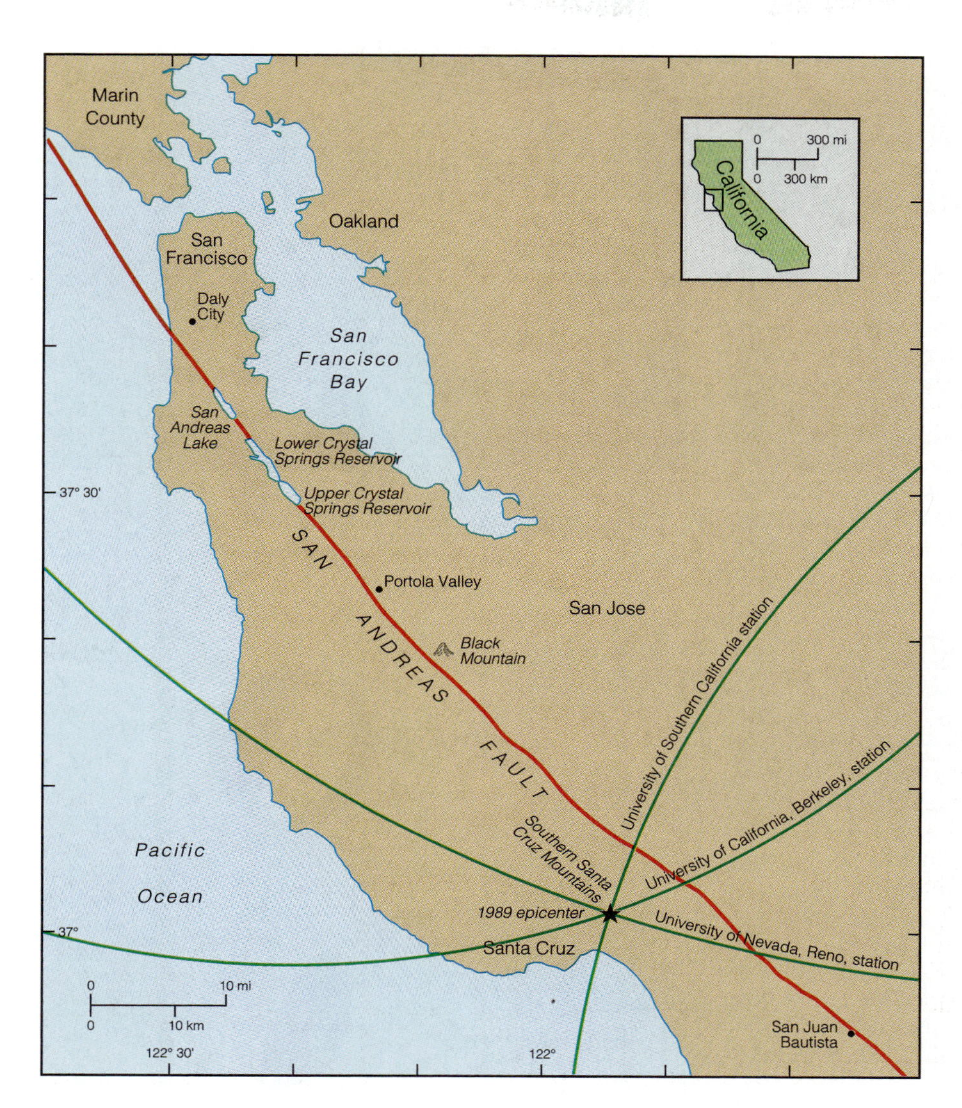

➢ **FIGURE 4.9 Method of locating an earthquake epicenter by triangulation from three seismic stations.**

ing the shapes of the P- and S-waves recorded on seismograms. Case Study 4.1 explains this.

Earthquake Measurement

The reactions of people (geologists included) to an earthquake, or an earthquake prediction, typically range from mild curiosity to outright panic. However, a sampling of the reactions of people who have been subjected to an earthquake can be put to good use. Numerical values can be assigned to the individuals' perceptions of earthquake shaking and local damage, which can then be contoured upon a map. One **intensity scale** developed for measuring these perceptions is the **Modified Mercalli Scale** (MM). The scale's values range from MM = I (denoting not felt at all) to MM = XII (denoting widespread destruction), and they are keyed to specific U.S. architectural and building specifications. (See Case Study 4.2.) People's perceptions and responses are compiled from returned questionnaires, and then lines of earthquake intensity, called **isoseismals,** are plotted on maps. Isoseismals enclose areas of equal earthquake damage and may indicate areas of weak rock or soil or of substandard construction techniques. Such maps have proved useful to planners and building officials in developing building codes and standards (➢ Figure 4.10). It is difficult to compare earthquakes in different regions using this scale because of differences in construction practices, population density, and sociological factors.

The most widely recognized measure of earthquake strength is the **Richter Magnitude Scale,** which was introduced in 1935 by Charles Richter and Beno Gutenberg at the California Institute of Technology. It is a scale of the energy released by an earthquake and thus, in contrast to the intensity scale, may be used to compare earthquakes in widely separated geographic areas. It is calculated by

CASE STUDY 4.1

Beach Balls and Fault Motion

Seismologists can determine whether a given earthquake resulted from strike-slip, normal, or reverse faulting or a combination of slip on a fault. Knowing how the fault moved at the focus of the quake—that is, its *focal mechanism*—is important to the understanding of earthquakes and the ground motions they produce. The upward or downward direction of an earthquake's first P-wave on a seismogram places the particular reporting station in a quadrant of compression or a quadrant of dilation (stretching), respectively, relative to the affected fault. Investigations of these initial P-waves are referred to as **first-motion studies.** When first-motion data are available from a large number of reporting stations, seismologists are able to locate and describe the fault very precisely by plotting the data on an imaginary sphere around the focus. This produces what might be called a "beach-ball" solution for explaining the cause of an earthquake.

Beach ball a of ➢ Figure 1 is a portrayal of a vertically oriented right-lateral strike-slip fault produced by this means. Note the regular appearance of the two compressional quadrants and the two dilational quadrants resulting from plotting the direction of the first P-waves reported from various locations. The quadrants are defined by two perpendicular planes, one of which is the fault plane.

In the case of nonvertical faults, such as normal and reverse faults, the fault plane is tilted toward the circumference of the projection sphere—or the "beach ball." With a normal fault, the areas of compression and dilation are distributed much like wedges cut from an orange (beach ball b). The same is true for a reverse fault (beach ball c), except that the stress fields are reversed.

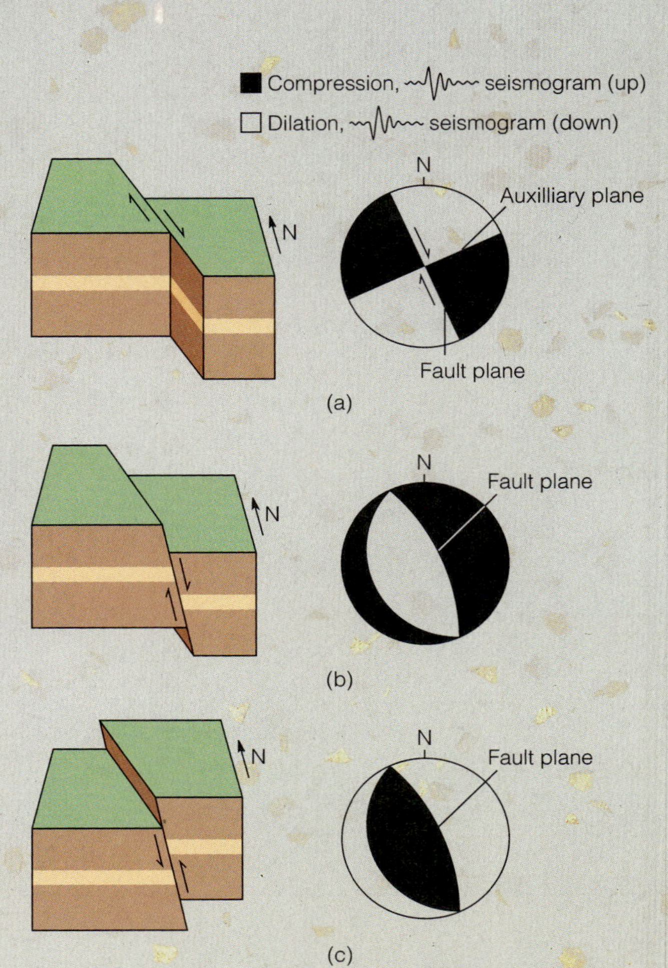

➢ **FIGURE 1 "Beach-ball" fault-plane solutions derived from first-motion studies of P-waves. Regions of compression (first seismogram motion is up) and dilation (first motion is down) as recorded by seismographs are shown. (*a*) Fault-plane solution for a vertical right-lateral strike-slip fault, (*b*) for an east-dipping normal fault, and (*c*) for an east-dipping reverse fault.**

measuring the maximum amplitude of the ground motion as shown on the seismogram using a specified seismic wave, usually the P-wave or S-wave or both. Next, the seismologist "corrects" the measured amplitude (in microns) to what a "standard" seismograph would record at the station. After an additional correction for distance from the epicenter, the Richter **magnitude** is the common logarithm of that ground motion in microns. For example, a magnitude-4 earthquake is specified as having a corrected ground motion of 10,000 microns ($\log_{10}$ of $10{,}000 = 4$) and thus can be compared to any other earthquake for which the same corrections have been made. It should be noted that because the scale is logarithmic, each whole number represents a ground shaking (at the seismograph site) 10 times greater than the next-lower number. Thus a magnitude-7 produces 10 times

CASE STUDY 4.2

Upstairs–Downstairs Earthquake Intensity

The Rossi-Forel Intensity Scale was the first scale developed for evaluating the effects of an earthquake on structures and humans. It was developed in Europe in 1878 and assigned Roman numerals from *I* (for *barely felt*) to *X* (for *total destruction*) for varying levels of intensity. This scale was first modified by Father Giuseppi Mercalli in Italy and later in 1931 by Frank Neumann and Harry O. Wood in the United States. The Modified Mercalli Scale is the scale most widely used today for evaluating the effects of earthquakes in the field. An abbreviated version is given here.

I. Not felt except by a very few persons under especially favorable circumstances.
II. Felt by a few persons at rest, especially by persons on upper floors of multistory buildings, and by nervous or sensitive persons on lower floors.
III. Felt quite noticeably indoors, especially on upper floors, but many people do not recognize it as an earthquake; vibrations resemble those made by a passing truck.
IV. Felt indoors by many persons and outdoors by only a few. Dishes, windows, and doors are disturbed, and walls make creaking sounds as though a heavy truck had struck the building. Standing cars are rocked noticeably.
V. Felt by nearly everyone. Many sleeping people are awakened. Some dishes and windows are broken, and some plaster is cracked. Disturbances of trees, poles, and other tall objects may be noticed. Some persons run outdoors.
VI. Felt by all. Many people are frightened and run outdoors. Some heavy furniture is moved, and some plaster and chimneys fall. Damage is slight, but humans are disturbed.
VII. General fright and alarm. Everyone runs outdoors. Damage is negligible in buildings of good design and construction, considerable in those that are poorly built. Noticeable in moving cars.
VIII. General fright approaching panic. Damage is slight in specially designed structures; considerable in ordinary buildings, with partial collapse; great in older or poorly built structures. Chimneys, smokestacks, columns, and walls fall. Sand and mud are ejected from ground openings (liquefaction).
IX. General panic. Damage is considerable in specially designed structures, well-designed frame structures being deformed; great in substantial buildings, with partial collapse. Ground is cracked; underground pipes are broken.
X. General panic. Some well-built wooden structures and most masonry and frame structures are destroyed. Ground is badly cracked; railway rails are bent; underground pipes are torn apart. Considerable landsliding on riverbanks and steep slopes.
XI. General panic. Ground is greatly disturbed, with cracks and landslides common. Sea waves of significant height may be seen. Few if any structures remain standing; bridges are destroyed.
XII. Total panic. Total damage to human engineering works. Waves seen on the ground surface. Lines of sight and level are distorted. Objects are thrown upward into the air.

greater shaking than a magnitude-6; 100 times that of a magnitude-5; and 1,000 times that of a magnitude-4. Total energy released, on the other hand, varies logarithmically as some exponent of 30. Compared to the energy released by a magnitude-5 earthquake, a M = 6 releases 30 times (30^1) more energy; a M = 7 releases 900 times (30^2) more; and a M = 8 releases 27,000 (30^3) times more energy (➢ Figure 4.11a and b).

The Richter Scale is open-ended; that is, theoretically it has no upper limit. However, rocks in nature do have a limited ability to store strain energy without rupturing, and no earthquake has been observed with a Richter magnitude greater than 8.9. One may approximate Richter magnitude with maximum intensity for a nearby earthquake as follows:

MAGNITUDE	INTENSITY
2.5	Felt locally; MM = II–III
4.5	Local damage; MM = V–VI
6.0	*Destructive* earthquake; MM = VII–VIII
7.0	*Major* earthquake; MM = IX–X
8.0	*Great* earthquake; MM = XI–XII

The magnitude scale does not accurately measure large earthquakes on faults with a great rupture length. This is because the seismic waves used to measure the magnitude of a large earthquake come from only a small part of the fault rupture and hence cannot accurately measure the total seismic energy released along the fault. For this reason seismologists now favor a new scale, **moment magnitude**

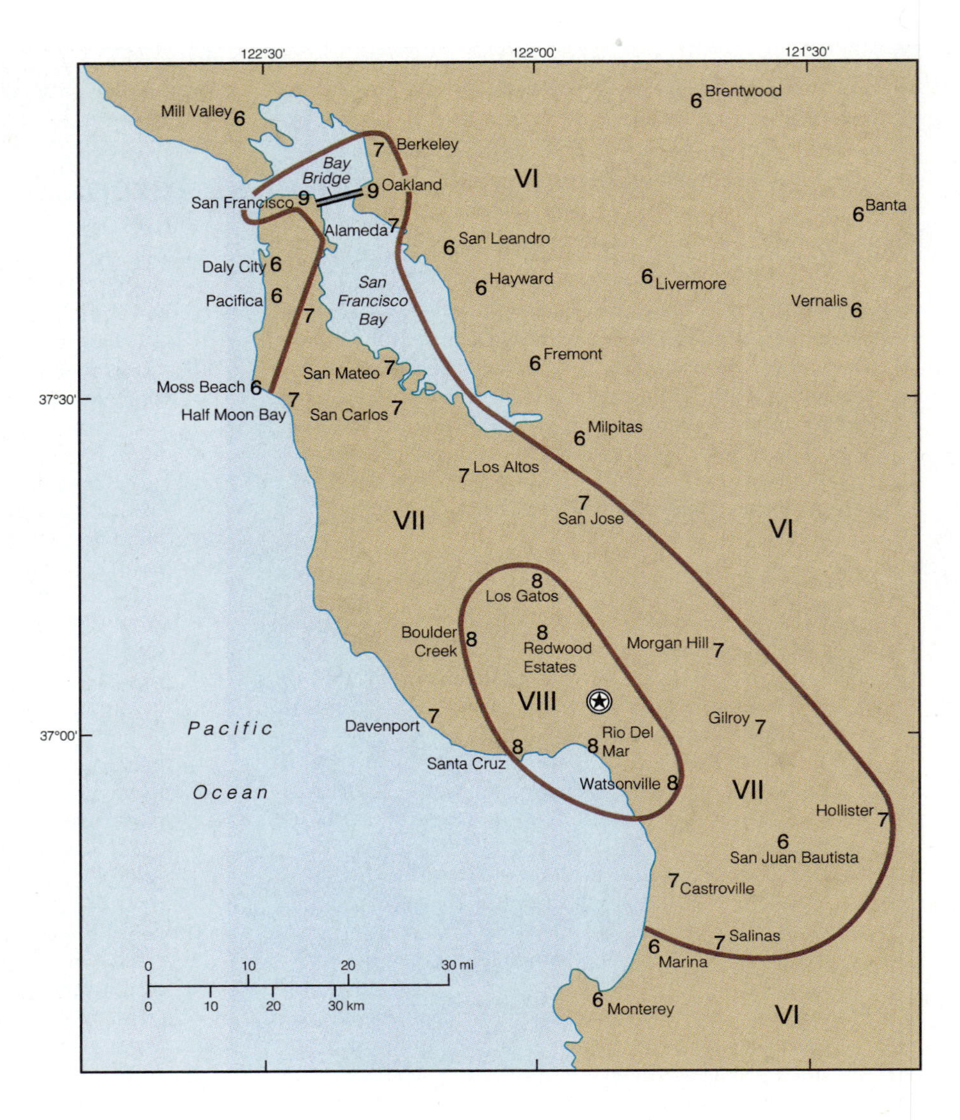

➢ **FIGURE 4.10 Isoseismal map of the Loma Prieta earthquake. Arabic numbers denote reported intensity levels at specific locations; Roman numerals, the general intensity levels of particular areas. Note the two small areas of intense damage (intensity 9) in the Oakland and San Francisco areas that were developed on bay muds.**

(**M**), which equals ($\frac{2}{3} \log M_o - 10.7$). Seismic moment ($M_o$ in dyne cm) is the product of rock rigidity *times* the area of movement along the fault surface *times* the amount of slip. Moment magnitude is favored by seismologists, because the amount of seismic energy (in ergs) released from the ruptured fault surface is linearly related to seismic moment (M_o) by a simple factor, whereas the Richter scale is logarithmically related to energy. The practice now is to assign large earthquakes both Richter and moment magnitudes, again because the latter can be related directly to the physical properties of the fault and energy release. For comparison, the equivalent energy released by other phenomena can be related to earthquakes by using this scale (➢ Figure 4.11c). Seismic moment magnitudes, derived from the seismic moments, are 7.9 for the 1906 San Francisco earthquake, 9.2 for the 1964 Alaskan earthquake, and 9.5 for the 1960 Chilean earthquake. With a Richter magnitude of 8.3, the Chilean earthquake has the largest seismic moment magnitude ever recorded.

Fault Creep, the "Non-Earthquake"

Some faults move almost continuously or in short spurts that do not produce detectable earthquakes. This type of movement, called **fault creep,** is well known but poorly understood. The Hayward fault, which is part of the San Andreas fault system, is an example of a creeping fault. It is just east of the San Andreas fault and runs through the cities of Hayward and Berkeley, California. Creep along this fault causes displacements of millimeters per year, and is one way that this plate boundary accommodates motion between the Pacific and North American plates. An alternative "accommodation" of plate motion on the San Andreas system is the storage of the strain energy and periodic release as a large displacement on a fault that

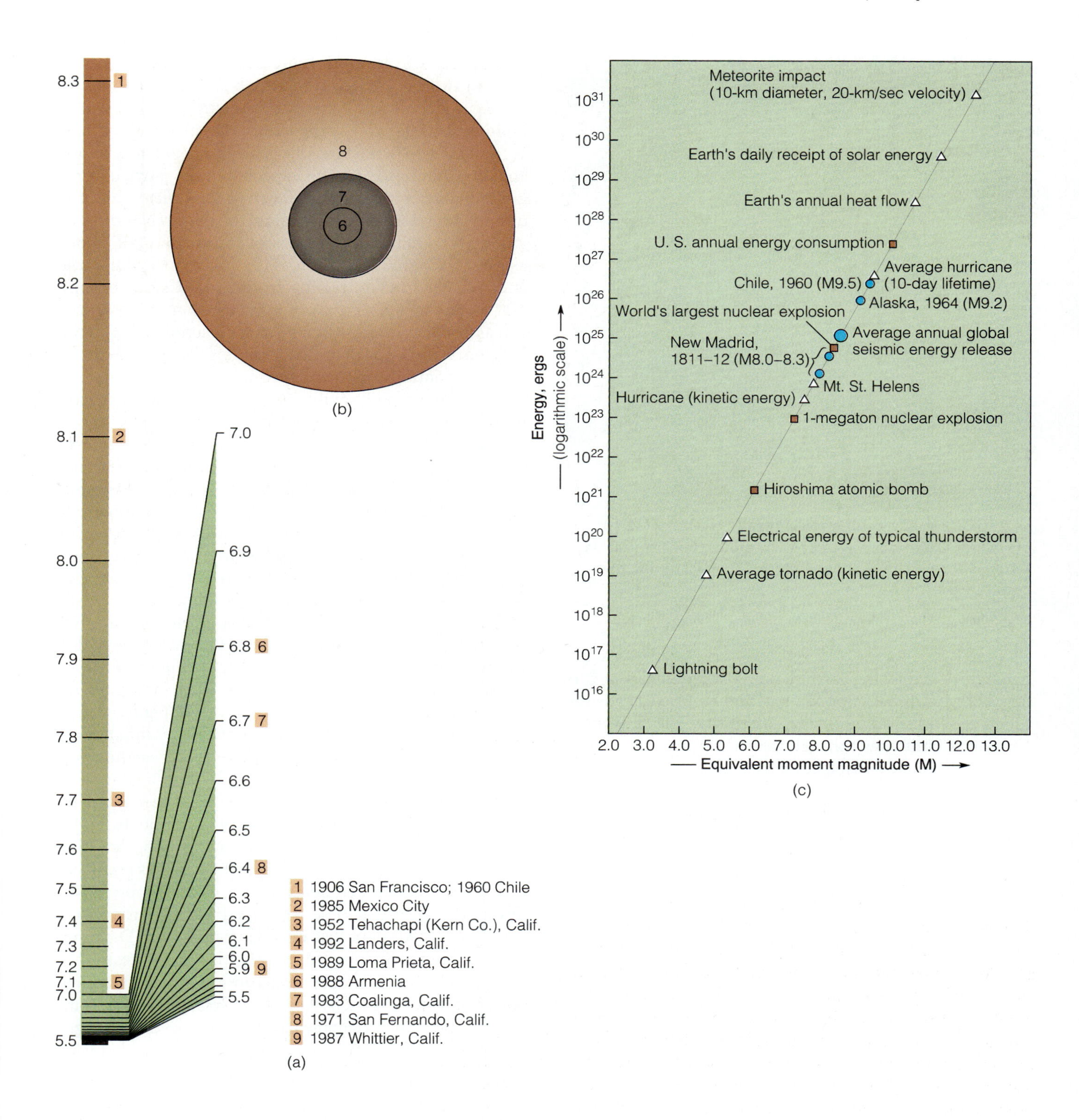

➢ FIGURE 4.11 Earthquake magnitude and energy. (*a*) Richter magnitudes of ten selected earthquakes, 1906–1992. (*b*) Relationship between magnitude and energy released. The volumes of the spheres are roughly proportional to the energy released for the given magnitudes. (*c*) Equivalent moment magnitudes of energetic human-caused events (*squares*), natural events (*triangles*), and large earthquakes (*circles*). The erg is a unit of energy or work in the metric (cgs) system. To lift a one-pound weight one foot requires 1.4×10^7 ergs.

➢ **FIGURE 4.12 Fault creep along the Hayward fault at Hayward, California. Creep of a few millimeters per year has resulted in about 18 centimeters (7 in) of offset during this curb's lifetime.**

generates a damaging earthquake. The obvious benefit of a creeping fault does not come without a price, as the residents of Hollister, California, can testify. The "creeping" Calaveras fault runs through their town and, as in Hayward, it gradually displaces curbs, sidewalks, and even residences (➢ Figure 4.12).

A good strategy in areas subject to fault creep is to map the surface trace of the fault precisely so that it can be avoided in future construction. Land adjacent to such known faults can be designated as limited-use areas. It may provide recreation space for local residents, for example.

An interesting lack of geological foresight is seen at the University of California–Berkeley Memorial Stadium. The huge structure was built on the Hayward fault before the hazard of fault creep was recognized, and creep-caused damage to the stadium's drainage system requires periodic maintenance. It is ironic that creep damage occurs at an institution that is highly regarded for seismological research and that installed the first seismometer in the United States.

Seismic Design Considerations

Ground Shaking

Movements, particularly rapid back-and-forth displacements, are the real "killers" during an earthquake. Shear and surface waves are the culprits, and the potential for them must be evaluated when establishing design specifications. The design objective for earthquake-resistant buildings is relatively straightforward: structures should be designed to withstand the maximum potential horizontal ground acceleration expected in the particular region. Engineers call this acceleration *base shear,* and it is usually expressed as a percentage of the *acceleration of gravity* (g). On earth, g is the acceleration of a falling object in a vacuum (980 cm/sec^2) (32 ft/sec^2). In your car it is equivalent to accelerating from a dead stop through 100 meters in 4.5 seconds. An acceleration of 1 g downward produces weightlessness, and a fraction of 1 g in the horizontal direction can cause buildings to separate from their foundations or to collapse completely. An analogy is to imagine rapidly pulling a carpet on which a person is standing; most assuredly the person will topple.

The effect of high horizontal acceleration on poorly constructed buildings is twofold. Flexible-frame structures may be deformed from cube-shaped to rhomb-shaped or they may be knocked off their foundations (➢ Figure 4.13). More rigid multistory buildings may suffer "story shift" if floors and walls are not adequately tied together (➢ Figure 4.14). The result is a shifting of floor levels and the collapse of one floor upon another like a stack of pancakes. The terrible death toll (over 25,000 fatalities) in Armenia in December 1988 was largely due to this kind of collapse. There, prefabricated multistory buildings simply came apart because their design could not accommodate the shearing force (➢ Figure 4.15). Such structural failures are not survivable by inhabitants and they clearly illustrate the adage that "earthquakes don't kill people; buildings do."

Damage due to shearing forces can be mitigated by bolting frame houses to their foundation and by *shear walls.* An example of a shear wall is plywood sheeting nailed in place over a wood frame, which makes the structure highly resistant to deformation. Wall framing, usually two-by-fours, should be nailed very securely to a wooden sill that is bolted into the foundation. Diagonal bracing and blocking also provide shear resistance (➢ Figure 4.16). L-shaped structures may suffer damage where they join, as each wing of the structure vibrates independently. Such damage can be minimized by designing *seismic joints* between the building wings or between adjacent buildings of different heights. These joints are filled with a compressible substance such as felt, plastic, or rubber that will accommodate any movement between the structures (Figure 4.16).

Wave period is the time interval between arrivals of successive wave crests, or of equivalent points of waves, and it is expressed as T in seconds. It is an important consideration when assessing a structure's potential for seismic damage, because if a building's natural period of vibration is equal to that of seismic waves, a condition of

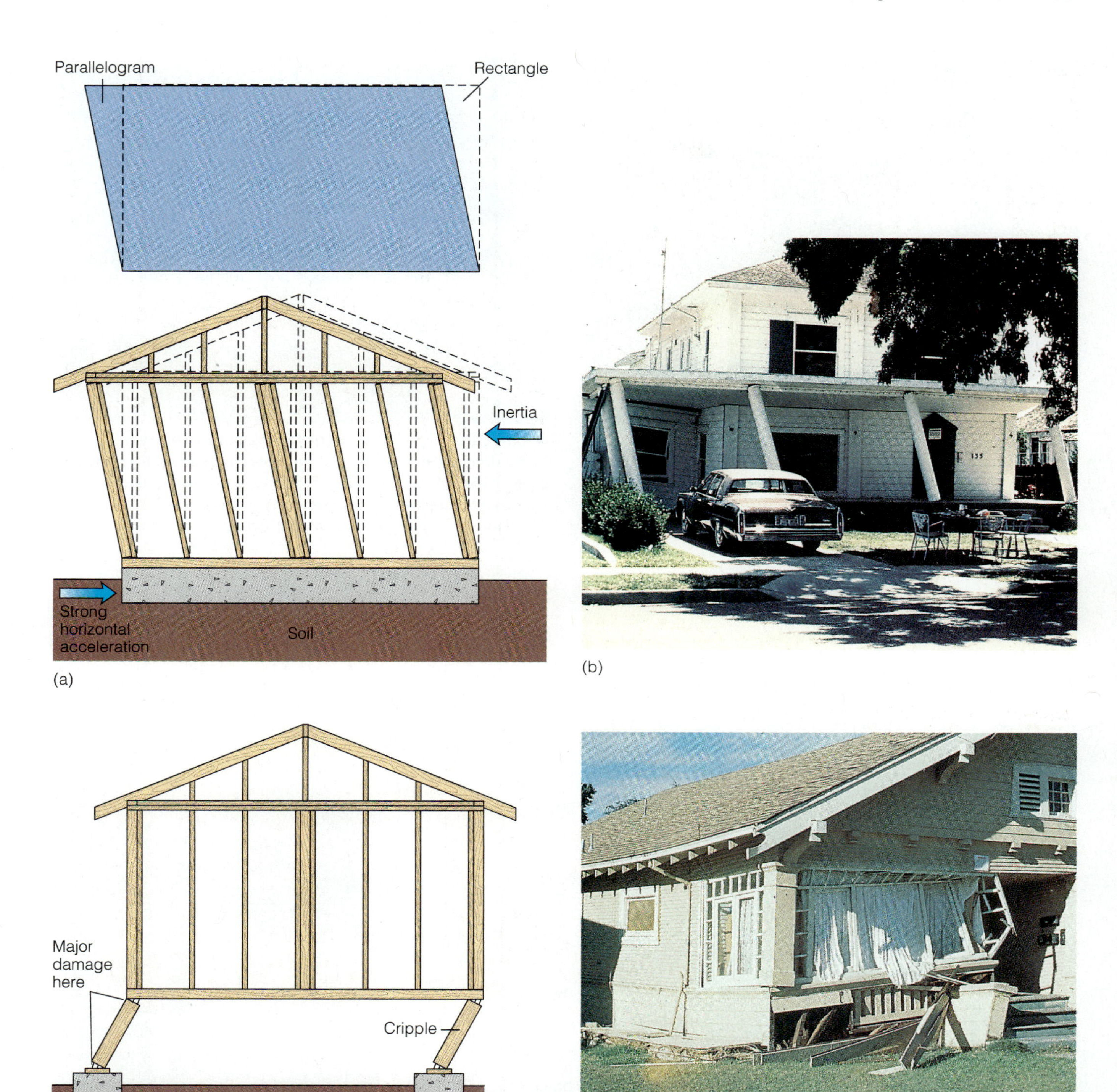

➢ **FIGURE 4.13** (*a*) **Strong horizontal motion may deform a house from a cube to a rhomboid or knock it from its foundation completely.** (*b*) **A Coalinga, California, frame house that was deformed by the magnitude-6.3 earthquake in 1983.** (*c*) **A cripple-wall consists of short vertical members that connect the floor of the house to the foundation. Cripple-walls are common in older construction.** (*d*) **A Watsonville, California, house that was knocked off its cripple-wall base during the Loma Prieta earthquake of 1989. Without exception, cripple-walls bent or folded over to the north relative to their foundations, which moved south.**

➢ **FIGURE 4.14** Total vertical collapse as a result of "story shift"; Mexico City, 1985. Such structural failures are not survivable.

➢ **FIGURE 4.15** Wreckage of an older three-story building; Leninakan, Armenia, 1988. A collapsed nine-story building is at the right. Membrane (floor) failures in precast-concrete-frame structures were common here; 132 of these buildings "pancaked," leaving little space for occupant survival.

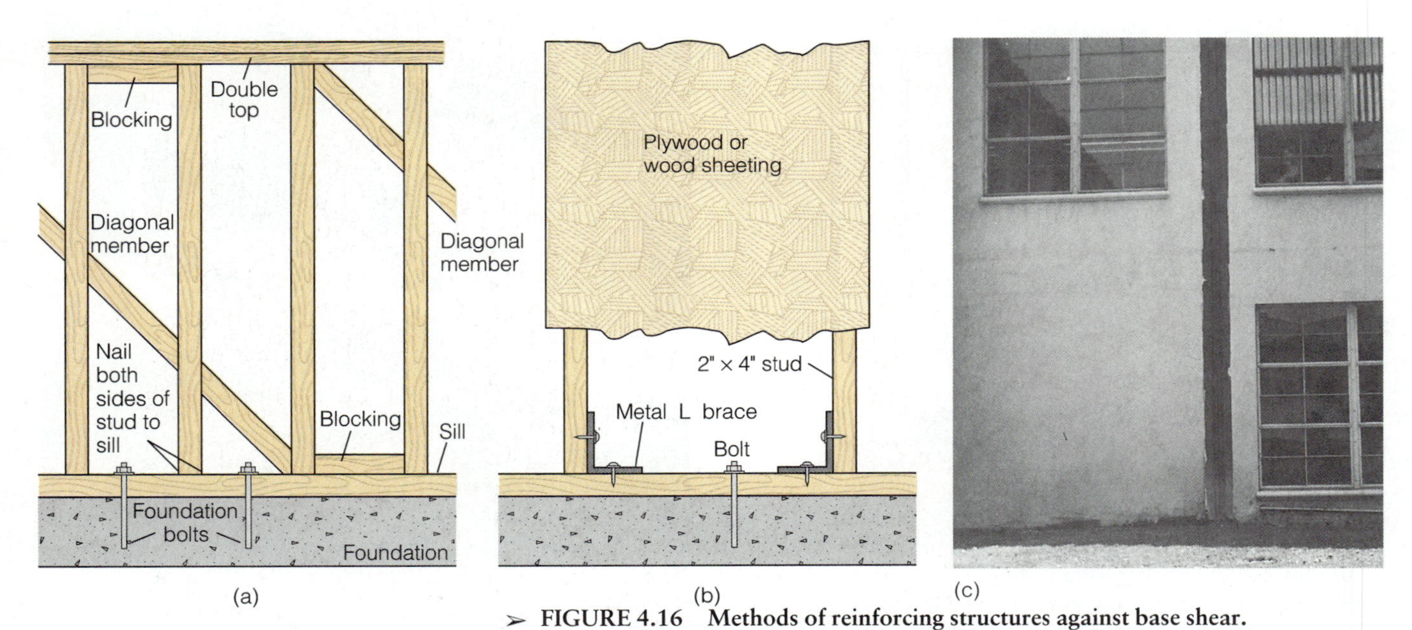

➢ **FIGURE 4.16** Methods of reinforcing structures against base shear. (*a*) Diagonal cross-members and blocks resist horizontal earthquake motion (shear). (*b*) Plywood sheeting forms a competent shear wall, and metal "L" braces and bolts tie the structure to the foundation. (*c*) Seismic joint between two wings of a classroom building; San Bernardino Valley College, California. The joint is the vertical, dark path in the center of the photo. The active San Jacinto fault runs through the campus of this community college.

resonance exists. **Resonance** occurs when a building sways in step with an oscillatory seismic wave. As a structure sways back and forth under resonant conditions, it gets a push in its direction of sway with the passage of each seismic wave. This causes the sway to increase, just as pushing a child's swing at the proper moments makes it go higher with each push. Resonance also may cause a glass tumbler to shatter when an operatic soprano sings just the right note or frequency.

Low-rise buildings have short natural wave periods (0.05–0.1 seconds), and high-rise buildings have long natural periods (1–2 seconds). Therefore, high-frequency (short-period) waves affect single-family dwellings and low-rise buildings, and long-period (low-frequency) waves affect tall structures. Close to an earthquake epicenter, high-frequency waves dominate, and thus more extensive home and low-rise damage can be expected. With distance from the epicenter, the short-period wave energy is absorbed or dissipated, resulting in the domination of longer-period waves. (The differential effects of long and short wave periods were evident after the 1985 Mexico City earthquake, as discussed and illustrated in the next section. See also Figure 4.32.)

Landslides

Literally thousands of landslides were triggered by ground shaking during the San Fernando and Loma Prieta earthquakes. Landsliding from the latter earthquake occurred in parts of the Monterey Bay and Santa Cruz areas, and also in adjacent parts of the California Coast Range. These areas are slide-prone, even under the best of conditions, and sometimes need only an earthquake to trigger a slide. In the greater San Francisco bay area there was at least $10 million damage to houses, utilities, and transportation systems.

One of the worst earthquake-triggered tragedies occurred in Peru in 1970. An earthquake-initiated avalanche consisting of a mixture of snow, ice, and rock gave way at an elevation of 6,000 meters on the western slopes of Nevado Huascarán. A kilometer wide and 1.5 kilometers long, it accumulated water as it plummeted downhill at more than 160 km/hr, jumping a natural barrier, and burying 25,000 residents and the villages of Yungay and Ranrahirca. The inundated areas remain completely covered with mud and boulders, leaving little indication of what lies below (➢ Figure 4.17).

➢ FIGURE 4.17 (*a*) and (*b*). **A 1970 earthquake triggered a rock and snow avalanche down the slopes of Nevado Huascarán in Peru that buried the towns of Yungay and Ranrahirca. Depth of mud at Yungay is as much as 4.6 meters (15 ft).**

Ground or Foundation Failure

Liquefaction is the sudden loss of strength of water-saturated sandy soils resulting from shaking during an earthquake. Sometimes called **spontaneous liquefaction,** it can cause large ground cracks to open, lending support to the ancient myth that the earth opens up to swallow people and animals during earthquakes. Shaking can cause saturated sands to consolidate and thus to occupy a smaller volume. If the water is slow in draining from the consolidated material, the overlying soil comes to be supported only by pore-water, which has no resistance to shearing. This may cause buildings to settle, earth dams to fail, and sand below the ground surface to blow out through openings at the surface much like small volcanoes (➢ Figure 4.18). Liquefaction at shallow depth may result in extensive lateral movement or spreading of the ground, leaving great cracks and openings.

Ground areas most susceptible to liquefaction are those that are underlain at shallow depth—usually less than 30 feet—by layers of water-saturated fine sand. With subsurface geologic data obtained from water wells and foundation borings, liquefaction-susceptibility maps have been prepared for many seismically active areas in the United States.

Similar failures occur in certain clays that lose their strength when they are shaken or remolded. Such clays are called *quick clays* and they are natural aggregations of fine-grained clays and water. They have the peculiar property of turning from a solid (actually a gel-like state) to a liquid when they are agitated by an earthquake, an explosion, or even vibrations from pile driving. They occur in deposits of glacial-marine or glacial-lake origin and are therefore found mostly in northern latitudes, particularly in Scandinavia, Canada, and the New England states. Failure of quick clays underlying Anchorage, Alaska, produced extensive lateral spreading throughout the city in the 1964 earthquake (➢ Figure 4.19).

The physics of failure is similar in spontaneous liquefaction and quick clays. When the earth materials are water-saturated and the earth shakes, the loosely packed sand consolidates or the clay collapses like a house of cards. The pore-water pressure pushing the grains apart becomes greater than the grain-to-grain friction, and the material becomes "quick" or "liquefies" (➢ Figure 4.20). The potential for such geologic conditions is not easily recognized. In many cases it can be determined only by information gained from bore holes drilled to depths of 100 feet. Because of this expense, many site investigations do not include deep drilling, and the condition goes unsuspected until a major earthquake occurs.

(b)

➢ FIGURE 4.18 **(*a*) These apartment buildings tilted as a result of soil liquefaction in Niigata, Japan, in 1964. Many residents of the building in the center exited by walking down the side of the structure. (*b*) "Sand blows" in a lettuce field caused by liquefaction during the Loma Prieta earthquake near Castroville, California.**

➢ FIGURE 4.19 **"House of cards" collapse of quick clay structure; Turnagain Heights, Anchorage, Alaska. Total destruction occurred within the slide area, which is now an earthquake park.**

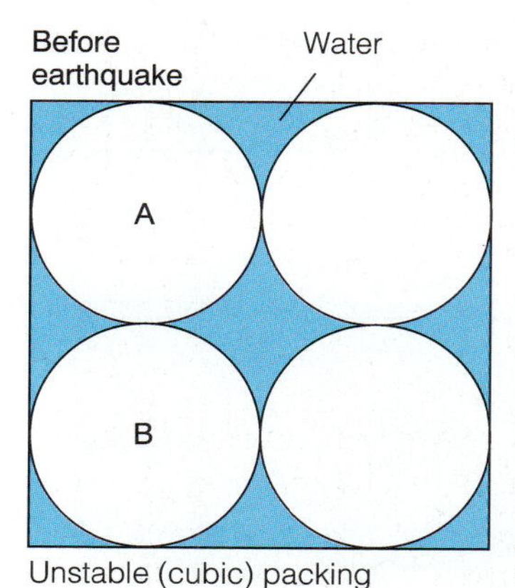

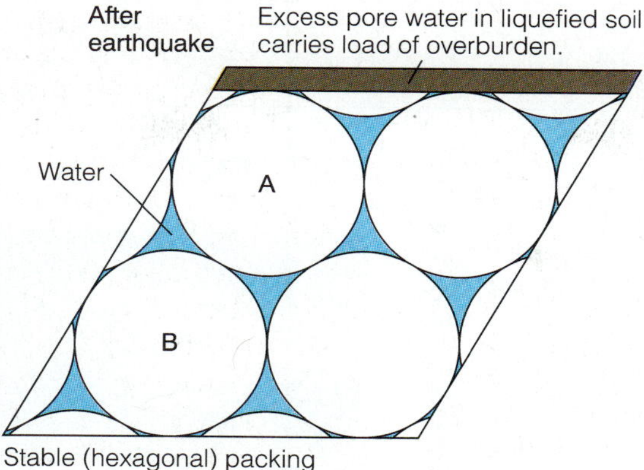

➢ **FIGURE 4.20 Liquefaction (lateral spreading) due to repacking of spheres (idealized grains of sand) during an earthquake. The earthquake's shaking causes the solids to become packed more efficiently and thus to occupy less volume. A part of the overburden load is supported by water, which has no resistance to lateral motion.**

Ground Rupture and Changes in Ground Level

Structures that straddle an active fault may be destroyed by actual ground shifting and the formation of a fault scarp. Although this kind of structural damage is not common, the hazard exists and some examples of scarp formation in urban areas can be shown (➢ Figure 4.21a). By excavating trenches across fault zones, geologists can usually locate potentially active rupture surfaces, thereby reducing the possibility of construction across an active fault trace. The Alquist–Priolo bill of 1970 was enacted in California specifically to mandate that active faults be located and mapped. Under the act the lands adjacent to active faults become Special Study Zones. This designation requires geological studies to locate the latest fault rupture and mandates local government to limit land use within the zone. The faults responsible for the 1992 Landers, California earthquake were designated Alquist-Priolo Special Study Zone faults just prior to the June 28 event (See Case Study 4.3).

Changes in ground level as a result of faulting may have an impact, particularly in coastal areas that are uplifted or down-dropped. For instance, parts of the Gulf of Alaska thrust upward 11 meters (36 ft) exposing vast tracts of former tidelands on island and mainland coasts (➢ Figure 4.21b). In fact, the 1964 Alaskan event produced the greatest tectonic landmass elevation changes in recorded history.

Fires

Fires caused by ruptured gas mains or fallen electrical power lines can add considerably to the damage caused

(a)

(b)

➢ **FIGURE 4.21** **(*a*) Geology students sitting on a fault scarp that formed in a San Gabriel Mountains, California, housing development in 1971. Fault displacement converted this lot to a split-level-home site in one day. In the background at the left is a seismograph installation for measuring aftershocks. (*b*) Fault scarp on Montague Island in Prince William Sound, Alaska. In places, the scarp is 4.8 meters (16 ft) high between the white, shell-covered uplifted sea floor (*right*) and the darker brown sand landward.**

CASE STUDY 4.3

Has the Big One Hit Southern California?

A great earthquake is forecast for Southern California. Ominously called the "Big One," it is expected to be centered within the "locked" segment of the San Andreas fault that extends from San Bernardino on the south to Parkfield on the north (see text Figure 4.43). On the morning of 28 June 1992, Southern Californians received a rude "quake up" call at 4:58 A.M., when a magnitude-7.5 earthquake shook an area encompassing San Diego, Los Angeles, Reno (Nevada), and Phoenix (Arizona) and disturbed swimming pools as far away as Denver and Boise, Idaho! This event is known as the *Landers earthquake,* named after a small town near the epicenter, 6 miles north of Yucca Valley in central San Bernardino County. Because of the long duration of shaking, many people in the greater Los Angeles area were convinced that this was the Big One. A magnitude-6.5 aftershock occurred 3 hours later in the mountain resort of Big Bear, 30 kilometers (18 mi) west of the main-shock epicenter. First-motion studies of the main shock indicate pure right-lateral strike-slip on the Camp Rock, Emerson, and Johnson Valley faults, oriented slightly west (10°) of north. The faults exhibited surface rupture for more than 70 kilometers (42 mi) with a maximum displacement of 6.7 meters (22 ft)—almost as great as that of the 1906 San Francisco earthquake (➢ Figure 1). The Big Bear aftershock occurred on a separate northeast-trending fault at depth with no surface offset. The Big Bear aftershock sequence intersects the Landers sequence, forming two legs of a triangle whose third leg is the San Andreas fault (➢ Figure 2).

More than 3,000 **aftershocks** were recorded in four days, with their magnitudes and distribution in time and space being about normal for an earthquake of this magnitude. The aftershock epicenters extended southward to the San Andreas fault, forming a map pattern similar to the Greek letter lambda, leading local seismologists to call them the "lambda sequence" (see Figure 2).

➢ FIGURE 1 Scarp formed on the Johnson Valley fault during the 1992 Landers earthquake in the Mojave Desert, California. Right-lateral strike-slip offset of 4.27 meters (14 ft) occurred here, resulting in a 2 meter (6.5 ft) vertical scarp due to lateral offsetting of a ridge across an adjacent arroyo. Geologist, looking like a rock hammer, is nearly 2 meters tall.

by an earthquake. In fact, most damage attributed to the San Francisco earthquake of 1906 was due to the uncontrolled fire that followed the event. Many gas pipelines that cross active faults in California today are above ground in order to facilitate repair in case of fault rupture. Fire is always a threat in urban areas after an earthquake. One of the principal "do's" for citizens immediately after a quake is to shut off the gas supply to homes and other buildings in order to prevent gas leaks into the structure from damaged lines. This bit of advice in itself saves lives.

Tsunami

The most myth-ridden hazard associated with earthquakes are so-called *tidal waves,* which are properly known as **tsunami.** Japanese art and literature are filled with references to them, American literature commonly includes narratives of great waves, and Hollywood has perpetuated people's misconceptions with exaggerated special effects in movies. Tsunami are impulsively generated waves that are produced when the sea floor is disrupted by faulting, a volcanic eruption, or a landslide.

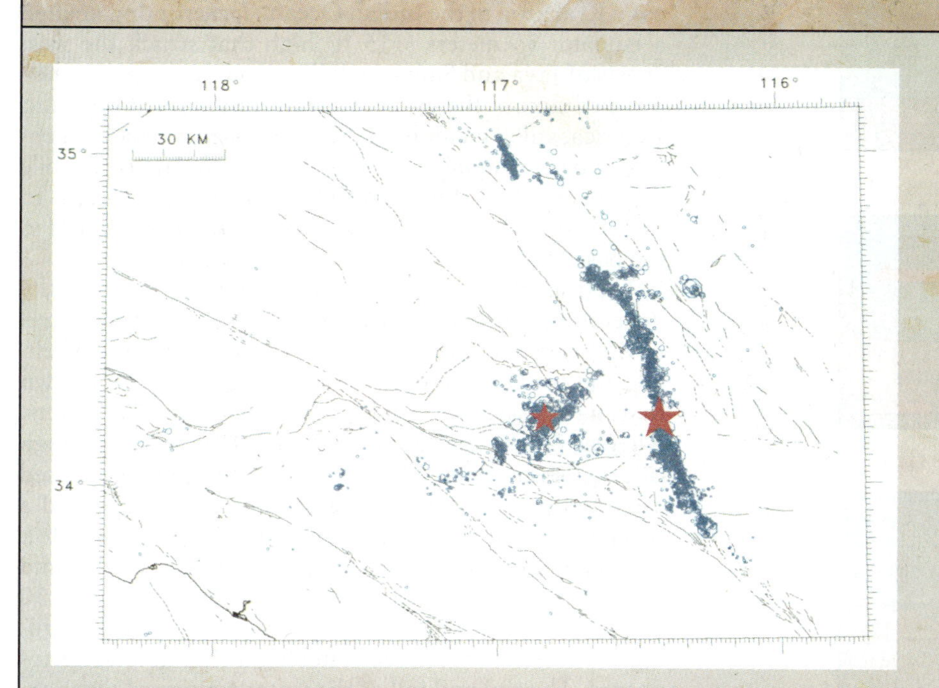

➢ **FIGURE 2 Computer-generated map of the 28 June 1992 Landers and Big Bear, California, earthquakes. The epicenters are denoted, respectively, by the large and small stars. The circles represent the aftershocks that occurred during the three weeks following the main events, with the sizes of the circles indicating their relative strengths. The map distribution of the aftershocks led seismologists to dub them the "lambda" sequence for their resemblance to the Greek letter.**

Miraculously, only one death and about 350 injuries resulted, because of the sparse population in eastern San Bernardino and Riverside Counties. The main shock and strong aftershocks destroyed a total of almost 600 homes and damaged another 5,700. Some homes damaged by the main shock were totally destroyed by the strong aftershocks. Yucca Valley and Landers, both small desert communities, sustained damage to mobile homes, chimneys, roadways, and power- and water-delivery systems. No water was available for fire fighting in Landers, because water mains were ruptured. In Big Bear, high in the San Bernardino Mountains, 268 homes were destroyed, 677 sustained major damage, and a thousand more sustained minor damage. Many of the fallen chimneys and damaged homes there had been constructed of rounded river boulders without reinforcing. Multiple landslides blocked the highways into the resort. High-rise buildings swayed in Los Angeles, and the Disneyland Hotel in Anaheim was evacuated—to avoid subjecting the patrons to an unexpected fun-ride.

In spite of this being a major earthquake, there were few casualties and there was relatively little damage because of the location of the epicenter. It would be a comfort to Southern Californians if the Landers event was indeed the forecasted Big One, but it does not meet the criteria. The energy released by the Landers events was less than 10 percent of what is expected from the Big One, and the motion of crustal blocks that produced these earthquakes served to reduce normal stress across the San Andreas fault. Thus, the probability of an earthquake on the fault was increased (by an unknown amount) by the events at Landers and Big Bear.

This motion displaces the overlying water column, giving rise to waves that move outward from the source in all directions—waves similar to those produced by throwing a rock into a pond. Tsunami may have wavelengths in excess of 160 kilometers (100 mi) and velocities on the order of 800 kilometers per hour (480 mph), but curiously, the waves are only a few feet high in the open ocean (➢ Figure 4.22). As the wave moves into shallow water, a transformation takes place as the velocity and wavelength decrease. This in turn causes the wave height to increase, creating a higher and more dangerous wave. Tsunami resulting from major (magnitude > 7) earthquakes may be 15 meters (50 ft) high and travel 50–60 kilometers per hour (30–36 mph) on land, like the one that struck Hawaii in 1946. The time between crests is about 20 minutes, and the first crest is not always the largest. The trough of the initial wave may arrive first, and many lives have been lost by curious persons who wandered into the tidal regions exposed as water withdrew. In Lisbon, Portugal, in 1755, many people were in church when an earthquake struck on All Soul's Day. The worshipers ran outside to escape falling debris and fire,

(a)

(b)

➢ **FIGURE 4.24** (*a*) **Photograph of Valdez, Alaska, three weeks after the 1964 earthquake, showing the extent of the tsunami runup.** (*b*) **Graphic depiction of the landslide that caused the wave and the fatalities at Valdez.**

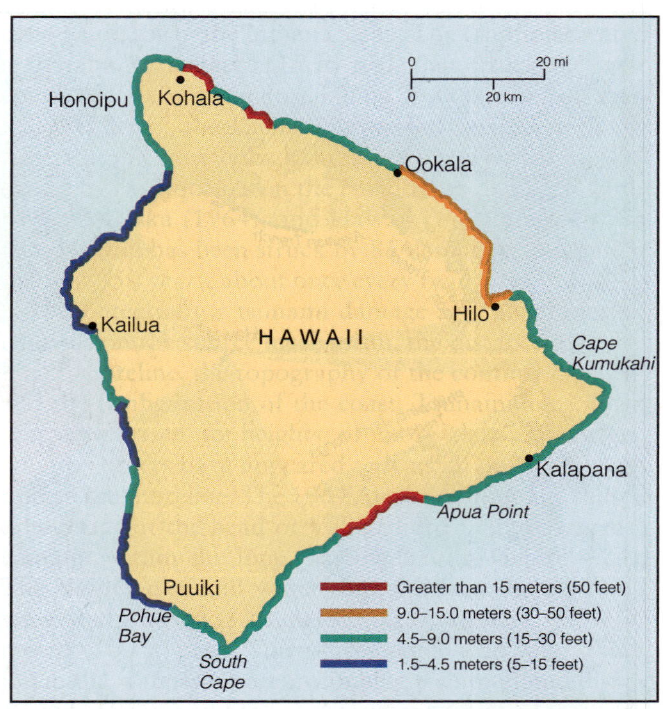

➢ **FIGURE 4.25 Forecast tsunami heights along the coasts of the island of Hawaii. Note that the greatest wave heights occur along the northeast shore. This shore faces the Aleutian trench, a common source of Hawaiian tsunami.**

tsunami that struck Hilo resulted in some recommendations for mitigating tsunami damage in that city. Parking meters throughout the runup area were bent flat in the direction of wave travel (➢ Figure 4.26), with the direction varying throughout the affected area. It is now recommended that new structures be built with their narrowest dimension facing the wave. Several waterfront structures had open fronts that allowed water to pass through them which demolished the solid bearing walls at the rear holding up the roof (➢ Figure 4.27). Waterfront buildings are now designed with open parking or glassed-in space on the lower level, so much damage from water impact can be avoided as wave runup flows through the open part of the structure.

Finally, people living in a coastal area subject to tsunami should heed civil defense officials and their tsunami warnings. When a warning was issued for the U.S. West Coast after the 1960 Chilean earthquake, Los Angeles Harbor was lined with people wanting to witness a possible "tidal wave." Although it is probably human nature to want to witness such an event, be forewarned that "the big one" could ruin your day. For obvious reasons, few photographs have been taken of tsunami breaking directly on shore. Thus, the series presented in ➢ Figure 4.28 is remarkable.

➢ **FIGURE 4.26 Parking meters bent by the 1960 tsunami at Hilo, Hawaii. The meters resemble arrows aligned with the wave runup.**

➢ **FIGURE 4.27 The lines painted on the window of a hotel near the Hilo waterfront mark the levels of several tsunami that have flowed through the structure.**

HISTORIC EARTHQUAKES

On average each year in the world there are at least two earthquakes of magnitude 8.0 or greater, and there are 20 earthquakes in the magnitude 7.0 to 7.9 range. Release of seismic energy in the form of earthquakes has occurred throughout geologic time, and recorded history contains many references to strong earthquakes. Over 3,000 years of seismicity is documented in China, and Strabo's *Geography* mentions an earthquake in 373 B.C. in Greece. Thus humans have always been subject to earthquakes, and noteworthy examples from the 14th century until the present may be found in Table 4.1 Two recent events have been chosen for discussion: the Mexico City earthquake of 1985 and the less damaging but spectacular San Francisco earthquake in 1989.

Mexico City, 1985

On 19 September 1985 a magnitude-8.1 earthquake shook southwestern Mexico (➢ Figure 4.29). Had you been in Acapulco at the time, 300 kilometers (180 mi) south, you may not have even felt the earthquake. Nearer the epicenter, at Ciudad Guzman, you would have been rudely awakened from your sleep, whereas at Mexico City, 350 kilometers (220 mi) distant, you would have felt like the victim of a World War II bombardment. From afar, you would have seen Mexico City's high-rise buildings begin to shake, and then sway in increasingly larger arcs, and then 3,000 structures would collapse or topple before your eyes (➢ Figure 4.30).

To North American scientists, this earthquake was not a surprise. Historically, the Cocos plate is the most active in the Western Hemisphere, having generated 42 earth-

(a)

(b)

(c)

➢ **FIGURE 4.28 (*a–c*) Sequential photographs of the 1957 tsunami at Laie Point, Oahu. The photos show the arrival of a major wave and its progressive runup. This tsunami caused 57 deaths and $300,000 worth of damage.**

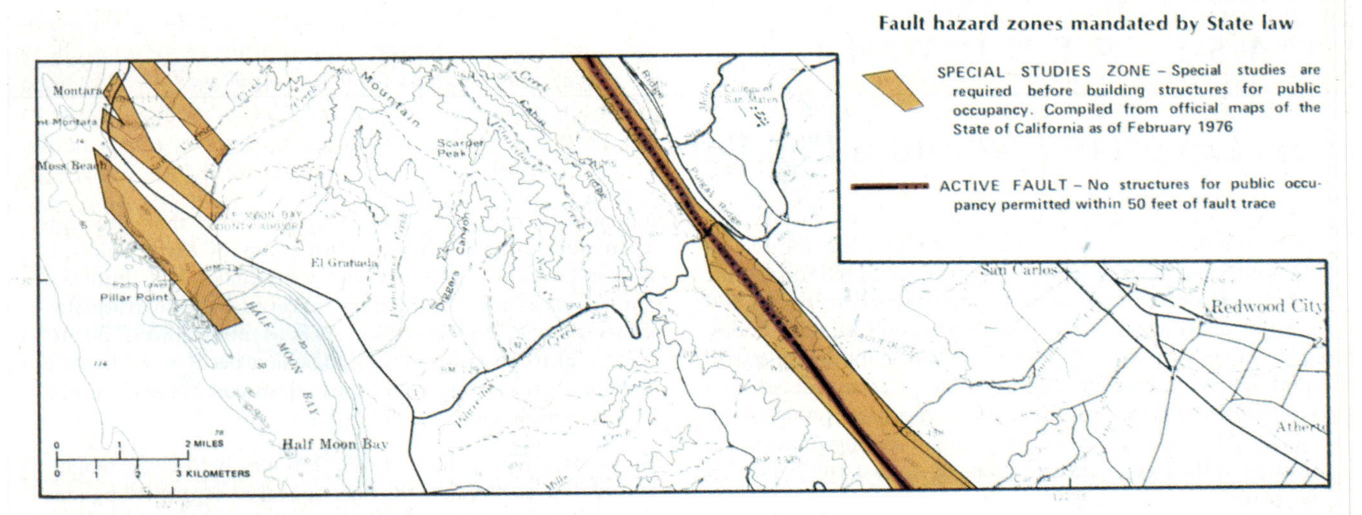

➢ **FIGURE 4.45 Alquist–Priolo Special Study Zone map of Half Moon Bay, California. The fault shown is the San Andreas.**

about 40 percent of the acceleration of gravity in the horizontal direction (0.4 g), and single-family dwellings are built to withstand about 15 percent (0.15 g). As one might expect, California is a leader in earthquake-hazard legislation. Fault-hazard rupture zones were targeted in 1972 and written into law as the Alquist–Priolo Special Studies Zones Act in 1973. Under the act, the state geologist is required to identify potential rupture zones along active faults. The affected cities and counties must regulate development within those zones and must not issue building permits within them until extensive geologic investigations demonstrate that specific sites are not susceptible to surface rupture or displacement from the fault. Special Study Zone maps are sufficiently detailed that planners can use their data for zoning and permit-granting decisions (➢ Figure 4.45). For the purposes of this act, an "active" fault is defined as one where surface displacement has occurred within Holocene time (about the last 11,000 years).

Finally, because a good defense is generally regarded as the best offense, knowing what to do when the Big One comes is important. Listed here are some suggestions for personal protection in the event of an earthquake:

- Above all, remain calm and consider the potential consequences of any action you may take.
- If indoors, watch for falling objects such as high bookcases, shelves, plaster, or bricks. If in danger, get under a desk or bed, or in a strong doorway.
- Usually it is best not to run outside, especially from a high-rise building. Power elevators may fail, and exits may be jammed with people.
- If outside, stay away from buildings, walls, power poles, and other objects that could fall. Move into an open area away from hazards. If driving in a car, stop in an open area.
- If at home, after the motion stops, turn off gas at the meter, check for downed power lines, and listen to a radio for pertinent information.

GALLERY: SEISMIC ODDITIES

Sometimes the pain of tragedy and loss can be softened by humor or humorous situations. This collage of photos presents examples of humorous, puzzling, and even downright comical results of the trembling earth. Please view them in this light (➢ Figures 4.46 and 4.47).

➢ **FIGURE 4.46 Statue of Louis Agassiz, the great geologist, naturalist, and educator, at Stanford University after the 1906 earthquake. Strong horizontal motion knocked the statue off its pedestal, prompting a noted scientist of the day to observe, "I have read Agassiz in the abstract. Now I see him in the concrete."**

(a)

(b)

(c)

➢ **FIGURE 4.47 San Fernando earthquake, 9 February 1971, 6:01 A.M. (*a*) Compression of sidewalk panels makes for rough skateboarding. (*b*) The approach to an Interstate-210 overpass settled during the earthquake, providing an upramp that launched an unknown motorist and his auto 46 feet through the air as indicated by the skidmarks. The flight ended with a safe landing. (*c*) The sign *said* NO PARKING!**

SUMMARY

Earthquakes

CAUSE: Movements on fractures in the crust known as *faults* that result in three types of wave motion: P- and S-waves, which are generated at the focus of the earthquake and travel through the earth, and L-waves, which are surface waves.

DISTRIBUTION: Most (but not all) large earthquakes occur near plate boundaries, such as the San Andreas fault, and represent the release of stored elastic strain energy as plates slip past, over, or under each other. Intraplate earthquakes can occur at locations far from plate boundaries where deep crust has been weakened, probably at "failed" continental margins.

MEASUREMENT SCALES: Some of the scales used to measure earthquakes are the Modified Mercalli Intensity Scale (based on damage), the Richter Magnitude Scale (based on energy released as measured by maximum wave amplitude on a seismograph), and moment magnitude (based on the total seismic energy released as measured by the rock rigidity along the fault, the area of rupture on the fault plane, and displacement).

Earthquake-Related Hazards

(1) GROUND SHAKING: Damaging motion caused by shear and surface waves.
Ways to reduce effects—seismic zoning, building codes, and construction techniques such as shear walls, seismic joints, and bolting frames to foundations.

(2) LANDSLIDES: Hundreds of landslides may be triggered by an earthquake in a slide-prone area.
Ways to reduce effects—proper zoning in high-risk areas (see Chapter 7).

(3) GROUND FAILURE (SPONTANEOUS LIQUEFACTION): Horizontal (lateral) land movements caused by loss of strength of water-saturated sandy soils during shaking and by liquefaction of quick clays.
Ways to reduce effects—building codes that require deep-drilling to locate liquefiable soils or layers.

(4) GROUND RUPTURE/CHANGES IN GROUND LEVEL: Fault rupture and uplift or subsidence of land as a result of fault displacement.
Ways to reduce effects—geologic mapping to locate fault zones, trenching across fault zones, implementation of effective seismic zonation like the Alquist-Priolo zones in California.

(5) FIRE: In the large earthquakes in Tokyo (1923) and San Francisco (1906), fire was the biggest source of damage.
Ways to reduce effects—public education on what to do after a quake, such as shutting off utilities and gas flow at meters.

(6) TSUNAMI: Multidirectional sea waves generated by disruption of the underlying sea floor.
Ways to reduce effects—early warning systems, building design that lessens impact of wave runup in prone areas.

Earthquake Prediction

STATISTICAL METHODS: Historical information on earthquakes in a region is used to calculate recurrence intervals and probabilities of damaging earthquakes. *Upside/downside*—good for planning purposes, but not for short-term warnings.

GEOPHYSICAL METHODS: Measurements of physical changes in the earth are used to identify seismic precursors. *Upside/downside*—vigorous and promising research is active on many potential precursors; limited results so far.

GEOLOGICAL METHODS: Active faults are studied to determine the characteristic earthquake magnitudes and recurrence intervals of particular fault segments; sediments exposed in trenches may disclose historic large fault displacements (earthquakes) and, if they contain datable C-14 material, their recurrence intervals. *Upside/downside*—useful for long-range forecasting along fault segments and for identifying seismic gaps; not useful for short-term warnings.

Key Terms

aftershock
asperity
body wave
dilatancy–diffusion model
elastic rebound theory
epicenter
fault
fault creep
first-motion studies
focus (pl. *foci)*
intensity scale
isoseismal
liquefaction
longitudinal wave
L-wave
magnitude
Modified Mercalli Scale (MM)
moment magnitude (M)
paleoseismicity
precursor
P-wave
recurrence interval
resonance
Richter Magnitude Scale
seismic gap
seismograph
seismogram
spontaneous liquefaction
stress
strain
S-wave
transverse (shear) wave
tsunami
wave period (T)

Study Questions

1. What is meant by elastic rebound, and how does it relate to earthquake motion?
2. Distinguish between intensity and magnitude scales. How can the scales be used to improve the quality of life in earthquake-prone country?
3. How does the amplitude of wave motion portrayed on a seismogram for a magnitude-4 earthquake compare to that for a magnitude-5 event, assuming the events are equidistant from the seismograph?
4. Why should one be more concerned about the likelihood of an earthquake in Alaska than of one in Texas?
5. In light of plate tectonic theory, explain why devastating shallow-focus earthquakes occur in some areas and only moderate shallow-focus activity takes place in other areas.
6. What are some earthquake precursors, and how useful is each as a predictor?
7. What are the 6 major considerations in designing against the impact of a severe earthquake on a major offshore fault?

8. Describe the motion of the three types of earthquake waves and their effects on structures.
9. Why do wood-frame structures suffer less damage than unreinforced brick buildings in an earthquake?
10. What geologic conditions were responsible for the extreme damage and loss of life in the 1985 Mexico City earthquake?
11. Explain why structures close to the epicenter of the 1989 Loma Prieta earthquake suffered less damage than buildings tens of miles away in the San Francisco and Oakland Bay areas.
12. How can the exact manner in which a fault moved to produce an earthquake be determined?
13. Define these terms: normal fault, reverse (thrust) fault, strike-slip fault, seismograph, seismogram, epicenter, focus, earthquake magnitude, earthquake intensity.

FURTHER READING

Bolt, Bruce. 1988. *Earthquakes.* New York: W. H. Freeman.

Bolt, B. A.; W. L. Horn; G. McDonald; and R. F. Scott. 1975. *Geological hazards.* New York: Springer-Verlag.

Borchardt, Glenn. 1991. Preparation and use of earthquake planning scenarios. *California Geology* 44, no. 3: 195–203.

Gere, James. 1984. *Terra non firma.* New York: W. H. Freeman.

Hodgson, J. H. 1964. *Earthquakes and earth sructures.* Englewood Cliffs, N.J.: Prentice-Hall.

Iacopi, Robert. 1971. *Earthquake country.* Menlo Park, Calif.: Lane Books.

Nance, John. 1989. *On shaky ground: America's earthquake alert.* New York: Avon Books.

Richter, C. 1958. *Elementary seismology.* New York: W. H. Freeman.

Scientific American. 1980. Earthquakes and volcanoes. *Scientific American Readings.* New York: W. H. Freeman.

Sherburne, R. W., and Chris Cramer. 1984. Focal mechanism studies: An explanation. *California Geology,* March.

U.S. Geological Survey. *Earthquakes and volcanoes.* Bimonthly publication. Washington: Government Printing Office.

———. 1990. *Probabilities of large earthquakes in the San Francisco Bay region, California.* Circular 1053 (free). Washington: Government Printing Office.

Walker, Bryce. 1982. *Earthquakes.* Alexandria, Va.: Time-Life Books.

Video

Nova, with WGBH Boston. *Predictable disaster.* 60 minutes.

CHAPTER

5

VOLCANOES AND THE ENVIRONMENT

Earth exists by geological consent, subject to change without notice.

WILL DURANT, HISTORIAN (1885–1981)

Although people commonly regard volcanoes and volcanic activity as frightening, environmentally dangerous geological phenomena, volcanoes are not all bad. They are attractive in a variety of ways: many of them are quite beautiful, they offer some of the best skiing and hiking in the world, and volcanic soils are highly productive. So why worry? Certainly the wealthy Romans vacationing on the Bay of Naples during the hot summer of A.D. 79 weren't concerned. Not until August 24th, that is, when Mount Vesuvius suddenly erupted and buried the cities of Pompeii and Herculaneum. Vesuvius had been quiet for hundreds of years and was thought to be extinct. Vegetation covered its slopes, and its central crater showed no signs of recent activity. In two days Pompeii was buried so completely that its ruins were not uncovered for almost 17 centuries. Although Pompeii was rediscovered in the sixteenth century during excavation for a water line, systematic work on the buried ruins did not begin until 1738. This date marks the beginning of the discipline we know as archaeology (➢ Figure 5.1). To date about half of Pompeii's 160 acres (65 hectares) have been exhumed. Herculaneum, settled by the Greeks and named after Hercules, is less well exposed because of its thick volcanic covering and because Portici, a new city, was built on top of its grave.

Incandescent spatter lights the sky above Paricutín Volcano, state of Michoacán, west-central Mexico. Paricutín is one of the youngest volcanoes on earth, breaking through to the surface in a farmer's field on February 20th, 1943. Its well documented eruptive phases ended in 1952.

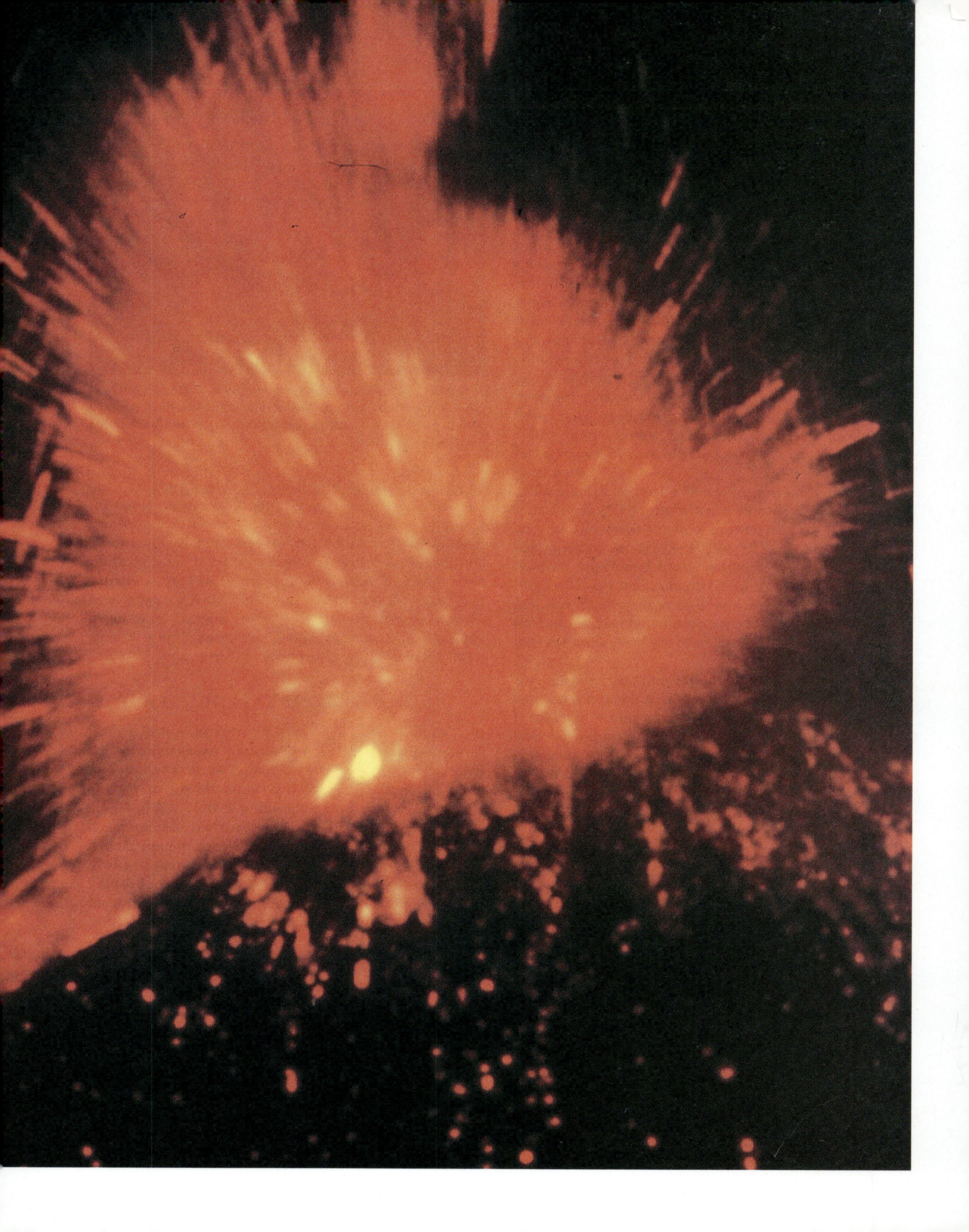

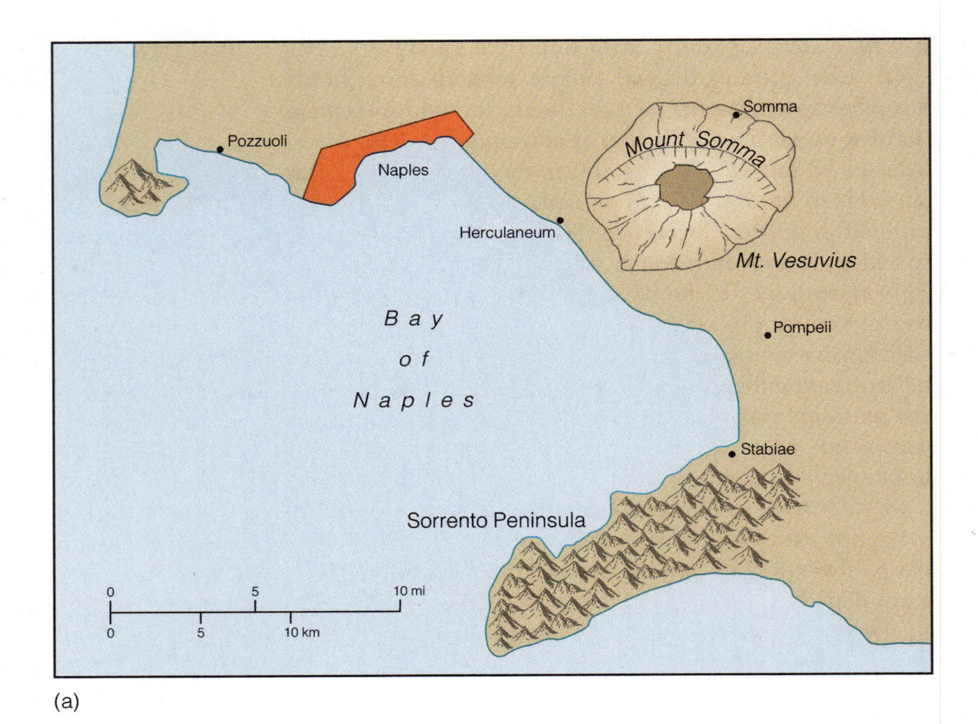

(a)

(b)

(c)

➢ **FIGURE 5.1** ***(a)* Map showing the locations of Mount Vesuvius, Herculaneum, and Pompeii on the Bay of Naples. *(b)* Air view of exhumed Pompeii as it looks today. *(c)* Naples with sleeping Vesuvius in the background. Note the remnants of an older, higher cone to the left of Vesuvius. Known as Mount Somma, it was obliterated during the A.D. 79 eruption.**

The heavy fall of ash at Pompeii suffocated more than 3,000 inhabitants and eventually buried the city under 6 meters (20 ft) of *pyroclastic* ("fire-broken") material. The story is preserved as molds in stone. Some people died attempting to flee the heavy fallout, whereas others, such as gladiators and slaves who were behind locked doors, had no chance to escape. An aristocratic woman was found in the gladiators' barracks; another died while attempting to retrieve her jewels. Imagine the desolation in late August A.D. 79 caused by the sudden blanketing with volcanic ash and cinders of hundreds of square kilometers, covering everything but an occasional spire or other tall structure.

Herculaneum, to the north and away from the wind-driven plume of ash that destroyed Pompeii, suffered a different fate. Water vapor emitted from the volcano precipitated as rain on the flanks of the mountain and mixed with the loose ash on its northern slopes to become great mudflows. Mudflows covered Herculaneum to a depth of 20 meters (65 ft). Although such flows usually move rapidly, the residents may have been able to evacuate the city, as only 30 bodies have been discovered to date. It is doubtful that the death toll at Herculaneum will be known because of the great depth of burial.

The A.D. 79 eruption of Vesuvius illustrates the nature of the volcano problem. Volcanoes may be intermittently active for a million years, and some continental ones are known to have erupted sporadically over 10-million-year periods. Thus, if dormant periods are sufficiently long, settlements may grow on or near a volcanic cone that later erupts. However, knowing that a potential volcanic problem exists allows policy makers, scientists, and others to plan for various eruption—and evacuation—scenarios, thus reducing casualties should an eruption occur. If successful, this can enable people to live in harmony with a sleeping giant.

WHO SHOULD WORRY

Most geologic activity that is dangerous to humans occurs along plate boundaries. This is where we find earthquakes and active volcanoes. Explosive volcanic activity, which presents the greatest challenge to life and property, occurs mostly at convergent plate boundaries. Inspection of the Pacific Ocean basin shows that it is surrounded by trenches that form at convergent plate boundaries. Associated with these subduction zones is the so-called Ring of Fire, the location of two thirds of the world's active violent volcanoes. A recent catalog of the 1,350 volcanoes that have been active within the last 10,000 years (*Holocene time* to the geologist) shows 900 of them around the Pacific Rim. The largest number are found in New Zealand, Japan, Alaska, Mexico, Central America, and Chile (➢ Figure 5.2). Mount Erebus in Antarctica is the southernmost active volcano in this belt. About 250 active volcanoes are found in the Mediterranean, including the famous volcanoes Vesuvius, Etna, and Stromboli. Stromboli has been called the "lighthouse of the Mediterranean" because of the nearly continuous activity in its crater for more than a century.

The most productive volcanic centers, which exhibit less spectacular, comparatively mild activity, occur along divergent plate boundaries at mid-ocean ridges. Hot-spot volcanism may be explosive or mild and may occur on the

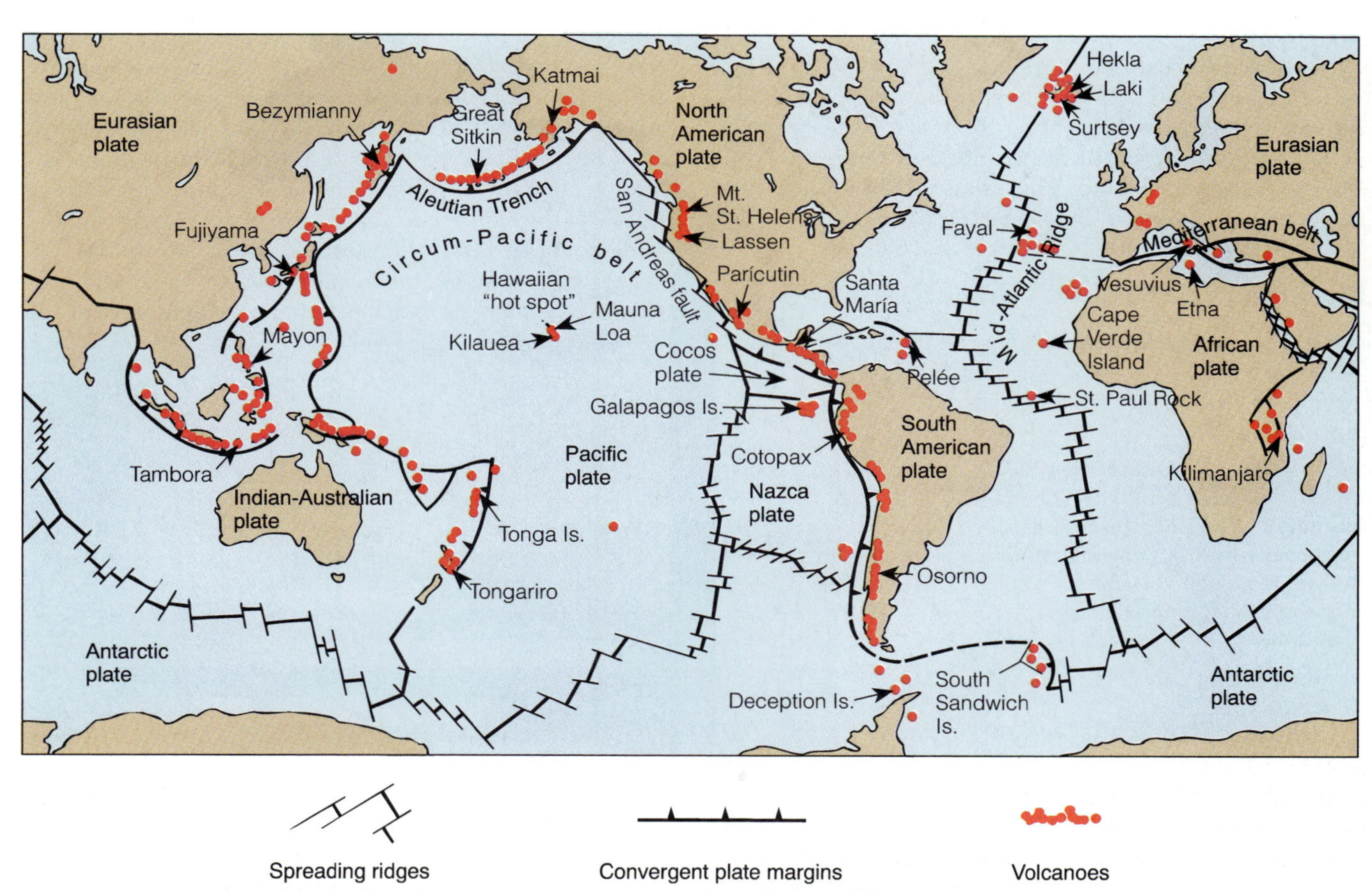

➢ **FIGURE 5.2 Distribution of earth's active volcanoes at plate boundaries and hot spots. Note the prominent "Ring of Fire" around the Pacific Ocean, which contains 900 (66%) of the world's active volcanoes. The remaining 450 are in the Mediterranean belt (subduction zones) and at mid-ocean ridge spreading centers (divergent boundaries). A few important volcanic centers are related to hot spots, such as those in the Hawaiian and Galápagos Islands.**

continents or in ocean basins (see also Chapter 3).

The distribution of the earth's active volcanoes according to latitude is shown in ➢ Figure 5.3. This information is particularly interesting to meteorologists, because large amounts of volcanic ash in the stratosphere result in a net heat loss for the earth, which affects climate. (High-flying jets are also affected; see Case Study 5.1). Any concentration of active volcanoes at a particular latitude must be considered as a potential agent of change in global climate. Although most climatic variability is related to causes other than volcanism, historical examples of volcanic eruptions affecting weather are those of Tambora (Indonesia, 1815), El Chichón (Mexico, 1982), and most recently, Mount Pinatubo (the Philippines, 1991). The figure shows that two thirds of known active volcanoes are found in the Northern Hemisphere, as is two thirds of the earth's land area. For this reason the Northern Hemisphere is more vulnerable to climatic impact by volcanic eruptions than is the southern half of the world.

Potentially dangerous volcanoes and volcanic centers in the contiguous 48 states are found in Washington (6), Oregon (21), and California (16) (➢ Figure 5.4). These are related to the Cascadia subduction zone, along which the volcanoes of the Cascade range occur. The zone of active volcanoes extends 1,100 kilometers (700 mi) from Mount Lassen in California northward into British Columbia. Alaska, our 49th state, ranks as the second-most active volcanic region in the world, and the 50th state, Hawaii, is not far behind. Thus the study of volcanoes is important to citizens of the United States.

The Nature of the Problem

How a volcano erupts determines its impact on humankind. A cataclysmic eruption, such as that of Mount St. Helens in 1980, has serious results. Simple outpourings of lava such as at Kilauea in Hawaii, on the other hand, can be good for tourism and business (except agriculture). Unfortunately, there are no measurement scales for describing the "bigness" of volcanic eruptions as there are for earthquakes. Table 5.1 shows some of the criteria commonly used to describe an eruption's "explosivity." The **Volcanic Explosivity Index (VEI)** values range from 0 to 8 according to the volume of material ejected, the height to which the material rises, and the duration of the eruption. The "Classification" scale associates the particular eruption with a well-known volcano that exhibited the same kind of activity. The "Description" scale employs adjectives such as those used in newspaper headlines to describe the eruption. The 1980 eruption of Mount St. Helens, for example, could be indexed as a *4* and appropriately described as an *explosive-to-cataclysmic* eruption.

A volcano's potential explosivity has been found to increase as the time since the start of its previous eruption increases (➢ see Figure 5.5 p. 108). This fact allows a region's risk of volcanic eruptions to be assessed when the date of the volcano's last eruption is known.

A lava's *viscosity*—its resistance to flow—is another determinant of a volcano's explosivity, and viscosity is a function of lava's temperature and composition. Viscosity varies inversely with temperature for most fluids, including lava: the higher the temperature, the less viscous; the lower the temperature, the more viscous, other factors being equal. A lava's composition, most importantly its silica (SiO_2) content, also influences its viscosity. The silicon–oxygen bond is very strong; considerable heat en-

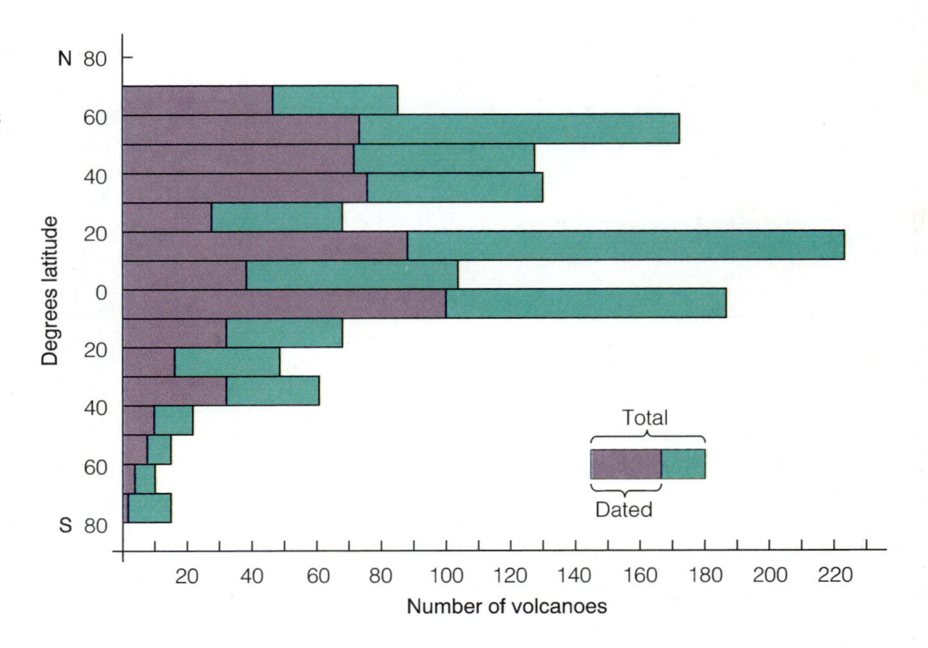

➢ FIGURE 5.3 Distribution of known volcanoes per 10 degrees of latitude. The purple color indicates volcanoes whose last date of eruption is known. Note that volcanoes are much more numerous in the Northern Hemisphere.

➢ **FIGURE 5.4 Active volcanoes and volcanic centers in the Western United States. The volcanoes of the Cascades are geologically related to the Cascadia subduction zone.**

ergy is required to break it. The fluidity of siliceous (containing silica) lavas depends on the continual breaking and remaking of this bond. Thus, the higher the SiO_2 content, the more viscous is the lava—again, other factors being equal.

Viscous lavas with a high gas content can build up such high gas pressures as to be explosive, whereas gases escape readily from fluid lavas, and hence they are much less dangerous. This fact gives us another method of categorizing large volcanoes and their activity: the nature of

TABLE 5.1 Modified Volcanic Explosivity Index (VEI)

VEI	0	1	2	3	4	5	6	7	8
GENERAL DESCRIPTION	NON-EXPLOSIVE	SMALL	MODERATE	MODERATE LARGE	LARGE	VERY LARGE			
Cloud column height (km)	<0.1	0.1–1	1–5	3–15	10–25	25			
Qualitative description	Gentle	Effusive	Explosive		Cataclysmic, paroxysmal, colossal				
Classification		Strombolian				Plinian			
		Hawaiian		Volcanian				Ultra-Plinian	
Total historic eruptions	487	623	3176	733	119	19	5	1	0

SOURCE: Smithsonian Institution's Volcano Reference File (*Volcanoes of the World*, Hutchinson Ross, 1981) with some fatality data from R. J. Blong (*Volcanic Hazards*, Academic Press, 1984). VEI descriptions in *Volcanoes of the World* and Newhall and Selt, *Journal of Geophysical Research (Oceans & Atmospheres)* v. 87, 1982.

CASE STUDY 5.1

Volcanic Ash and the Friendly Skies

One night in 1982 a British Airways Boeing 747 was on a routine flight from Kuala Lumpur, Malaysia, to Australia, cruising at an altitude of 12,000 meters (37,000 ft). Just before midnight, sleeping passengers were awakened by a pungent odor filling the cockpit. Looking out the windows, they saw that the huge plane's wings were lit by an eerie blue glow. Suddenly the number-4 engine flamed out. Practically immediately the other three did as well, and the plane began to lose altitude. At 4,500 meters (14,500 ft), the number-4 engine was restarted, followed by numbers 2, 1, and 3. Nonetheless, an emergency was declared, and the plane landed in Jakarta, Indonesia, with only three engines operating.

This near-death experience gained the attention of the world's airline passengers and pilots. Before this, such a failure had seemed virtually impossible in modern aircraft, which have redundant (backup) systems for almost every contingency. Unfortunately, flying air-gulping jet engines through clouds of volcanic ash is not one of these.

In December 1989, a KLM Boeing 747 encountered airborne ash from Redoubt Volcano at about 8,500 meters (28,000 ft) during a descent to land at Anchorage, Alaska (➢ Figure 1). All four engines flamed-out, causing the large aircraft to suddenly become a glider. Thousands of feet lower, the engines were restarted, and the aircraft continued on to make what was described as an "uneventful" landing. All four engines required replacement, as did the windshield and the leading edges of the wings, flaps, and vertical stabilizer, all of which had been "sandblasted." One may wonder what is required to make a landing "eventful." The interior of the plane was so filled with ash that the seats and avionic equipment had to be removed and cleaned. The total cost of returning the aircraft to service was $80 million.

Volcanic eruptions constitute a significant hazard to civil air transportation. Numerous encounters with ash clouds have been reported by aircraft since the late 1970s, most recently from the Mount Pinatubo eruptions of June 1991, as shown in the table. No fewer than 11 commercial aircraft declared emergencies in the air during the Mount Pinatubo eruptions and required extensive maintenance after landing (➢ Figure 2). Similar incidents have been reported over Central America, Chile, and Malaysia, going back as far as World War II, when Allied bombers encountered ash in the skies over Italy.

The U.S. Weather Service has established procedures for issuing in-flight Aviation Weather Advisories Services (AWAS) regarding conditions associated with volcanic eruptions. Such advisories state that an eruption has occurred and give information regarding the extent and movement of the airborne volcanic material. These advisories are issued as Significant Meteorological Advisories (SIGMETS), which are the most urgent weather warnings the service issues. A SIGMET was issued in August 1992 when Mount Spurr in Alaska erupted, shutting down the Anchorage airport.

Selected Aircraft Encounters with Mount Pinatubo Ash, June 1991

DAY	LOCATION	COMMENTS
14	Not reported	Two engines had to be replaced; cockpit and cabin were contaminated.
15	10.5-km altitude	Temperature of all four engines rose; sparks flew from the windows.
15	Approach to Manila	Ash in engines; sulfur odor; electrostatic discharge; abrasion.
15	1100 km west, 9-km altitude	All 4 engines were damaged.
18	11.0-km altitude	Engine #1 stalled, # 4 lost power; descended to 9.0 km to restart; engine #1 had to be replaced.

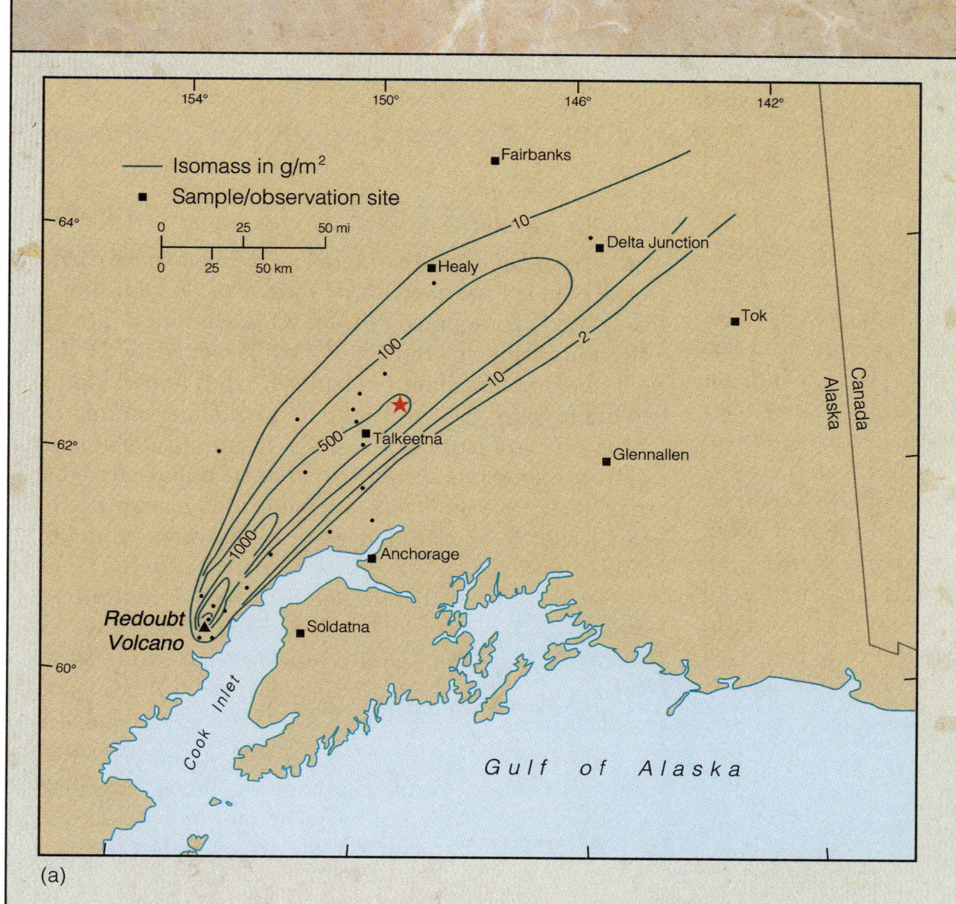

(a)

(b)

➢ FIGURE 1 (*a*) Deposits of tephra from Mount Redoubt's eruption of 15 December 1989. (The star indicates the KLM 747's position en route from Fairbanks to Anchorage when it encountered the ash cloud and flamed out.) Isomass contours enclose areas of equal tephra weight. Note that the weight of tephra near the vent was greater than a kilogram per square meter. (*b*) A worker removes ash from a Boeing 747 at Moses Lake, Washington, following the 1980 eruption of Mount St. Helens.

➢ FIGURE 2 Heavy ashfall from Mount Pinatubo made this World Airways DC-10 very tail heavy; Cubi Point Naval Air Station.

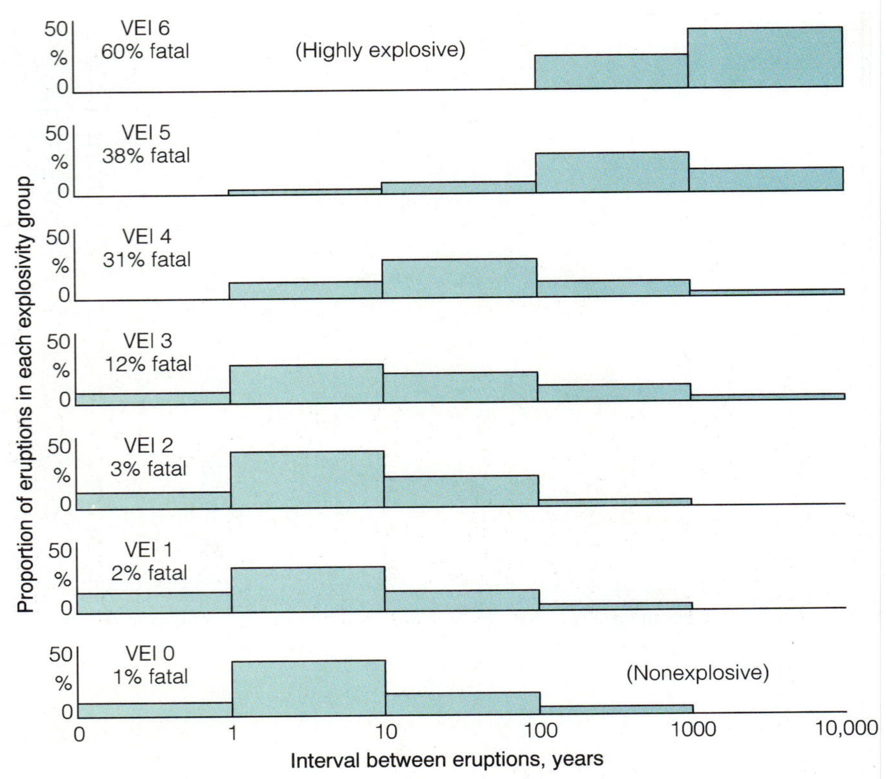

➢ **FIGURE 5.5 The longer the time interval since an eruption, the greater is the potential explosivity of the next eruption. These data, for 4,320 historic eruptions, relate known intervals between eruptions to the Volcanic Explosivity Index (Table 5.1). Also shown is the percentage of eruptions in each VEI that have caused fatalities.**

the magmas they tap. Magmas form at depths of 50 to 250 kilometers (30–150 mi), where temperatures are sufficiently high to melt rocks completely or partially. At mid-ocean ridges and hot spots, volcanoes draw from magmas in the upper mantle that are high in iron and magnesium and low in silica (50% or less). These are **mafic** magmas (*ma-* for magnesium and *-f-* for the Latin word for iron, *ferrum,* plus *-ic*), which yield fluid (low-viscosity) lavas that retain little gas and hence do not erupt violently. Volcanoes that are adjacent to subduction zones, on the other hand, tap magmas that are a mixture of oceanic crust and sediment, upper-mantle material, and melted continental rocks. These are **felsic** magmas (*fel-* for feldspar and *-s-* for silica, plus *-ic*), which yield thick, pasty lavas with SiO_2 contents up to 70 percent. Even though hot lavas have lower viscosities and flow more readily than do cool ones, just as with honey and various other common fluids, the major determinant of lava fluidity is silica content.

High-silica, high-viscosity lavas retain gases, which leads to violent, explosive-type eruptions. The geologically important boundary between mild oceanic eruptions and the more explosive continental ones of the Pacific basin is called the **andesite line** (after the rock andesite from the Andes Mountains). The line is generally drawn southward from Alaska to east of New Zealand by way of Japan, and along the west coasts of North and South America. The andesite line is also a petrologic boundary between mafic magmas that yield basalt, and felsic magmas that yield andesite or some other viscous high-silica rock. Thus, explosive volcanoes are always found on the continental side of the andesite line, and gentle activity characterizes volcanoes within the Pacific Ocean basin.

Types of Eruptions and Volcanic Cones

The type of volcanic eruption determines the shape of the structure or cone that is built, and by appearance alone, one can get an idea of a volcano's hazard potential (➢ Figure 5.6).

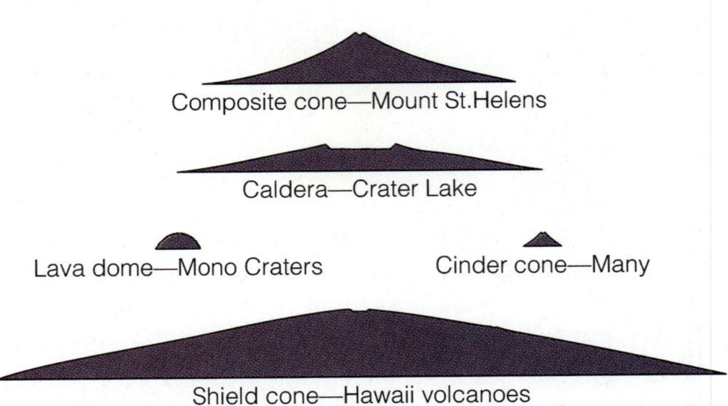

➢ **FIGURE 5.6 Shape, relative size, and example of each type of volcanic cone. Note the large diameter of the Crater Lake caldera and its resemblance to a truncated composite cone.**

➢ **FIGURE 5.7 Shield volcano Fernandina in the Galápagos Islands. This active volcano is related to the Galápagos "hot spot." Its convex profile is similar to those of the volcanoes of Hawaii.**

Quiet Eruptions

SHIELD CONES. **Shield volcanoes** are built by gentle outpourings of fluid lavas from a central vent or conduit, and the lavas cool to form basalt, the most common volcanic rock. The name is due to the fact that a shield volcano's profile is gently convex upward like that of a shield laid on the ground. Mostly oceanic in origin, shield volcanoes are found in Iceland (where the *shield* name was first applied), the Galápagos Islands, and the Hawaiian Islands (➢ Figure 5.7). They are built up from the sea floor, layer upon layer, to elevations thousands of meters above sea level. Mauna Loa ("long mountain") and Mauna Kea ("white mountain") volcanoes on the island of Hawaii are shield volcanoes that project 4.5 kilometers (2.8 mi) above sea level, and their bases are in water about 5 kilometers (3 mi) deep. With a total height of 9.5 kilometers (31,000 ft), these are the highest mountains on earth, exceeding Mount Everest by about 650 meters (2,150 ft). The Island of Hawaii is composed of 5 separate volcanoes (➢ Figure 5.8), the most active of which is Kilauea ("much spreading"), the most easterly of the group. This is exactly what plate movement would predict as the sea floor moves northwesterly over the Hawaiian "hot spot" (see Chapter 3, Plate Tectonics). Loihi, a new volcano, is forming on the sea floor southeast of Kilauea, providing further evidence that the Pacific plate is moving northwest over the plume.

It is not uncommon for a shield volcano to erupt from a *fissure,* or crack, on its flanks, rather than from a central vent. This is typical of Kilauea's east fissure zone, where the countryside has been flooded under a sea of lava. Pu'u o'o ("hill of the o'o bird") is a large crater built upon a vent in the east fissure zone. Pu'u o'o's crater has reached impressive proportions as a result of scores of eruptions since 1983. During Pu'u o'o's eruptions, Kilauea's summit deflates slightly, rising again or reinflating between eruptive episodes. This indicates that the fissure zone and Kilauea's vent plumbing are connected with the main lava reservoir. Lava flows associated with Pu'u o'o caused significant damage in the Royal Gardens subdivision near

➢ **FIGURE 5.8 Satellite photograph of the Island of Hawaii. North is approximately at the left of the photo. The two most active volcanoes visible in the photograph are the older, snow-capped Mauna Kea in the north and Mauna Loa, conspicuous for the abundant, fresh-looking basalt flows on its flanks in the south-central part of the island. The east fissure zone and Kilauea, the youngest active volcano, are to the southeast of Mauna Loa. The Pacific plate is moving from southeast to northwest.**

(a)

(c)

(b)

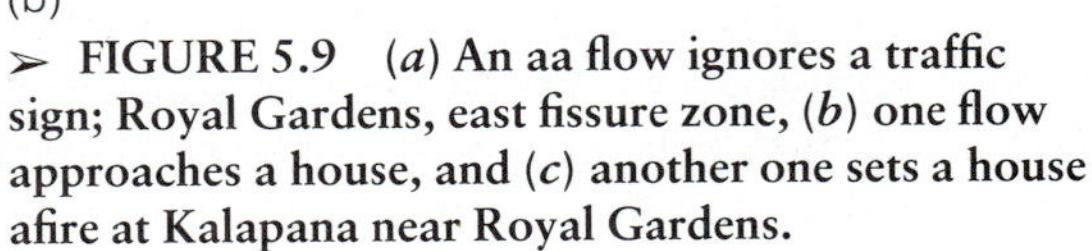
➢ **FIGURE 5.9** ***(a)* An aa flow ignores a traffic sign; Royal Gardens, east fissure zone, *(b)* one flow approaches a house, and *(c)* another one sets a house afire at Kalapana near Royal Gardens.**

Kalapana about 8 kilometers (5 mi) from the vent. Flows destroyed at least 75 homes and covered 10 kilometers (6 mi) of residential streets (➢ Figure 5.9). In addition, lava has built out hundreds of meters into the sea, creating new lands.

Spectacular lava fountains 400 (1300 ft) meters high have occurred with fissure eruptions at Pu'u o'o (➢ Figure 5.10). Some of the more fluid droplets set into glassy, tear-shaped blobs known as "Pele's tears," named after the Hawaiian goddess of volcanoes. Trailing behind these droplets may be tiny threads of lava, some more than a meter in length, that detach and fall to earth as "Pele's hair" (➢ Figure 5.11). Some of the detached fibers are carried many kilometers by the wind. Birds have been known to make nests from some of the more pliable threads of volcanic glass. Volcanic glass, which has a disordered or random atomic structure, forms when rapid cooling or quenching prevents the formation of orderly crystals.

➢ **FIGURE 5.10** ***(Left)* Lava fountaining at Pu'u- o'o, east fissure zone, Kilauea volcano, Island of Hawaii.**

➢ **FIGURE 5.11 "Pele's hair," thin filaments of glassy lava that have been carried by wind.**

FISSURE ERUPTIONS. In some shield volcano areas, lava flows *only* from long fissures, forming broad **lava plateaus** (➢ Figure 5.12a). Lava plateaus are found in Iceland, India, and the Columbia River Plateau in Washington, Oregon, and Idaho in the United States. The Deccan lava plateau of India (➢ Figure 5.12b) is enormous. Its massive outflow with degassing of toxic sulfurous fumes during Cretaceous time has been linked by some paleontologists to the extinction of the dinosaurs. In addition, many submarine eruptions at oceanic spreading centers are fissure eruptions.

Explosive Eruptions

COMPOSITE CONES. Explosive volcanic activity builds **composite cones.** These cones are typically thousands of feet high with a concave upward profile and a central vent. Composite cones are stratified, consisting of alternating layers of ash, cinders, and lava. The upper, steep slopes are mostly formed of **pyroclastic** ("fire-broken") volcanic ejecta, and the less steep, lower slopes are composed of alternating layers of lava and pyroclastics. These cones are some of the most beautiful tourist attractions in the world. Noteworthy examples are Mount Vesuvius in Italy, Mount Fujiyama in Japan, Mount Hood in Oregon, Mount Rainier in Washington, and Agung in Bali (➢ Figure 5.13). They occur on the landward side of subduction zones, where melting of oceanic crust and mantle forms magmas that rise because they are expanded and thus more buoyant than the surrounding lithosphere. As bodies of magma rise in the lithosphere, they mix with continental rocks and form thick, viscous, gas-charged lavas. These lavas do not flow easily; rather, they congeal in the volcano's central conduit (vent), which permits gas pressures to build to explosive proportions. For this reason, composite-cone volcanoes present the most immediate threat to humans. Some eruptions have been so great that large cones have simply disappeared in the explosion. Crater Lake in Oregon represents the stump of a very

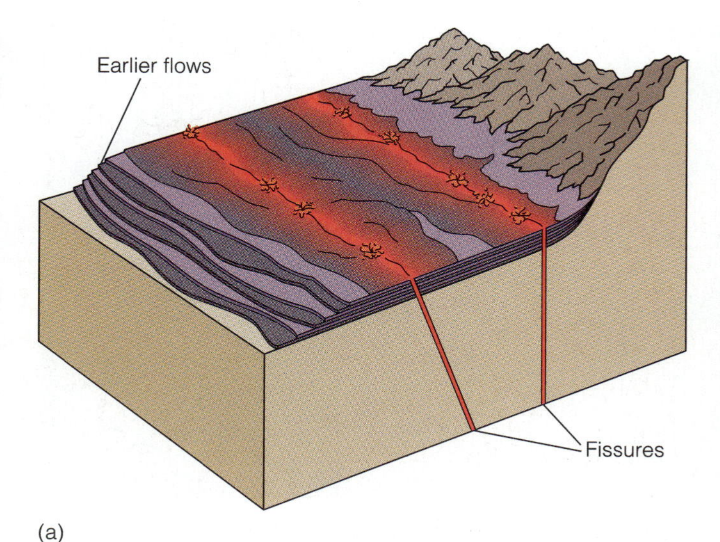

(a)

(b)

➢ **FIGURE 5.12** (*a*) **Fissure eruptions forming a lava plateau.** (*b*) **Antoja caves, Buddhist shrines excavated into the basalts of the Deccan lava plateau, India.**

large composite cone, called Mount Mazama; it was probably about the size of Mount Rainier. About 6,900 years ago, 70 cubic kilometers (17 mi^3) of Mount Mazama disappeared, due in part to explosive activity, but mostly due to the collapse of the remaining cone into the deflated magma chamber beneath it. The "crater" of Crater Lake is a **caldera,** defined as a volcanic crater that is many times larger than the vent that feeds it (➢ Figure 5.14). Calderas may form by explosive disintegration of the top of a volcano, by collapse into the magma chamber, or by both mechanisms. Yellowstone National Park occupies several huge calderas, as does Long Valley in the eastern Sierra Nevada, both of them lying above shallow bodies of magma that are capable of producing destructive eruptions.

LAVA DOMES. Lava domes are formed when bulbous masses of lava pile up around the vent because the lava is

(a)

(b)

➢ **FIGURE 5.13 Two beautiful examples of composite cones from widely separated areas. (*a*) Mount Rainier, Washington, with active glaciers on its flanks. (*b*) Agung Volcano, Bali.**

(a)

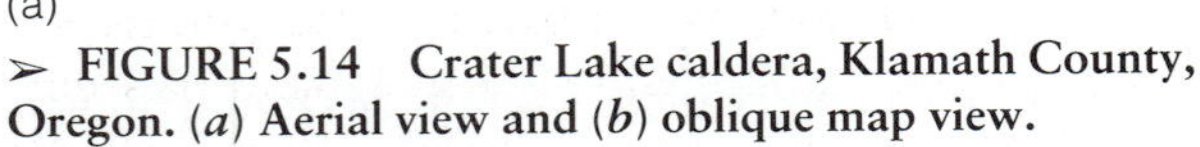

➢ **FIGURE 5.14 Crater Lake caldera, Klamath County, Oregon. (*a*) Aerial view and (*b*) oblique map view.**

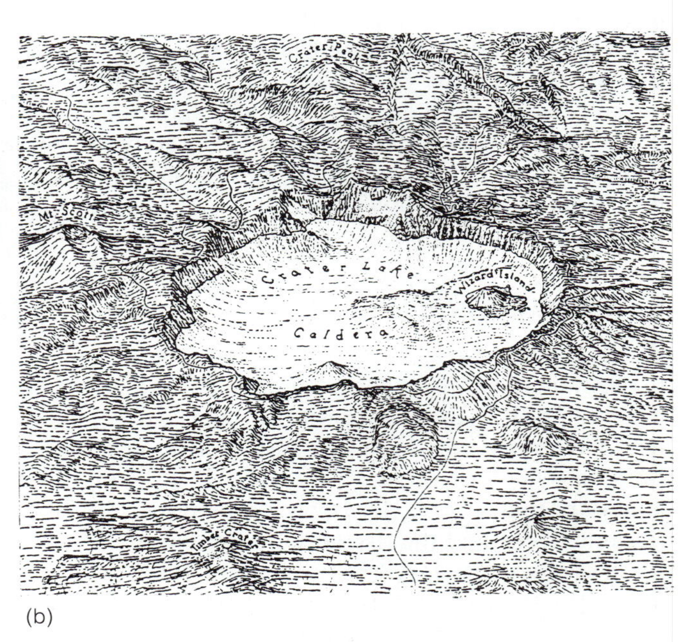

(b)

too thick and viscous to flow any significant distance from its source. They usually grow by expansion from within. As the outer surface cools, the brittle crust breaks and tumbles down the sides, as at the Mono Craters on the eastern Sierra Nevada of California (➢ Figure 5.15). Relatively small domes may form in the crater of a larger composite cone, such as the one that formed in the crater of Mount St. Helens after the 1980 eruption (➢ Figure 5.16). The source vents of large domes may be substantially plugged, which offers the potential for explosive eruptions, particularly where the lavas have access to underground water or seawater. Mount Pelée on Martinique in the West Indies, Mount Lassen, Mono Craters, and Mammoth Mountain in California (one of the largest ski areas in the United States) are all dormant lava domes. They extrude lavas with silica contents of 65–75 percent, such as rhyolite, and they also extrude glassy rocks that cool quickly, such as obsidian and pumice. Although pumice is largely SiO_2, it does not look like glass as does obsidian. This is because it is derived from gas-charged magmas and is spongy and full of gas holes, in contrast to smooth, glasslike obsidian. It should be noted that not all volcanic glasses are silica-rich; basalts that are chilled by extrusion underwater or by flowing into the sea sometimes form a glass.

➢ FIGURE 5.15 Lava domes; Mono Craters, east-central California. The Sierra Nevada range is in the background.

➢ FIGURE 5.17 Cinder cones; Owens Valley, California.

(a)

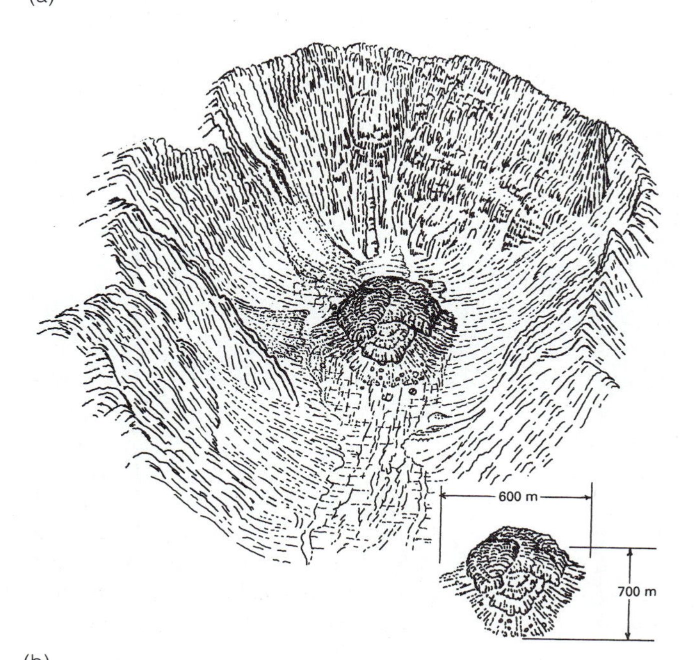

(b)

➢ FIGURE 5.16 (*a*) Steaming lava dome in crater of Mount St. Helens, October 1981. (*b*) Oblique map of same lava dome, 8 October 1981.

CINDER CONES. **Cinder cones** are the smallest, and most numerous of volcanic cones (➢ Figure 5.17). They are built of pyroclastic material of all sizes—from blocks and bombs to the finest ash (Table 5.2). Pyroclastic material of all shapes and sizes is collectively known as **tephra.** *Bombs* are blobs of still-molten lava that assume an aerodynamically induced spindle shape and solidify before striking the ground (➢ Figure 5.18). They can be found in great numbers around certain cinder cones and are prized by collectors and geologists. Cinder cone activity is local, within a few kilometers of the source vent, and is usually short-lived. The only new volcano to form in historic time in North America erupted in a farm field near the village of Paricutín in the state of Michoacán, Mexico, in 1943 (see Chapter opening photo). Eruptions caused by frothing gases hurled blobs of lava into the air that became cinders and then fell back around the vent, eventually building a cone nearly 400 meters (1,312 ft) high. An observatory was established on a nearby hill, and the volcano's every burp and belch was recorded for nine years. After the cone-building stage had passed, basaltic lava flowed from the crater, inundating the nearby pueblo of San Juan and temporarily removing about 160 square kilometers (100 mi^2) from agriculture. Fortu-

TABLE 5.2 Classification of Pyroclastic Ejecta

NAME	SIZE	CONDITION WHEN EJECTED
Blocks	>32 mm	Cold, solid
Bombs	>32 mm	Hot, plastic
Lapilli (cinders)	4–32 mm	Molten or solid
Ash	¼–4 mm	Molten or solid
Dust	<¼ mm	Molten or solid

➢ **FIGURE 5.18 A volcanic bomb; Galápagos Islands, Ecuador. Note the spindle shape that the bomb acquired as it was flung through the air.**

nately, no lives were lost. Cinder cones are generally thought to be one-shot events; that is, once the eruption sequence ends, it is rarely reactivated. Nonetheless, exceptions are known, and one usually finds many cinder cones within a volcanic area.

VOLCANIC PRODUCTS

Pyroclastic Materials

Volcanic activity provides products that the earth's peoples utilize in many ways. The final polish your teeth receive when you get them cleaned involves volcanism. Dental pumice is refined and flavored volcanic pumice with a hardness just slightly less than tooth enamel. "Lava" brand hand soap, a rough, abrasive bar soap, contains the same powdered rock. Light-weight bricks, cinder blocks, and many road-foundation and decorative-stone products originated deep within the earth. Where volcanic cinders are mined for road base, they may also be mixed with oil to form asphalt pavement. The reddish pavement of state highways in Nevada and Arizona contains basalt cinders that are rich in oxidized iron, which provides the color. Pieces of rock pumice may be purchased in drugstores, supermarkets, and hardware stores to use as a mild abrasive for removing skin calluses and unsightly mineral deposits from sinks and toilets. Powdered, it is used in abrasive cleaners and in furniture finishing.

Glassy volcanic rock, such as obsidian, is easy to chip and form. Hence, it was used to make tools and arrowheads by native Americans and stone-age people in many parts of the world. Today obsidian and many similar volcanic rocks are the raw material of "rock hounds" and artisans for producing polished pieces and decorative art.

Geothermal Energy

By far the most important and beneficial volcanic product is also the most hazardous; it is heat. The same heat energy that causes eruptions also drives geysers and hot springs, and when controlled, it can be converted to other uses. Not surprisingly, the prospects for geothermal energy are best at or near plate boundaries where active volcanoes and high heat flow are found. The Pacific Rim (Ring of Fire), Iceland on the Mid-Atlantic Ridge, and the Mediterranean belt offer the most promise. The first facility to utilize geothermal energy was built in 1905 at Larderello, Italy. Today, the United States, Japan, New Zealand, Mexico, and countries of the former USSR are utilizing energy from earth heat. Iceland uses geothermal energy directly for space heating (one geothermal well there actually began spewing lava), and many other countries have geothermal potential (➢ Figure 5.19). The versatility of earth heat ranges from using it to grow mushrooms to driving steam-turbine generators with it (➢ Figure 5.20).

Geothermal energy fields may be found and exploited where magma exists at shallow depths and where there is sufficient underground water to form steam. Where the water is heated to 200°–250°C, it can be used to produce electricity. Geothermal fields are classified as either *steam-dominated* systems or *hot-water-dominated* systems. Deep, insulated reservoirs may produce live steam, whereas shallower reservoirs contain hot water, some of which flashes to steam at the surface. Yellowstone National Park and Wairaki, New Zealand, are examples of

➢ **FIGURE 5.19 Wairaki, one of several large volcanic centers on the North Island of New Zealand. It produces significant amounts of geothermal energy in a country that is not energy-rich.**

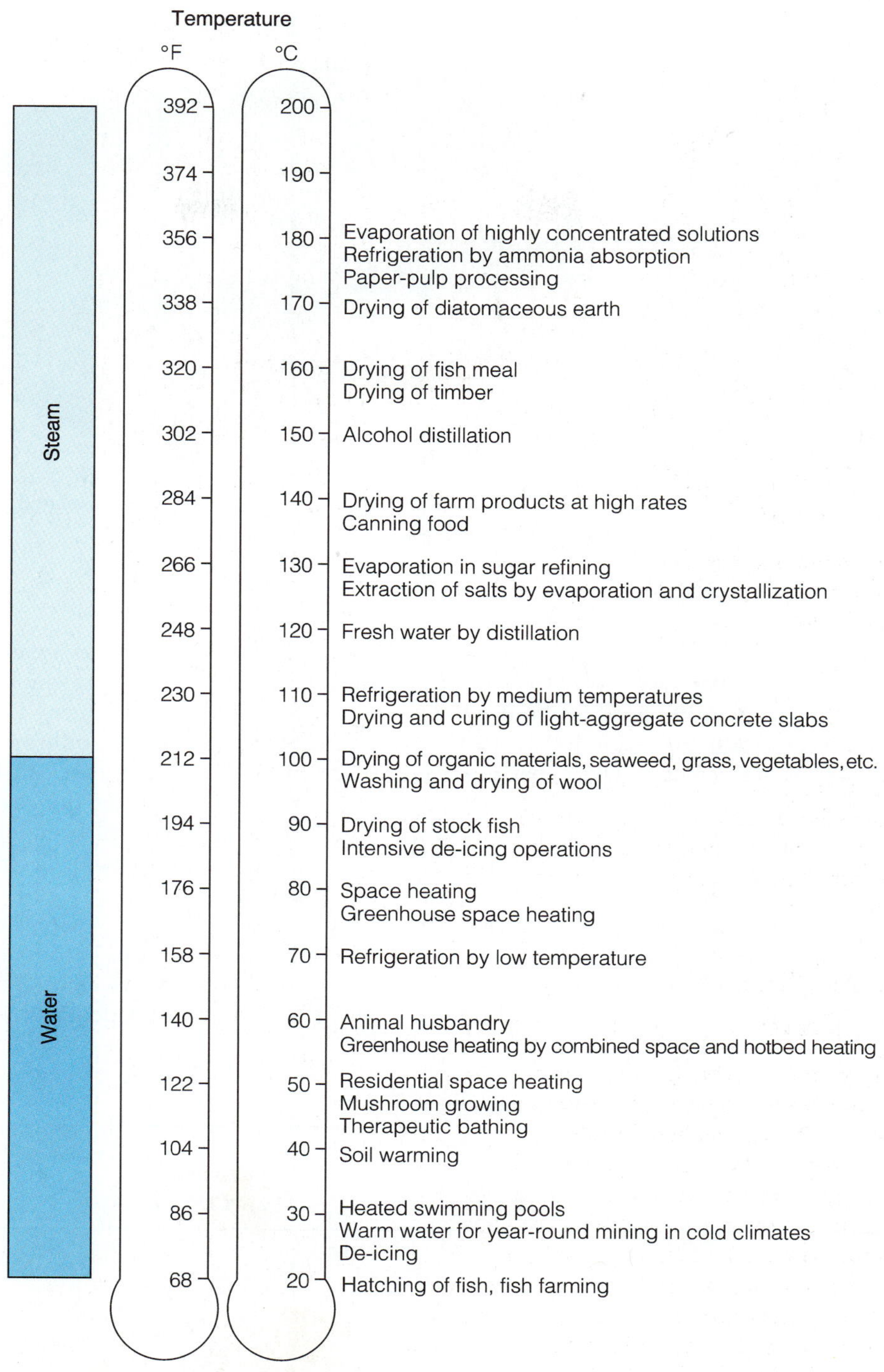

➢ **FIGURE 5.20 Established commercial uses of geothermal energy.**

hot-water reservoirs (➢ Figure 5.21). Larderello, Italy, and The Geysers near Napa Valley in northern California are examples of steam-dominated fields. The latter type are the rarest and most efficient energy producers, and they occur where temperatures are high and water discharge is low so that steam forms. They are much like an oil reservoir in that porous rocks that hold steam or very hot water are overlain by an impermeable layer that prevents the upward escape of the steam or water. Table 5.3 shows the electrical power produced at selected geothermal fields in 1982. Note that the fields are rated in megawatts (MW); a megawatt is 1,000 kilowatts (kw) or one million watts. The Geysers in California is the largest producer of geothermal power in the world (➢ Figure

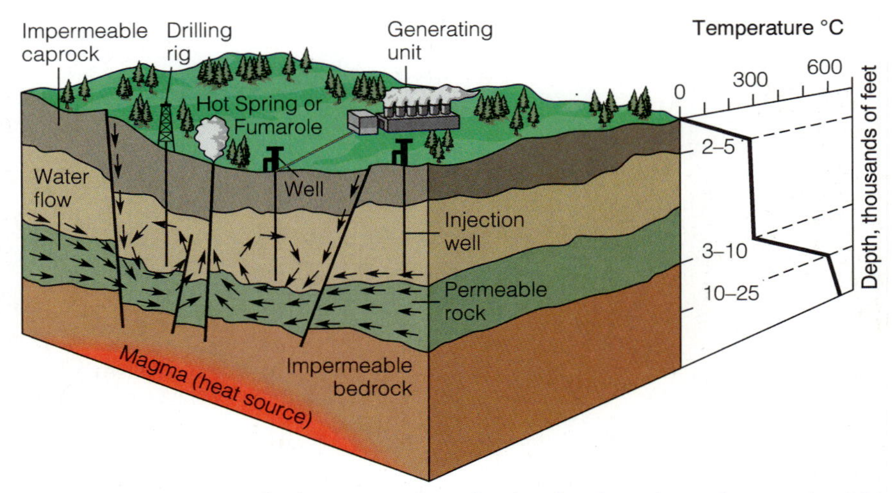

➢ **FIGURE 5.21 Geologic cross section of an insulated geothermal reservoir with little or no leakage.**

5.22); its production is sufficient to supply electricity to a good part of the city of San Francisco. To give a feel for the scale of this production, 1,000 MW is about what a modern nuclear reactor produces, and in 1981, the Geysers produced 960 MW of electricity. This capacity is expected to double by the year 2000. Lesser but important amounts of electricity are produced at Mammoth Lakes, Coso Springs, and the Salton Sea in California, all of which are underlain by cooling magma.

The problems associated with geothermal energy production are related to water withdrawal and water quality. A major problem in the Salton Sea area of southeastern California is the salinity of the geothermal waters. Several times as salty as seawater, it contains such toxic elements as arsenic and selenium and such heavy metals as silver, gold, and copper. Technology is in place to transfer the heat of the hot brine deep within the wells to a clean, working fluid that can be brought safely to the surface. This eliminates the easier but environmentally less desirable process of bringing salty water to the surface, using its heat to make electricity, and then disposing of it there.

In geothermal areas where there is clean hot water or steam, removal of this resource from underground can cause surface subsidence and perhaps even earthquakes. At The Geysers, subsidence of 13 centimeters (5 in) has

TABLE 5.3 Selected Geothermal Field Electrical Generating Capacity

COUNTRY	FIELD	GENERATING CAPCITY (MW)*	ADDITIONAL CAPACITY (MW)** YEAR 2000
United States	The Geysers, Sonoma Co., California	960	1350
	Casa Diablo (Mammoth Lakes), Mono Co., California	12	
	Heber, Imperial County, California	70	
Italy	Larderello (since 1905)	169	44
New Zealand	Wairaki	192	80
Japan	Hatchobaru	50	3275
	Matsukawa	22	
	Otake	12.5	
Iceland	Namafjall, Krafla, Svartsengi	3.5	36
El Salvador	Ahuachapán	110	0
Mexico	Cerro Prieto (Sonora)	150	470
Philippine Islands	Makaling Banahaw	220	550

*Production in place 1982 (Heber, California, 1993 production)

**Definite planned geothermal energy production for entire country

SOURCE: United Nations Conference on New and Renewable Sources of Energy, 1981; Armistead, H.C.H., 1983, *Geothermal Energy;* E. and F.N. Spon, London, 2nd Edition.

➢ **FIGURE 5.22 The Geysers geothermal field in Sonoma County, California, is steam-dominated. It is the largest geothermal-energy producer in the world.**

been measured without noticeable impact. Reinjecting the cooled water back into the geothermal reservoir decreases subsidence risk and ensures a continued supply of steam to the electrical generators. All geothermal fields in California are being monitored for any possible seismic activity related to steam production. Small quakes have been reported from The Geysers, but none of significant magnitude. Despite technical and environmental problems, it is anticipated that geothermal energy will supply at least 1 percent of U.S. energy needs by the year 2000.

VOLCANIC HAZARDS

Sicily's Mount Etna (from the Greek *aitho,* "I burn"), in the Mediterranean volcanic belt, exhibits almost continuous activity in its crater. In 1992 eruptions and lava flows were threatening several villages. Empedocles (circa 490–430 B.C.), Greek philosopher and teacher, is perhaps most notable for throwing himself into the crater of Mount Etna to convince his followers of his divinity. His dramatic suicide inspired Matthew Arnold's epic poem "Empedocles on Etna," which bears no resemblence to the little rhyme attributed to Bertrand Russell:

> Empedocles that ardent soul—
> Fell into Etna and was roasted whole!

Volcanoes comprise the third-most dangerous natural hazard in terms of loss of life, after coastal flooding (hurricanes and typhoons) and earthquakes. After the Mount St. Helens eruption of 1980, U.S. civil defense agencies published suggestions for what to do when a nearby volcano erupts. The Federal Emergency Management Agency's suggestions appear in the next section. For now, however, let us examine the various types of hazards associated with eruptions, several of which might occur during a single eruptive phase.

Lava Flows

Most hazards, natural and human-made, decrease in severity with distance from the point of origin. This is true of earthquakes, tornadoes, and falls of volcanic ash, for example. Lava flows may be the exception to this, because they generally burn or bury everything in their path, even to their farthest limit. Holocene lava flows in Queensland, Australia, traveled 100 kilometers (62 mi) from their vents down riverbeds with very low gradients. Knowledge of the paths flows might take from given vents makes it possible to delineate low- and high-risk volcanic-hazard areas. Basaltic lavas, such as those from Mount Etna in 1992, may emerge from a vent or fissure at temperatures in excess of 1,100°C and flow rapidly downslope like a river because of their low viscosity. As a flow cools, its viscosity increases, and then it may only "chug" along slowly. In either case, the end product is a layer of rock of variable thickness that covers everything in its path.

Two kinds of basalt flows are recognized. **Pahoehoe** ("pa-hoy-hoy") forms when a skin develops on a flow and buckles up as the flow moves, making a smooth or ropy-looking surface (➢ Figure 5.23). Other basalt flows develop upper and lower surfaces that are rough and blocky. This rough, angular lava is called **aa** ("ah-ah") and it is characterized by a very slow advance in which the top, cold rubble rolls over the front of the flow—moving much like the track of a Caterpillar tractor (see Figure 5.9*a*).

A volcanic eruption cannot be prevented, but in some cases, it is possible to divert or chill a flow so as to keep it from encroaching onto croplands or structures. Flows have been diverted by constructing earthen dikes in Hawaii and Italy. Icelandic fire fighters have successfully chilled flow fronts by spraying them with large volumes of seawater from fire hoses. During World War II, the

➢ **FIGURE 5.23 Smooth, ropy pahoehoe lava; Galápagos Islands, Ecuador.**

U.S. Army Air Corps tried, unsuccessfully, to divert lava flows by bombing them.

Ashfalls and Ash Flows

Roman naturalist and historian Pliny the Elder (born A.D. 23) died during the eruption of Vesuvius in A.D. 79 from complications brought on by inhaling noxious fumes and overexertion. He is revered as geology's first martyr, having been quoted by his nephew, Pliny the Younger, as saying, "Fortune favors the brave." Bravery was to be his undoing, as corpulent Pliny died in an area of heavy fallout from the eruption. Fortunately, however, Pliny had enjoyed life: he is credited with coining the phrase *In vino veritas* ("In wine there is truth"). Before Gaius Plinius (Pliny the Elder) died, he may have seen a great vertical plume of ash with a mushroom- or anvil-shaped head rising from Vesuvius's crater. This high-velocity column of steam, gas, and fragmented lava is called, for Pliny, a **Plinian eruption.** Plinian eruptions have been seen at Mount Lassen, California, in 1915, at Mount St. Helens in 1980, and at Mount Pinatubo in 1992 (➢ Figure 5.24 and Case Study 5.2). Such an ash cloud, consisting of tephra (jagged pumice, rock fragments, and bits of glass), may be carried away from the crater by prevailing winds and deposited as an **ashfall.** Ashfalls, like the ones at Pompeii and Mount St. Helens, bury objects close to their source. At Mount St. Helens, westerly winds carried ash eastward, causing difficulties in Yakima and Spokane by depositing several inches of ash there. At Butte, Montana, about 750 kilometers (465 mi) east, the fine ash and dust in the atmosphere forced the closure of the airport. People in Yakima wore masks to filter out the choking dust. Bankers politely requested that all patrons remove their masks before entering the banks because of potential identification problems. ➢ Figure 5.25 is a satellite photograph of the ash cloud being carried eastward across several states.

➢ **FIGURE 5.24 Plinian ash cloud over Mount Pinatubo.**

Ash flows are turbulent mixtures of hot gases and pyroclastic material that travel across the landscape with great velocity. They are generated quickly and with such force that obstructions in their path may be blown down or carried away. The term ***nuée ardente,*** French for "glowing cloud," is commonly applied to this kind of eruption because of the intense heat in the flow's interior. A nuée ardente caused most of the fatalities at Mount St. Helens and devastated an area in excess of 160 square kilometers (100 mi^2). ➢ Figure 5.26 is a graphic representation of the sequence of events at Mount St. Helens that culminated in the devastating ash-flow eruption of May 1980.

No other ash-flow eruption is as famous as that at Mount Pelée on the Island of Martinique in the West Indies on 8 May, 1902. It caused the deaths of at least 30,000 people in Saint-Pierre at the foot of the volcano and initiated the study of this type of eruption, and the unique volcanic deposits it forms. Like Mount St. Helens,

➢ **FIGURE 5.25 Satellite photo of an ash cloud from Mount St. Helens (two days after the eruption) superimposed on an outline map. Note that the cloud covered western Washington and parts of Idaho and Montana. Ash fell as far away as New York State.**

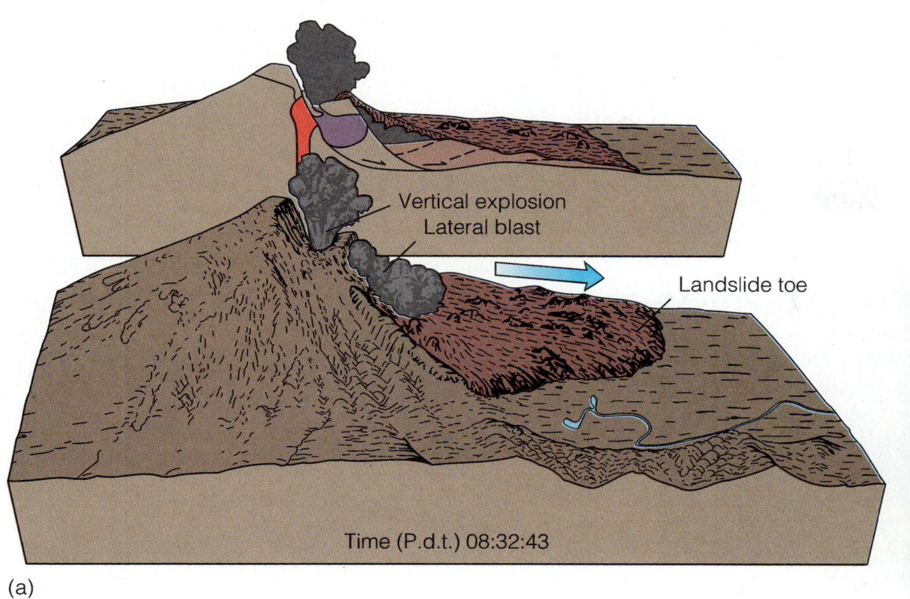

(a)

(b)

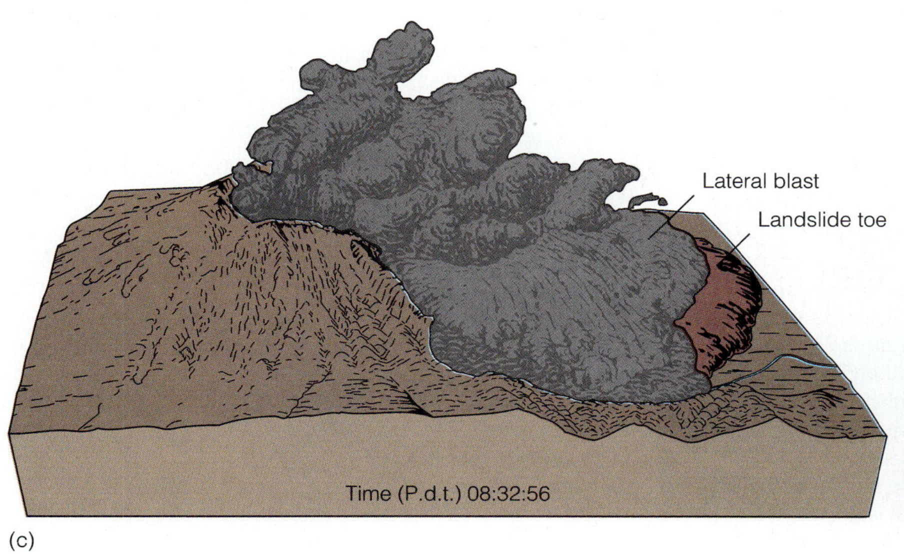

(c)

(d)

(e)

➢ FIGURE 5.26 Sequence of events at Mount St. Helens on Sunday morning, 18 May 1980. (*a* and *b*) A landslide opens the vent, and a skyward eruption begins. The cross section shows the slide plane. (*c* and *d*) Lateral blast begins 13 seconds later. The toe of the landslide is still exposed. (Both (*a*) and (*c*) are drawn as viewed toward the west, with no vertical exaggeration.) The eruption culminated in a powerful ash flow (nuée ardente). (*e*) The force of the ash flow is evidenced by the downed trees in the so-called blow-down area. This is the area where most of the fatalities occurred.

CASE STUDY 5.2

Mount Pinatubo—the Biggest Eruption of the Century?

Mount Pinatubo, with a summit elevation of 1,745 meters (5,670 ft), is a volcano 100 kilometers northwest of Manila on Luzon Island, the Philippines (➢ Figure 1). Before 2 April 1991, its dome had been dormant for 600 years. Throughout April minor eruptions and low-level seismicity were observed and recorded. These were followed by stronger and more frequent tremors and by heavy SO_2 emissions (as much as 500 metric tons per day) in late May and early June. Finally, things culminated in spectacular cataclysmic eruptions on June 15–16 (see Figure 5-24). This "finale" lasted 15 hours and sent tephra more than 30 kilometers (18 mi) into the atmosphere. Ash flows 200 meters (650 ft) thick and 10 kilometers (6 mi) long filled all low spots in the topography, and the volcano lost 400 meters (1,300 ft) of summit elevation, leaving a crater 3 cubic kilometers (0.73 mi^3) in volume. Following the pyroclastic flows, heavy monsoon rains and winds triggered lahars that traveled down major drainage systems, adding to the Filipinos' misery by destroying more homes and causing more fatalities. It is estimated that there were no survivors within 10 kilometers (6 mi) of the volcano, and only a few within 15 kilometers (9 mi). Estimates of deaths directly attributable to the eruption, volcanic mudflows, and disease in evacuation camps range from 323 to 435.

➢ **FIGURE 1 Map of the Mount Pinatubo area, the Philippines.**

Precursors to the main event—harmonic tremors and SO_2 and ash emissions—increased up to the time of the main eruptions, giving notice of the impending disaster. The 2½-month warning time allowed areas considered to be in the path of ash flows or lahars to be evacuated, which saved many lives. When Clark Air Force Base, a major U.S. installation 15 kilometers (9 mi) east of the volcano, was ordered evacuated on June 10, 15,000 military personnel and their families were moved to Subic Bay Naval Base in a matter of hours. An estimated 250,000 villagers were evacuated, but others remained in their homes to face the imminent danger of falling ash and mudflows. Tragically, mudflows continued well into July due to Typhoon Amy. By July 26, 100,000 homes had been destroyed or severely damaged, and 90,000 people remained in evacuation camps.

The cloud of ash and SO_2 from Pinatubo expanded rapidly to the west and southwest, reaching Bangkok, 2,000 ki-

Mount Pelée rumbled for some time before its main eruption. Because of this precursor activity, the population of Saint-Pierre swelled from 19,700 to at least 30,000 as refugees fled there from outlying villages. The concentration of sulfurous gases in the air at Saint-Pierre became so great that horses, suffocating from the fumes, dropped in their tracks. Citizens were forced to cover their faces with wet cloths as ash fell in the avenues and streets. Many inhabitants wanted to evacuate the city, but officials encouraged them to stay, saying there was little to fear from Mount Pelée. On the fateful day there was a tremendous explosion and lateral blast from Pelée that boiled and rolled down the slopes of the mountain at a velocity estimated at between 100 and 150 km/h (60–90 mph). The searing cloud engulfed the town in seconds and burned or suffocated all its inhabitants except two survivors, one of them a condemned prisoner who had been in a dungeon (➢ Figure 5.27).

Observations at Pelée in the 1930s by F. A. Perret of the Carnegie Institution led to the first accurate description of an ash flow:

> First of all it should be realized that the horizontal movement, three kilometers in three minutes, is due to an avalanche of an exceedingly dense mass of hot, highly gas-charged, and constantly gas-emitting fragmental lava, much of it finely divided, extraordinarily mobile, and practically frictionless, because each particle is separated from its neigh-

➢ FIGURE 2 Ash-filled sky and parking lot at Clark Air Force Base.

➢ FIGURE 3 Satellite images that show thickness of aerosols, including SO_2, before (May 28–June 6) and after (July 4–July 10) Mount Pinatubo's eruption. By about 20 days after the major eruptions, the aerosol cloud had completely encircled the earth.

lometers (1,240 mi) away, by June 17th. Pumice the size of apricots fell at Clark Air Force Base (➢ Figure 2), and the size of marbles at Olongapo, south of Pinatubo, where the ash thickness reached 30 centimeters (1.1 ft). Satellite data (Nimbus-7) showed very high concentrations of SO_2 over a large area, with a mass twice that of the sulfurous eruption of El Chichón in Mexico in 1982. By mid-July the ash cloud had encircled the earth (➢ Figure 3), and by September, atmospheric ash was creating spectacular sunsets off the west coast of North America.

The Pinatubo eruptive sequence was a major volcanic event. It may have put 15–20 million tons of sulfur dioxide into the atmosphere, which could make it the largest eruption of the twentieth century. The stratospheric dust and aerosols (SO_2) exerted a cooling influence of about 1°C on the earth during 1992–93, partly offsetting the greenhouse warming of the atmosphere. Also, gases from the volcano may have damaged the stratospheric ozone layer, which serves to shield the earth from ultraviolet radiation (see Chapter 12). Great Britain's Faraday Research Station in Antarctica recorded the lowest amount of stratospheric ozone in its short history in early 1993. The eruption was 150 times larger than Mount St. Helens, but it was not as large as the eruption of Krakatoa in 1883. (The subsection Weather and Climate later in the chapter provides some historic data that might serve to predict how long the climatic impact of Pinatubo will last.)

bors by a cushion of compressed gas. For this reason, too, its onward rush is almost noiseless.

Suffocating ash flows are a major volcanic hazard, and they have been observed in historic times in Alaska, Washington, and California. Prehistoric but geologically young ash-flow deposits are found in all the Western states. They are known as **welded tuffs,** because the minerals and glass composing them were fused together by intense heat.

➢ FIGURE 5.27 (*Right*) Saint-Pierre, Martinique, after the 1902 ash-flow eruption of Mount Pelée in which 30,000 to 40,000 people were killed.

CASE STUDY 5.3

The Exploding Lake

Lake Nyos is contained within a volcanic crater, one of 40 such craters that are roughly aligned southwest to northeast in Cameroon, West Africa (➢ Figure 1). Specifically, the lake is in a **maar**, a low-relief crater produced by explosive interaction of magma with underground water. Such a high-pressure steam explosion, called a **phreatic eruption**, does not produce much new magmatic material, which is why the resulting crater is relatively shallow. Lake Nyos and several nearby crater lakes and springs are known to be highly charged with CO_2, just as is a bottle of soda water.

On the evening of 21 August, 1986, almost noiselessly and without warning, the lake erupted, discharging 80 million cubic meters of CO_2 dissolved in its waters into the atmosphere. So much gas escaped from the lake that its level dropped 1 meter. Because carbon dioxide weighs about one and a half times as much as air, the suffocating gas descended downslope, following stream valleys and drainages for 16 kilometers (10 mi). The CO_2 cloud was 50 meters thick, and it killed all animal life in its path, including 1,200 residents of Nyos village and 500 persons in nearby communities (➢ Figure 2). The entire event took less than an hour.

Below its surface waters, Lake Nyos contains so much dissolved gas that a liter of water is charged with 1 to 5 liters of gas, 99 percent of which is carbon dioxide. (At a depth of 208 meters—680 ft—the lake's depth, water can hold 15 times its own volume in CO_2 at saturation.) The source of the CO_2 is believed to be volcanic, although none of the gases usually associated with volcanic activity—H_2S, CO, SO_2—is found in the waters. Carbon-14 study of the CO_2 indicates that it is more than 35,000 years old. This rules out a biogenic origin for the gas, as the lake is less than 1,000 years old. However, the isotopic ratios of He^3/He^4 and C^{13}/C^{12} in the gas are consistent with a magmatic source. Therefore, since evidence to the contrary has not been found, the CO_2 of Lake Nyos is believed to be escaping from magma at depth and seeping upward through a shattered volcanic vent into the bottom of the lake.

What happened the evening of the tragedy that caused the gas in the lake to explode catastrophically? Was it a small volcanic disturbance, or did some other physical phenomenon upset the lake? It is known that lakes periodically *overturn;* that is, their surface water sinks to the bottom and is replaced by bottom water rising to the surface. This usually occurs in temperate climates in the fall, because as the surface water cools, it becomes denser than the bottom water and displaces it. Overturn usually requires a trigger, such as strong winds, a landslide, or a drastic change in atmospheric pressure. Lake Nyos is only seven degrees north of the equator, however, so seasonal temperature change alone would not be sufficient to cause overturn. However, Cameroon lakes are known to be least stable during the rainy season of August and September, when cool fresh water is added to the lake surface. The best hypothesis appears to be that winds associated with the rainy season blew the lake's surface water to one side of the crater. This reduced the hydrostatic (confining) pressure on the bottom waters, which allowed

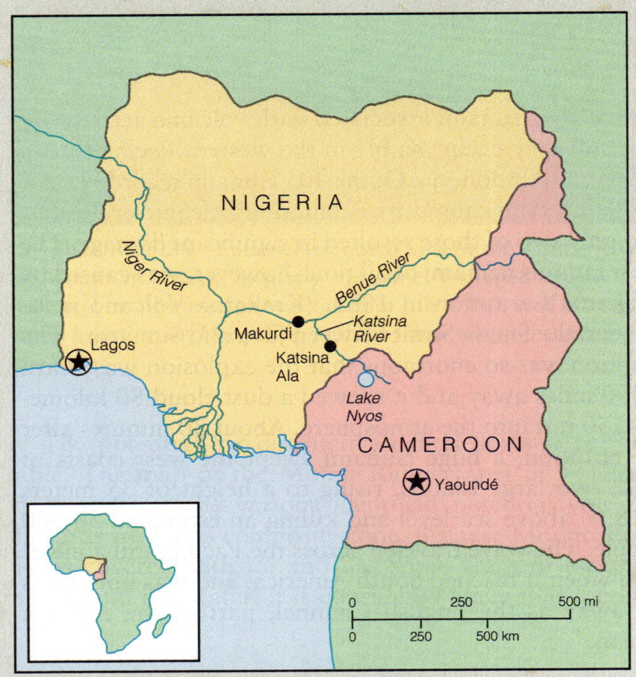

➢ **FIGURE 1 The location of Lake Nyos in Cameroon, West Africa.**

MITIGATION AND PREDICTION

Diversion

As mentioned earlier in the chapter, various tactics have been used in attempts to prevent lava flows from overwhelming the land. At various times, people have dammed them, diverted them, and even bombed them. Diversion barriers, high earth and rock embankments piled up by bulldozers, have been constructed on Mauna Loa to protect the city of Hilo, and also to divert potential flows away from the NOAA observatory at Mauna Loa (➢ Figure 5.30). Practically, diversion can be considered

the highly charged gas to come out of solution and form bubbles. The formation of bubbles further reduced the density of the water, and the gas in solution frothed and rose with dramatic speed. That this sequence of events indeed happened is supported by stripped vegetation along the lake shore. The leaves and other parts of plants were most likely removed by a surge of water associated with the explosion—just as a fountain of foam is associated with popping the top of a warm can of soda. Evidence to support the instability hypothesis, rather than a volcanic cause, for the Nyos tragedy is a similar explosion on 16 August 1984 at Lake Manoun, 95 kilometers (60 mi) south, with a loss of 37 lives. The occurrence of these two events during the season of minimum stability for Cameroon lakes suggests that such events may occur during this season in the future.

Bantu folklore is filled with stories of exploding lakes and "rains" of dead fish. According to legend, there have been at least three earlier episodes of catastrophic lake explosions. Dead fish found floating on the water surface during minor degassing events are believed by the local people to be gifts from their ancestors who live in the lake, and they believe that they cannot drown there without prior approval of their ancestors.

Since 1986 the gas content of the lake has increased by additions from subterranean springs and magmatic sources. Engineering studies indicate that Lake Nyos and other dangerous crater lakes in the region could be defused by installing pipes in them to bring charged water from the lake bottom to the surface as fountains. Once started, the fountains would be driven entirely by the lifting force of the released gas.

The danger is far from over at Lake Nyos. In addition to its high concentration of carbon dioxide, Lake Nyos has a very fragile dam of soft volcanic ash on its perimeter. Piping through this thin ash dam could cause failure and the release of a flood calculated at twice the flow of the 1899 Johnstown, Pennsylvania, flood (see Chapter 9). There are plans to lower the lake level and to improve the natural dam in order to minimize the flood threat, but the plans cannot be carried out before there is a better understanding of the potential impact of a lower lake level on the dissolved gas.

(a)

(b)

➢ **FIGURE 2** **(*a*) Deadly emission of carbon dioxide sinks to valley floors and other low areas at Lake Nyos, Cameroon, West Africa. (*b*) Cattle suffocated by carbon dioxide at Lake Nyos.**

only if the topography allows the diverted lava to flow onto unimproved lands whose owners do not object to it. Flows have been slowed and even stopped by spraying them with seawater at Heimaey Island, Iceland (➢ Figure 5.31). This was done in 1973 at the fishing village of Vestmannaeyjar. Both damming and diversion were used in Sicily in 1992 when eruptions of Mount Etna generated extensive lava flows.

Prediction

The ability to predict when and where a volcanic eruption

TABLE 5.5 Selected Notable Worldwide Volcanic Eruptions*

YEAR	VOLCANO NAME AND LOCATION	VEI	COMMENTS
4895B.C. ±	Crater Lake, Oregon	7	Post-eruption collapse forms caldera
1390B.C. ±	Santorini (Thera), Greece	6	Late Minoan civilizatin devastated
79	Vesuvius, Italy	5	Pompeii and Herculaneum buried, at least 3,000 killed
130 ±	Taupo, New Zealand	7	16,000 km^2 devastated
1631	Vesuvius, Italy	4	Modern Vesuvius eruptive cycle begins
1783	Laki, Iceland	4	Largest historic lava flows; 9350 killed
1792	Unzen, Japan	2	Debris avalanche and tsunami kill 14,500
1815	Tambora, Indonesia	7	Most explosive eruption in history, 92,000 die, weather changed
1883	Krakatoa, Indonesia	6	Caldera collapse; 36,000 killed, mostly by tsunami
1902	Mount Pelée, Martinique, West Indies	4	Saint Pierre destroyed, 30,000–40,000 killed, spine extruded from lava dome
1912	Katmai, Alaska	6	Maybe largest 20th-century eruption, 33 km^3 of tephra ejected
1943	Paricutiń, Mexico	3	New cone formed, observed and documented from first eruption
1914–17	Lassen Peak, California	3	California's last historic eruption
1959	Kilauea, Hawaii	2	Lava lake formed that is still cooling
1963	Agung, Bali	4	1100 killed, climatic effects
1968	Fernandina, Galápagos	4	Caldera floor drops 350 meters
1980	Mount St. Helens, Washington	5	Ash flow, 600 km^2 devastated
1982	El Chichón, Mexico	4	Ash flows kill 1877, climatic effects
1991	Mount Pinatubo, Philippine Islands	5†	Probably the second largest eruption of the 20th century, huge volume of SO_2 emitted
1991	Unzen, Japan	4†	Pyroclastic flows killed 41 people including 3 volcanologists, lava dome
1991–1993	Etna, Italy	?	Longest activity (473 days) in 300 years ended, 300 million m^3 lava extruded
1983–1993	Kilauea, Hawaii	?	Longest continuing eruption with over 50 eruption events

* Historic lava or tephra volume >2 km^3, Holocene >100 km^3, fatalities >1500.
† Estimated VEI by author.
SOURCE: Data adapted from Smithsonian Institution, Global Volcanism Program.

➢ **FIGURE 5.30 Lava diversion dike; Mauna Loa, Hawaii.**

➢ **FIGURE 5.31 Lava flows on Heimaey Island off the south coast of Iceland were successfully chilled by spraying them with seawater in 1973.**

will occur has attained a high degree of refinement at the U.S. Geological Survey's Hawaiian Volcano Observatory. It is located at Kilauea crater in Hawaii Volcanoes National Park on the east slope of Mauna Loa, Island of Hawaii. Kilauea and Mauna Loa are two of the most active volcanoes in the world. These volcanoes and the magma reservoir that feeds them can be viewed as balloons buried below thin layers of sand and clay. As magma enters the reservoir prior to an eruption, the volcano's surface layers are pushed upward and outward—just as if gas were being blown into the balloon. This expansion is measurable with tiltmeters. As magma works its way to the volcano surface, earthquakes so numerous that they are referred to as *swarms* occur. Earthquake swarms have been found to be quite reliable precursors of an impending eruption. ➢ Figure 5.32 illustrates the three stages of the eruption and the accompanying changes in tilt and seismicity. Increasing tilt indicates that something is about to happen, and the locations of earthquake epicenters and foci indicate where the outbreak is

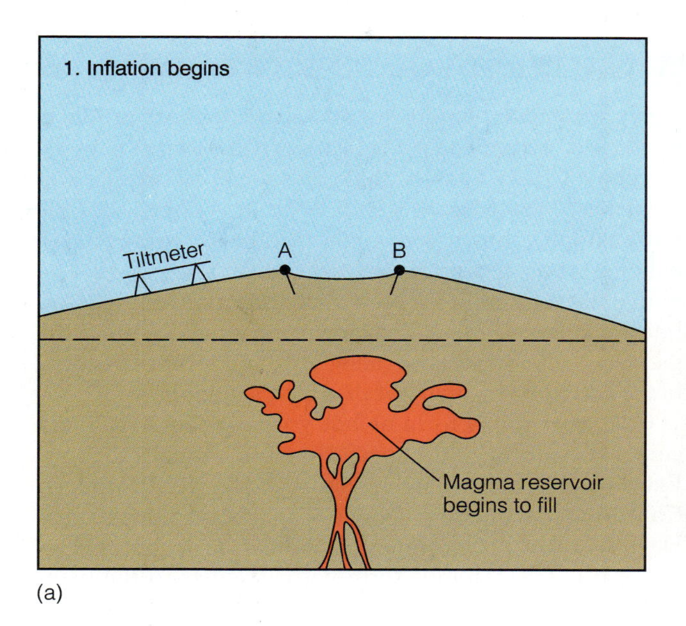

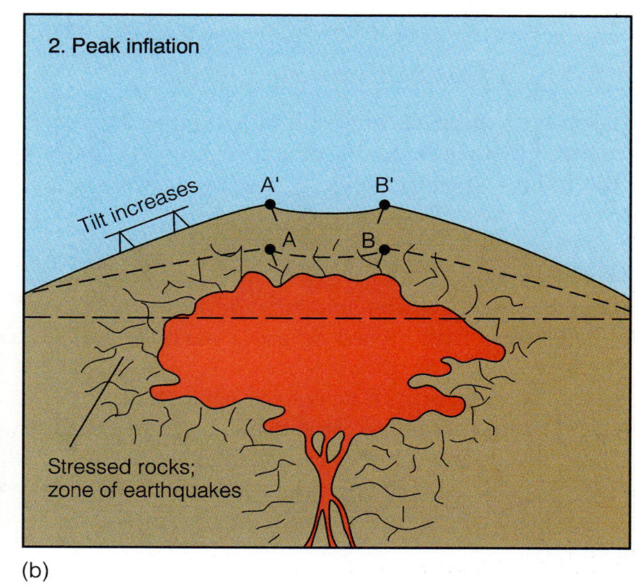

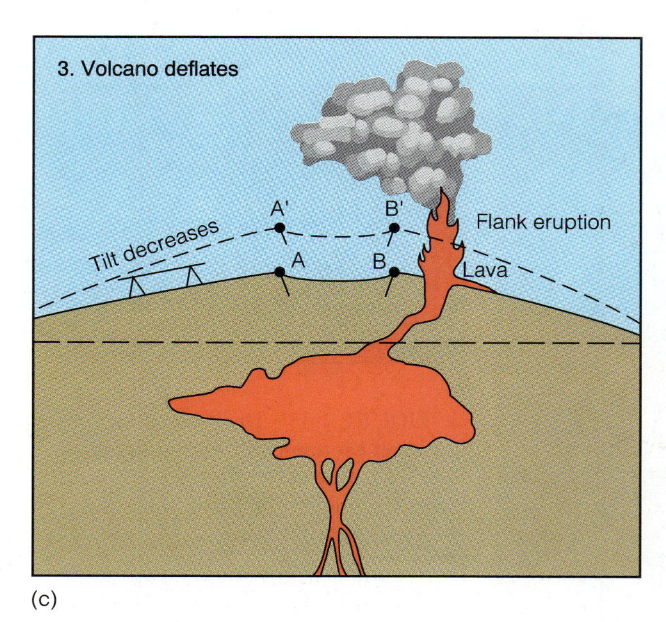

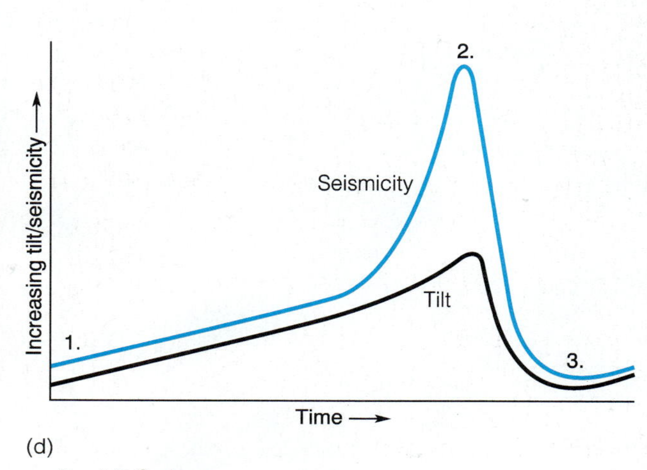

➢ **FIGURE 5.32 Hawaiian volcanoes' typical eruption sequence. (*a–c*) Inflation and deflation of the cone are accompanied by (*d*) measurable changes in slope and seismicity. Three stages are apparent.**

likely to occur.

These same phenomena were observed at Mount St. Helens, leading geologists to expect an eruption—though not such a cataclysmic one. The swelling there produced a noticeable change in the shape of the cone, and the harmonic tremors (caused by magma moving continuously within the reservoir) caused such surface motion on the cone that many persons were uncomfortable. Unlike earthquakes, the eruptions of active volcanoes can be predicted with fair accuracy as to time and place; the magnitude of the pending eruption, however, is more difficult to assess. Predicting the behavior of dormant volcanoes is much more difficult, and much remains to be learned about predicting their future activity.

What to Do When an Eruption Occurs

Based upon the experience of people at Mount St. Helens, the Federal Emergency Management Agency (FEMA) made the following suggestions to ease the trauma of a volcanic eruption.

At Home

- Stay calm and get children and pets indoors.
- If outside, keep eyes closed or use a mask and seek shelter.
- If indoors, stay there until heavy ash has settled.
- Close doors and windows, and place damp towels at door thresholds and other draft openings.
- Do not run exhaust fans or clothes dryers.
- Remove ash accumulations from low-pitched roofs and gutters.

In Your Car

- Drive slowly because of limited visibility.
- Change oil and air filters every 50–100 miles (80–160 km) in heavy dust, or 500–1,000 miles (800–1600 km) in light dust (visibility up to 60 m).
- Use both windshield washer and wipers.
- Volcanic ash is abrasive and can ruin engines and paint.

Pets

- Keep pets indoors.
- Brush or vacuum them if they are covered with ash.
- Do not let them get wet, and do not try to bathe them.

GALLERY: VOLCANIC WONDERS

Many spectacular landforms and attractions are products of volcanic action and earth heat—Yellowstone's geysers, Oregon's Crater Lake, California's Devil's Postpile, and northwest New Mexico's Ship Rock. It is hoped that the few examples shown here might inspire you to travel and enhance your appreciation of the earth's wonders. The native Maori people of New Zealand say "Man passes, but the land endures." This adage originated around the great volcanic centers of New Zealand's North Island, but it is echoed by many of the earth's inhabitants.

Columnar joints form as hot volcanic rocks cool and contract, splitting into spectacular polygonal columns. Excellent examples may be seen at the Devil's Postpile in California, an andesitic lava flow (➢ Figure 5.33), and at Devil's Tower in Wyoming, a shallow intrusion (➢ Figure 5.34). The columns form perpendicular to the main

(a)

(b)

➢ FIGURE 5.33 Devil's Postpile National Monument, Mammoth Lakes area, east-central California. (*a*) Beautiful columnar jointing in basalt. (*b*) Glacier-polished tops of the polygonal columns.

➢ **FIGURE 5.34 Devil's Tower National Monument in northeastern Wyoming is an exposed volcanic neck with spectacular columnar jointing. Also known as Mato Tepee, it rises 865 feet above the surrounding landscape.**

cooling surfaces of the flows, which is why they are sometimes curved. The 264-meter-tall (865 ft) Devil's Tower, called "Grizzly Bear Lodge" by early Native Americans, is surrounded by colorful Cheyenne and Lakota Sioux legends. According to one legend, a group of children were being chased by a grizzly bear. The Great Spirit saw their plight and raised the ground they were on. The bear scratched deep grooves into the rocks in his attempts to reach the frightened children. It is not difficult to visualize the tribal elders passing on this legend to wide-eyed children. It is a much more memorable explanation for the origin of the columns than the scientific one of contraction cracks. The columns at Devil's Tower are a challenge to rock climbers, but they have been scaled many times.

Diamond Head in the city of Honolulu (➢ Figure 5.35) is a tuff cone composed of glassy volcanic ash and pyroclastic material. Lava tubes and tunnels form when a lava flow crusts on top while continuing to flow internally. The ones at Crater of the Moon in Idaho are outstanding (➢ Figure 5.36). Dying stages of volcanism are often reflected by geysers, steam vents (*fumaroles*), and hot springs like those found at Yellowstone National Park and at Whakarewarewa Hot Springs near the city of Rotorua, New Zealand (➢ Figure 5.37). The springs lie in a giant caldera that is still hot, and the local people use the flowing waters directly for heating and bathing.

➢ **FIGURE 5.35 Diamond Head crater, a beloved landmark of Honolulu, Island of Oahu, Hawaii.**

➢ **FIGURE 5.36 Lava tunnel at Craters of the Moon National Monument, south-central Idaho.**

(a)

(b)

➢ **FIGURE 5.37 Whakarewarewa Hot Springs, Rotorua, North Island, New Zealand. (*a*) The earth-heated water is used directly for heating homes and as domestic hot water. (*b*) This geyser erupts at regular intervals near the homes shown in (*a*).**

SUMMARY

Volcanoes

DEFINED: A vent or series of vents that issue lava and pyroclastic material.

DISTRIBUTION: Adjacent to convergent plate boundaries around the Pacific Ocean (the Ring of Fire); in a west–east belt from the Mediterranean region to Asia; along the mid-ocean ridges; and in the interior of tectonic plates above hot spots in the mantle. In the U.S., a 1,100-kilometer (680 mi) belt extends northward from northern California through Oregon and Washington.

MEASUREMENT: The Volcanic Explosivity Index assigns values of 0 to 8 to eruptions of varying size.

EXPLOSIVITY: Depends upon the lava's viscosity and gas content. Viscosity, a function of temperature and composition, determines the type of eruption: felsic (high-SiO_2) lavas are viscous and potentially explosive; mafic (low in SiO_2, high in magnesium and iron) lavas are more fluid and less explosive.

Types of Volcanoes and Volcanic Landforms

SHIELD (HAWAIIAN OR "QUIET" TYPE): Gentle outpourings of lava produce a convex upward edifice resembling a shield.

FISSURE ERUPTION: Lava erupts from long cracks, building up broad lava plateaus such as those found in the Columbia Plateau (Oregon–Washington–Idaho), Iceland, and India.

COMPOSITE CONE (EXPLOSIVE TYPE): Characterized by concave upward cone thousands of meters high. Vesuvius (Italy), Fujiyama (Japan), and Mount Hood (Oregon) are examples. Crater Lake (Oregon) is the stump of a composite cone that exploded and collapsed inward, leaving a wide crater known as a *caldera*.

LAVA DOMES: Bulbous masses of high-silica, glassy lavas, such as found at Mono Craters, California. Domes may form in the crater of a composite cone after an eruption, such as at Mount St. Helens, Washington.

CINDER CONES: The smallest and most numerous of volcanic cones, composed almost entirely of tephra.

Volcanic Products

BUILDING MATERIALS: Cinder blocks, road-base materials, pumice, light-weight concrete, decorative stone.

GEOTHERMAL ENERGY: Cooling magma at depth heats water that is converted to steam to drive electrical generators. Steam- or hot-water-dominated geothermal energy is utilized in Italy, Iceland, New Zealand, California, Nevada, Japan, Mexico, and the former Soviet Union.

Hazards

LAVA FLOWS: Destroy or burn everything in their path, can travel a distance of 100 kilometers or more.

ASHFALLS AND ASH FLOWS: Heavy falls of volcanic ash from Mount Vesuvius buried Pompeii in A.D. 79. There were heavy ashfalls from Mount Pinatubo, the Philippines, in 1991. Vertical plumes of ash may rise many kilometers from a vent, which are then carried by the wind and blanket the terrain. Such plumes rising from a vent are known as *Plinian eruptions*. Ash flows are hot, fluid masses of steam and pyroclastic material that travel down the flanks of volcanoes at high speeds, blowing down or suffocating everything in their path. Such flows have killed thousands of people.

DEBRIS FLOWS (LAHARS): Catastrophic mud flows down the flanks of a volcano. They cause most volcanic fatalities, more than ash flows.

TSUNAMI: Submarine volcanic eruptions (such as that of Krakatoa in 1883) create enormous sea waves that may travel thousands of miles and still do damage along a shoreline.

WEATHER: Sulfur dioxide coatings on ash and dust particles increase reflectivity of the earth and cause cooling of weather, such as occurred after the eruption of Mount Pinatubo, 1991–1993. Large eruptions have been related to stormy conditions and El Niño. The effects of ash and dust in the stratosphere are short-term—a few years.

GASES: Corrosive gases emitted from a volcano can be injurious to health, structures, and crops. Carbon dioxide emitted from the bottom of volcanic Lake Nyos in Cameroon formed a cloud that sank to the ground and suffocated 1,700 people in 1986.

Mitigation and Prediction

DIVERSION AND CHILLING: Flows have been diverted in Hawaii and elsewhere and have been chilled with seawater in Iceland.

PREDICTION: Prediction in Hawaii, based upon seismic activity (earthquake swarms and harmonic tremors) and the tilt of the cone as magma works its way upward, has become quite accurate. Similar phenomena were observed at Mount St. Helens and Mount Pinatubo.

Key Terms

aa
andesite line
ashfall (air fall)
ash flow (*nuée ardente*)
basalt
caldera
cinder cone
composite cone
felsic
lahar
lava dome (volcanic dome)
lava plateau
maar
mafic
pahoehoe
phreatic eruption
Plinian eruption
pyroclastic
shield volcano (cone)
tephra
Volcanic Explosivity Index (VEI)
welded tuff

Study Questions

1. What are the four kinds of volcanoes, and what type of eruption can be expected from each kind? What are fissure eruptions, and what have they produced to shape the earth's landscape?
2. How would you expect the eruption of a volcano that taps mafic magma to differ from the eruption of one that taps felsic magma? What kind of rocks will result from each eruption?
3. Explain the distribution of the world's 1,350 active volcanoes in terms of plate tectonic theory.
4. List the five geologic hazards connected with volcanic activity in decreasing order of their threat to human life and limb.
5. Where in the U.S. are the most dangerous volcanic hazards encountered? Which geologic hazard related to volcanic activity presents the greatest danger to humans, and how can this hazard be mitigated, or at least minimized?
6. Why can eruptions on the Island of Hawaii be predicted more accurately than volcanic eruptions at subduction zones?
7. How may calderas be formed? What geologic evidence is found at Crater Lake, Oregon, that indicates it was formed by a special set of circumstances?
8. In what ways are volcanic activity and its products useful to humankind?
9. Name five volcanic areas of great scenic wonder, and indicate the geologic feature found there; for example, Crater Lake, caldera.
10. How do volcanic eruptions affect climate and weather? What impact does this have on populations, particularly Third World peoples?
11. In what ways do gases, including water vapor, emitted by a volcano impact the area and people living near an active volcano?
12. Define or sketch these volcanic features: aa, pahoehoe, volcanic neck, lahar, dust veil index, lapilli, tephra, columnar jointing, welded tuff.

FURTHER READING

Bullard, Fred M. 1962. *Volcanoes in history, in theory, in eruptions.* Austin Texas: University of Texas Press.

Casadevall, T. J., ed. 1991. *Volcanic ash and aviation safety: First international symposium,* U.S. Geological Survey Circular 1065. Washington, D.C.: Government Printing Office.

Decker, Robert, and Barbara Decker. 1981. *Volcanoes.* New York: W. H. Freeman.

Edwards, L. M. 1982. *Handbook of geothermal energy.* Houston, Texas: Gulf Publishing Company.

Francis, Peter. 1976. *Volcanoes.* New York: Penguin Books.

Hazlett, Richard W. 1987. *Kilauea volcano: Geological field guide.* Hilo, Hawaii: Hawaiian Natural History Association.

Kling, George W., and others. 1987. The 1986 Lake Nyos gas disaster in Cameroon, West Africa. *Science* 236: 169–174.

McClelland, Lindsay; D. Leschinsky; and K. Kivimaki. 1991. Global volcanism network. *Bulletin of the Smithsonian Institution,* March–July.

McDonald, G. A.; A. T. Abbott; and F. L. Peterson. 1983. *Volcanoes in the sea: the geology of Hawaii,* 2d ed. Honolulu, Hawaii: University of Hawaii Press.

McPhee, John. 1989. *The control of nature.* New York: Farrar, Straus, and Giroux.

Schuster, Robert L., and J. P. Lockwood. 1991. Geologic hazards at Lake Nyos, Cameroon, West Africa. *Association of Engineering Geologists News,* April.

Sharpton, V. L., and P. E. Ward. 1991. *Global catastrophes in earth history: An interdisciplinary conference on impacts, volcanism, and mass mortality.* Geological Society of America Special Paper 247.

Simkin, Tom, and others. 1981. *Volcanoes of the world.* Stroudsburg, Penn.: Hutchinson Ross Publishing Co.

Smith, A. L., and M. J. Roobol. 1991. *Mt. Peleé, Martinique: An active island-arc volcano.* Geological Society of America Memoir 175.

Stager, Curt. 1987. Killer lake. *National Geographic,* September, pp. 404–420.

Tilling, Robert I. 1984. *Monitoring active volcanoes.* U.S. Geological Survey, Box 25286, Federal Center, Denver, CO 80225.

Wright, Thomas L., and T. C. Pierson. 1992. *Living with volcanoes:* U.S. Geological Survey Circular 1073, U.S. Geological Survey, Box 25425, Federal Center, Denver, CO 80225.

Video

National Geographic Society, with WQED Pittsburgh. 1983. *Born of fire,* 60 minutes.

CHAPTER

6

WEATHERING AND SOILS

A few inches between humanity and starvation

ANONYMOUS

Soil is one of the solid earth's most valuable resources. Through it we produce much of our food, and it supports forest growth, which gives us essential products. Soil material, organisms within the soil, and vegetation constitute an ecological system in which the four ingredients needed for plant growth are recycled. These ingredients are water, air, humus (organic matter), and mineral matter (➢ Figure 6.1).

The multiple uses of **soil** are reflected in the many definitions and classification schemes that have been developed for soils. To the engineer, a soil is the loose material at the earth's surface; that is, material that can be moved about without first being dynamited and upon which structures can be built. The geologist and the soil scientist see soil as weathered rock and mineral grains that are capable of supporting plant life. A farmer, on the other hand, is mostly interested in which crops a soil can grow and in whether the soil is rich or depleted with respect to humus and minerals. Soils form by the weathering of **regolith,** the fragmental rock material at the earth's surface. By studying soils we can make inferences about their parent materials and the climate under which they formed and can determine their approximate age.

The carrying capacity of our planet—that is, the number of people the earth can sustain—depends on the availability and productivity of soil. This is why reducing soil

Bryce Canyon National Park, Utah. The park was created in 1928 to protect an area of 14,500 hectares (36,000 acres) of colorful, eroded sandstones and limestones. Its geologic story is closely related to those of Grand Canyon and Zion, except that the rock formations of the Bryce Canyon are much younger than those of the adjacent parks. The canyon was named for Ebenezer Bryce, an early settler in the area.

(a)

(b)

➢ **FIGURE 6.10 Volcanic ash (horizontal layers) overlying weathered basalt; Pahala, Hawaii. This volcanic soil grows rich crops of sugar cane, pineapple, and macadamia nuts.**

Soil Classification

ZONAL CLASSIFICATION. **Pedologists** (Greek *pedon,* "soil," and *logos,* "knowledge of") have long known that soils that form in tropical regions are different from those that form in arid or cold climates. Where rainfall is heavy, the soils are deep, acidic, and dominated by chemical weathering. Soluble salts and minerals are leached (removed) from the soil, iron and aluminum compounds accumulate in the B horizon, and the soil supports abundant plant life. In arid regions, on the other hand, soils are usually alkaline and thinner, coarse-textured or

➢ **FIGURE 6.11 Thin, dark soil developed on light-colored loess.**

rocky, and dominated by physical weathering. Because of low precipitation there, salts are not leached from soils, and when soil moisture evaporates, it leaves additional salts behind (Case Study 6.1). These salts may form crusts and lenses of caliche ($CaCO_3$ and other salts) in the B horizon.

These understandings led to the development of a "zonal" classification of soils based upon climate, which then evolved into a system of Great Soil Groups. These groups may be simplified into four major types (➢ Figure 6.12):

- **pedalfers,** soils that are high in aluminum (Al) and iron (Fe) and that are characteristic of humid regions (rainfall $>$ 50 cm/year, or $>$ 20 in/y)
- **pedocals,** soils that are high in calcium and found in desert or semiarid regions (rainfall $<$ 50 cm/y)
- **laterites,** the brick-red soils of the tropics that are enriched in hydrated iron oxides (rust)
- **tundra soils,** the soils of severe polar climates, found mostly in the Northern Hemisphere

Pedalfers are the rich soils of the prairies and steppes that support forest and grassland growth. Pedocals form where there is little rainfall, resulting in desert or scrub vegetation. Pedocals and pedalfers are mid-latitude, temperate-climate soils. In contrast, laterites (Latin *later,* "brick, tile") are the residual soils of the tropics, where

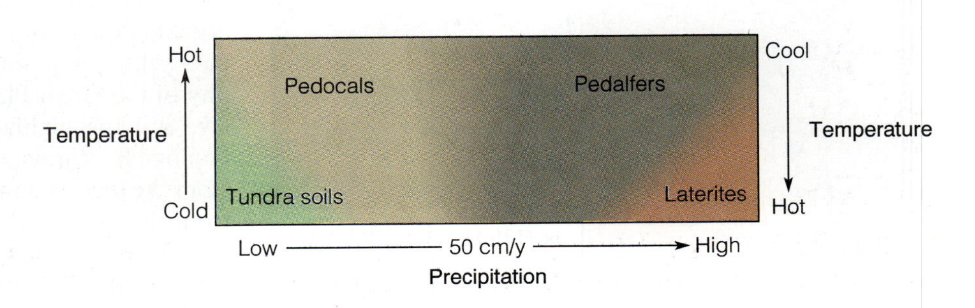

➢ **FIGURE 6.12 The Great Soil Groups and their climatic environments. Note that the temperature scale is reversed on the left and right sides of the diagram.**

CASE STUDY 6.1

Salinization and Waterlogging

Saline soils contain sufficient soluble salts to interfere with plant growth. Salinization is the oldest soil problem known to humans, dating back at least to the 4th century B.C. At that time a highly civilized culture was dependent upon irrigation agriculture in the southern Tigris–Euphrates Valley of Iraq, the "Cradle of Civilization." By the 2nd century B.C., the soils there were so saline that the area had to be abandoned. Soils in parts of California, Pakistan, the Ukraine, Australia, and Egypt are now suffering the same fate.

Salinization may be human-induced by intensive irrigation, which raises underground water levels very close to the soil. Capillary action then causes the underground water, containing dissolved salts, to rise in the soil, just as water is drawn into a paper towel or sugar cube. The soil moisture is then subject to surface evaporation, and as the water in the soil evaporates, it leaves behind salts that render the soil less productive—and eventually barren (see ➢ Figure 1). The productivity of 15 percent of all U.S. cropland is totally dependent upon irrigation, and 30 percent of U.S. crops are produced on this land.

➢ **FIGURE 1 Salts have formed at the surface of this soil from evaporating irrigation and ground water, Imperial Valley, California.**

For millenia, annual flooding of the Nile River added new layers of silt to the already rich soil and removed the salts. After completion of the Aswan Dam in 1970, however, the yearly flooding and flushing action stopped, and salts have now accumulated in the soils of the floodplain. Salinization is particularly evident in the arid Imperial Valley of California. Soils of the Colorado Desert, or any desert for that matter, can be made productive if enough water is applied to them. Because deserts have low rainfall, irrigation water must be imported, usually from a nearby river. The All-American Canal brings relatively high salinity water from the Colorado River to the Imperial Valley, where it is used to irrigate all manner of crops from cotton to lettuce. In places the water table has risen to such a degree that the cropland lies fallow, its surface covered by white crusts of salt. The Imperial Valley's Salton Sea is one of the largest saltwater lakes in the United States, and it grows saltier and larger every year by the addition of saline irrigation water. The return of degraded irrigation water in canals to the Colorado River in southern Arizona has required that a desalinization facility be established near Yuma to make the water acceptable for agricultural use in Mexico.

The good news is that soil salinization can be reversed. It can be done by lowering the water table by pumping and then applying heavy irrigation to flush the salts out of the soil. Salinization can also be mitigated, or at least delayed, by installing perforated plastic drain pipes in the soil to intercept excess irrigation water and carry it offsite. Neither remedy is an easy or cheap solution to this age-old problem.

heavy rainfall leaches out all soluble minerals, leaving behind the clay minerals and hydrated aluminum and iron oxides that give it a red color (➢ Figure 6.13). Laterite is not uniquely identified with any particular rock type, but it does require an iron-containing parent, abundant rainfall, and well-drained terrain. Laterites are currently forming in humid tropical and subtropical regions. Leaching of these soils may be so extensive that with the right parent rock, economically valuable aluminum-rich *bauxite* deposits are the end products of the soil-forming process (➢ Figure 6.14). Finally, tundra soils form in arctic and subarctic regions and support only such vegetation as mosses, sedges, lichens, and dwarf shrubs, making for bleak, forestless landscapes (➢ Figure 6.15).

➢ **FIGURE 6.13 Brick-red laterite in an urban setting; Brazil.**

➢ **FIGURE 6.14 Bauxite ore developed on aluminum-rich igneous rock; Jamaica.**

➢ **FIGURE 6.15 Treeless tundra landscape near Kotzebue, Alaska. Most of the soil in tundra regions is frozen much of the year.**

U.S. COMPREHENSIVE SOIL CLASSIFICATION SYSTEM. Because of dissatisfaction with existing classification schemes, the U.S. Soil Conservation Service instituted the first of seven trial classification systems in 1952. Each trial system was sent to a select group of international pedolgists for their comments. After incorporating the suggestions and findings of these experts, the "Seventh Approximation," as it is called, was officially adopted by the Soil Conservation Service in 1965. The distinguishing features of this system are that:

1. Soils are classified by their physical characteristics, rather than their origin; that is, the system is nongenetic.
2. Soils that have been modified by human activities are classified with natural soils.
3. The soil names convey information about the soils' physical characteristics.

The system subdivides soils into a hierarchy of six levels: orders, families, series, and so forth. There are 10 soil orders at the highest level in the system, which, with a practiced eye, can be easily identified (Table 6.1). Because the number of soil entries increases astronomically at lower levels (14,000 soil series are recognized in the U.S.), this classification is for the professional, not for the layperson. Nonetheless, one should be aware that it is the classification used by today's soil scientists.

Soil Problems

Soil Erosion

Although soil is continually being formed, it is for practical purposes a nonrenewable resource, because hundreds or even thousands of years are necessary for soil to develop. Ample soil is essential for providing food for the earth's burgeoning population, but throughout the world it is rapidly disappearing. Although the United States has one of the world's most advanced soil conservation programs, soil erosion remains a problem after 60 years of conservation efforts and expenditures of billions of dollars. Water erosion removes 3 billion tons of topsoil from U.S. farmlands annually, with the average loss estimated at 10–12 tons per hectare per year (4.0–4.8 tons/acre/year). Assuming soil forms at the estimated rate of 2–4 tons per hectare per year, 3–5 times as much soil is being lost as is being formed. Losses vary considerably with area, but about 70 percent of U.S. farmlands have experienced moderate to severe erosion in the recent past. Not only does soil erosion destroy fertility, several billion tons of soil settle each year in lakes and clog waterways and drainages with sediment, pesticides, and nutrients. Losses of only 2–3 centimeters (1 in) of topsoil represent about 200 tons per hectare, and such erosion cuts crop yields and may render the land unproductive.

On a global basis, an area the size of China and India combined has suffered *irreparable degradation* from agricultural activities and overgrazing—mostly in Asia, Africa, and Central and South America. Lightly damaged areas can still be farmed using the modern conservation techniques explained later in this chapter, but "fixing" serious erosion problems is beyond the resources of most Third World farmers and their governments. It is estimated that more than 9 million hectares (22 million acres) of farmland are currently beyond reclamation.

EROSION PROCESSES. The agents of soil erosion are wind and running water. Heavy rainfall and melting snow create running water, which removes soil by sheet, rill, and gully erosion. **Sheet erosion** is the removal of soil particles in thin layers more or less evenly from an area of gently sloping land. It goes almost unnoticed. **Rill erosion,** on the other hand, is quite visible in discrete stream-

TABLE 6.1 U.S. Comprehensive Soil Classification System

SOIL ORDER	DESCRIPTION AND AREAS WHERE FOUND
1. Oxisols	The intensely leached, oxidized red soils of the tropics (laterites).
2. Ultisols	Soils that have chemically weathered under warm, moist climatic conditions; support forest canopy. Dominant in southeastern U.S., China, Brazil, Australia, and West Africa.
3. Alfisols	Fine-textured, yellow-brown soils that are low in humus; clay-rich B horizon, moderate to high fertility. Texas, Ohio Valley, Mississippi Valley regions, Russian steppes, and Australia.
4. Spodosols	Acidic soils developed on quartz-rich sandy soils in cool climates. Low fertility and short growing season make these poor soils, except for root crops such as potatoes.
5. Mollisols	Thick humus, soft (*molli,* "soft") consistency when dry. Almost all are developed on transported parent material, loess, or alluvium. Prairie and steppe grasslands of mid-latitudes, the choice fertile soils of the Midwest (pedalfers).
6. Aridisols	Alkaline, with caliche layers (pedocals). Productive when water and good drainage available. Desert regions, areas that receive less than 25 cm rainfall per year.
7. Vertisols	Clay-rich soils that expand upon absorbing water and shrink upon drying. Dessication cracks are common, and their expansive nature requires special foundation design. Mostly in California, Colorado, and Texas.
8. Histosols	Organic soils (*histos,* "tissue") in bogs or swamps. Least important and least extensive soil order. Florida, the Everglades, and the Mississippi River and delta.
9. Entisols	Poorly developed, recent (*ent,* "recent") soils with little or no profile. Sandy deserts, tundra areas, and areas recently glaciated or covered by alluvium or volcanic ash.
10. Inceptisols	Immature soils (*incept,* "beginning") containing weatherable minerals; better developed than entisols. Tundra regions, steep mountain areas, and floodplains or deltas; very productive along the Mississippi, Nile, and Ganges Rivers.

lets carved into the soil (➢ Figure 6.16). If these rills become deeper than about 25–35 centimeters (10–14 in), they cannot be removed by plowing, and gullies form. Gullies are created in unconsolidated material by widening, deepening, and headward erosion of rills (➢ Figure 6.17). **Gullying** is a big problem in the easily eroded loess soils of the Palouse region of western Idaho and eastern Washington. Snow melting there saturates the loess soil (see Chapter 12), which becomes fluid and then flows like wet concrete across the underlying still-frozen ground. Cropland soil loss from sheet and rill erosion is shown by state in ➢ Figure 6.18a. Note that Tennessee heads the list, followed by Hawaii, Missouri, Mississippi, and Iowa (Table 6.2).

Wind erodes soils when tilled land lies fallow and dries out without vegetation and root systems to hold it together. The dry soil breaks apart, and wind removes the

➢ **FIGURE 6.16 Rill erosion in cultivated soil.**

➢ **FIGURE 6.17 Extreme gullying in soft sediments encroaching on tilled land near Lumpkin, Georgia. (Width of image area equals approximately 1¾ kilometers.)**

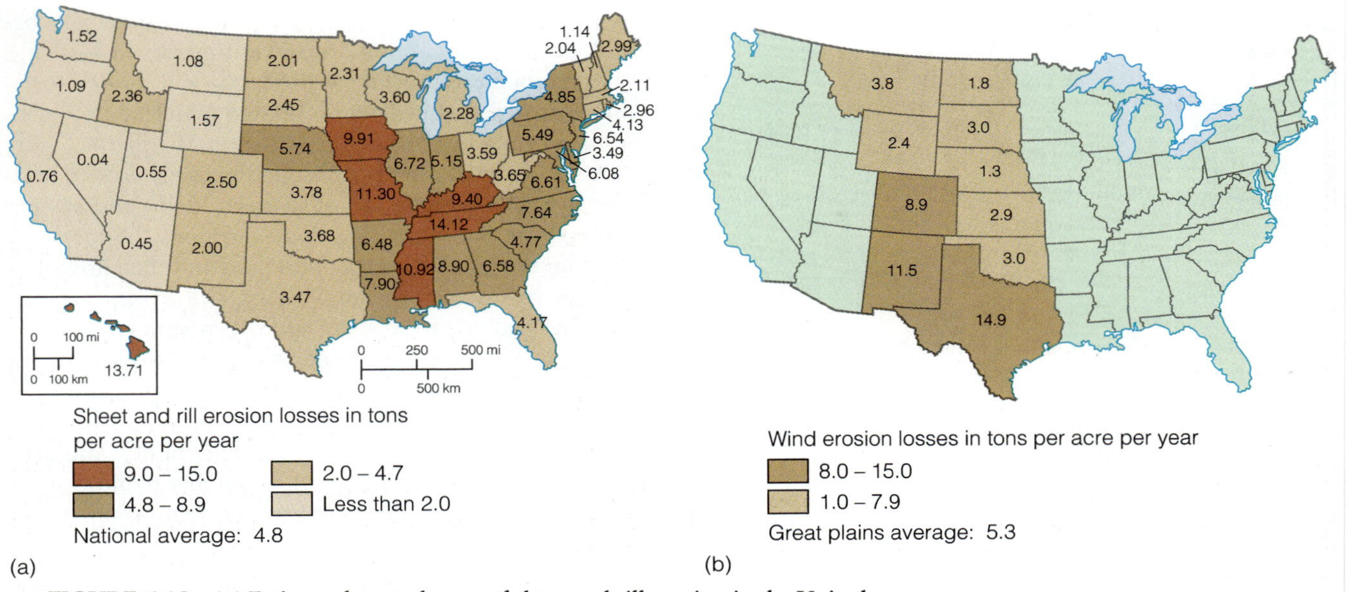

➢ **FIGURE 6.18** **(*a*) Estimated annual rates of sheet and rill erosion in the United States. (*b*) Estimated annual rates of wind erosion in the Great Plains.**

TABLE 6.2 Significant U.S. Soil Erosion by State

RANK	STATE	TONS/ACRE/YEAR
Sheet and Rill Erosion		
1	Tennessee	14.12
2	Hawaii	13.71
3	Missouri	11.30
4	Mississippi	10.92
5	Iowa	9.91
Wind Erosion		
1	Texas	14.9
2	New Mexico	11.5
3	Colorado	8.9
4	Montana	3.8
5	Oklahoma	3.0
*Total Soil Erosion, Water and Wind**		
1	Texas	18.4
2	Tennessee	14.12
3	Hawaii	13.71
4	New Mexico	13.50
5	Colorado	11.40

* The U.S. average annual rate of total soil erosion is estimated at 4.0–4.8 tons/acre (10–12 tons/hectare).

SOURCE: *Environmental Trends* (Washington, D.C.: Council on Environmental Quality, 1981), cited in Sandra Batie, *Soil Erosion: A Crisis in America's Cropland?* (Washington, D.C.: The Conservation Foundation, 1983).

lighter particles. If the wind is strong, a dust storm may occur. The great "black blizzards" of the 1930s created the "dust bowl" of the southern Great Plains during a severe drought (Case Study 6.2). Much of the reason for the dust bowl was the widespread conversion of marginal grasslands, which receive only 25–30 centimeters (10–23 in) of rainfall per year, to cropland. Unfortunately, this is still occurring in eastern Colorado. ➢ Figure 6.18b presents estimates of the amount of soil lost each year to wind erosion in the Great Plains states. When the 50 states' estimated wind and water erosion losses are combined, the state with the greatest cropland loss is Texas. Texas is followed by Tennessee, Hawaii, and New Mexico, respectively. Table 6.3 summarizes the recognized causes of accelerated soil erosion and assesses the relative impact of each major cause.

MITIGATION OF SOIL EROSION. Thomas Jefferson was among the first people in the United States to comment on the seriousness of soil erosion and he called for such soil conservation measures as crop rotation, contour plowing, and planting grasses during the fallow season. These practices are now common in industrialized nations, but unfortunately, only minimally employed in Third World countries. Proper choices of where to plant, what to plant, and how to plant are the first steps toward soil conservation.

Terracing—creating flat areas, terraces, on sloping ground—is one of the oldest and most efficient means of saving soil and water (➢ Figure 6.19). *Strip-cropping,* by which close-growing plants are alternated with widely

TABLE 6.3 Causes of Accelerated Soil Erosion and Deterioration

CAUSE	ESTIMATED % OF TOTAL
Overgrazing • reduced vegetation cover • trampling that leads to decreased water-holding capacity	35
Bad agricultural practices • salinization due to improper draining of irrigated land • fertilization that results in acidification • lack of terraces on tilled sloping land • wind erosion of fallow lands	28
Deforestation	30
Other causes, natural and human	7

SOURCE: *World Resources: A Guide to the Global Environment.* World Resources Institute, United Nations Environment and Development Program. (New York: Oxford University Press, 1992).

spaced ones, efficiently traps soil that is washed from bare areas and it also supplies some wind protection. An example of this is alternating strips of corn and alfalfa. *Crop rotation* is the yearly alternation of soil-depleting crops with soil-enriching crops. Rotating corn and wheat with clover in Missouri, for example, has been found to reduce soil runoff from 19.7 tons/acre for corn to 2.7 tons/acre, a huge loss reduction indeed. Groundcovers such as grasses, clover, and alfalfa tend to be soil-conserving, whereas rowcrops like corn and soybeans can leave soil vulnerable to runoff losses. Retaining crop residues, such as the stubble of corn or wheat, on the soil surface after harvest has been found to reduce erosion and to increase water retention by more than half.

Conservation-tillage practices eliminate unnecessary plowing in the fall and encourage the contour plowing of furrows perpendicular to the hill slope to catch runoff and increase water infiltration (➢ Figure 6.20). *No-till* and *minimum-till* practices also conserve soil. With no-till farming, seeds are planted directly into the soil through the previous crop's residue, and weeds are controlled with chemicals. Specialized no-till equipment is designed to implant seeds beneath crop stubble and plant debris. "No-till" has been successful in soybean farming. In 1993 farmers left 30 percent of their crop debris on 37 million cultivated hectares (92 million acres), nearly a third of all cultivated land in the United States. Ten percent of U.S. planted acreage was not plowed at all. No-till agriculture is most beneficial during times of drought, when plowed fields dry out and topsoil is exposed to wind erosion. During the Great Flood of 1993 in the Mississippi River Valley, no-till practices saved much of the agricultural soil that would have otherwise been eroded by floodwaters.

Wind-caused soil losses can be reduced by planting windbreaks of trees and shrubs near fields and by planting strip crops perpendicular to the prevailing wind direction. Wind losses are almost zero with cover crops such as sod and clover.

➢ FIGURE 6.19 Terracing on a relatively steep hillside prevents extreme erosion; China.

➢ FIGURE 6.20 Contour plowing. Runoff is intercepted in furrows that follow an elevation and that are perpendicular to the slope of the land. The practice retains water and reduces erosion.

CASE STUDY 6.2

The Wind Blew and the Soil Flew—The Dust Bowl Years

The stock market crash of 1929 ushered in the Great Depression of the 1930s, which brought widespread unemployment without unemployment benefits or entitlements such as welfare, social security, and free health care. It was a bad time, and to make things worse, there was protracted drought in the Great Plains, an area that was totally dependent upon rainfall for crop production at the time. Thousands of acres of marginally productive land, mostly grasslands, had been tilled because the government had begun guaranteeing wheat prices. Farmers thought they couldn't lose—but they were banking on rain. The entire United States dried up, except for the New England area, but between 1932 and 1940 the southern Great Plains was hit the worst (➢ Figure 1).

A longstanding agricultural practice is to plow up cropland after the fall harvest and allow the land to stand fallow through the winter. Plowing turns crop stubble into the soil, gets rid of weeds, enhances soil water and air uptake, and makes fields look better. This is a fine practice as long as it doesn't rain too much—or too little. If too much rain falls, the soil is subject to severe sheet and rill erosion. If the pulverized soil dries out and the wind blows, the result is the same: extreme soil erosion. The wind did blow in the southern Great Plains in the mid-1930s, and huge quantities of topsoil were simply blown away. There were many dust storms during this period, but the one of April 1935 was the worst. Soil silt, clay, and organic matter were lifted 5 kilometers (16,000 ft) into the air and carried as far eastward as Washington, D.C., and even 500 kilometers (300 mi) off the East Coast into the Atlantic Ocean. The source area of the dust storms became known as the "Dust Bowl." It is a zone

➢ **FIGURE 1 A dust storm hits a southwestern high plains village in October, 1937. Storms like this one destroyed crops and buried pasture lands. Minutes after this photograph was taken the village was completely engulfed in choking dust.**

roughly 900 kilometers long by 450 kilometers wide (540 mi × 270 mi) in Colorado, Kansas, New Mexico, Texas, and Oklahoma (➢ Figure 2).

The wind and dust made farming impossible, and farm families, burdened by debt for equipment, seed, and supplies, left their farms to become migrant workers. Many traveled to California, where their lot was hardly improved, and the entire country was soon referring to this huge unfortunate group, derisively, as *Okies* (➢ Figure 3). This subculture of the 1930s, innocent victims of severe soil erosion, was the subject of John Steinbeck's touching novel *The Grapes of Wrath*.

Even today wind erosion exceeds water erosion in Texas, Colorado, Montana, New Mexico, Oklahoma, and North Dakota. Mismanagement of plowed lands and overgrazing are making parts of these six states potential dust bowls.

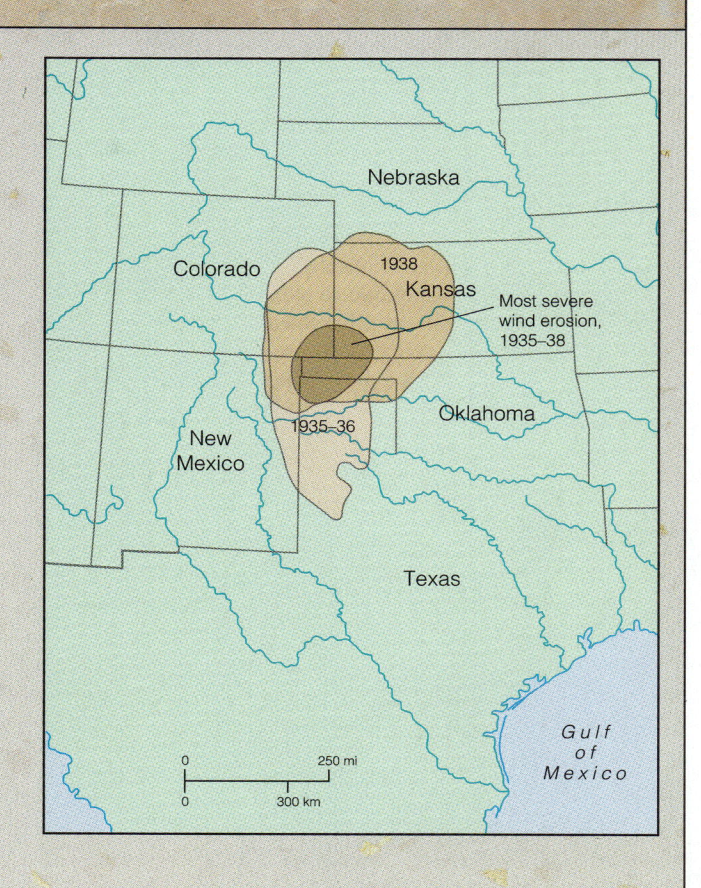

➢ FIGURE 2 (*Right*) Map of states most seriously affected by blowing dust during the Dust Bowl years 1935–1936, and 1938. Kansas was the state most devastated in terms of percentage cultivated area degraded by strong winds.

➢ FIGURE 3 Dust Bowl America. In an historic 1938 photograph by Dorothea Lange, who worked for the Farm Security Administration, a homeless migrant family of seven walks a highway in Pittsburgh county, Oklahoma. By 1939, half of the families in the state were on relief.

➢ **FIGURE 6.21 Damage to interior walls from expansive soil. Note the cracks around the door.**

Expansive Soils

Certain clay minerals have a layered structure that allows water molecules to be absorbed between the layers, which causes the soil to expand. This process differs from the *adsorption* of water to the surface of nonexpansive soil clays. Soils that are rich in these minerals are said to be **expansive soils.** Although this reaction is somewhat reversible, as the clays contract when they dry, soil expansion can exert extraordinary uplift pressures on foundations and concrete slabs with resulting structural and cosmetic damage (➢ Figure 6.21). Damage caused by expansive soils costs about $6 billion per year in the United States, mostly in the Rocky Mountain states, the Southwest, and Texas and the Gulf Coast states (➢ Figure 6.22). The swelling potential of a soil can now be identified by several standard tests, usually performed in a soils engineering laboratory. An obvious test is to compact a soil in a cylindrical container, soak it with water, and measure how much it swells against a certain load. Clays that expand more than 6 percent are considered highly expansive; those that expand 10 percent are considered critical. Treatments for expansive soil include (1) removing them, (2) mixing them with nonexpansive material or with chemicals that change the way the clay reacts with water, (3) keeping the soil moisture constant, and (4) using reinforced foundations that are designed to withstand soil volume changes. Identification and mitigation of damage from expansive soils is now standard practice for U.S. soil engineers.

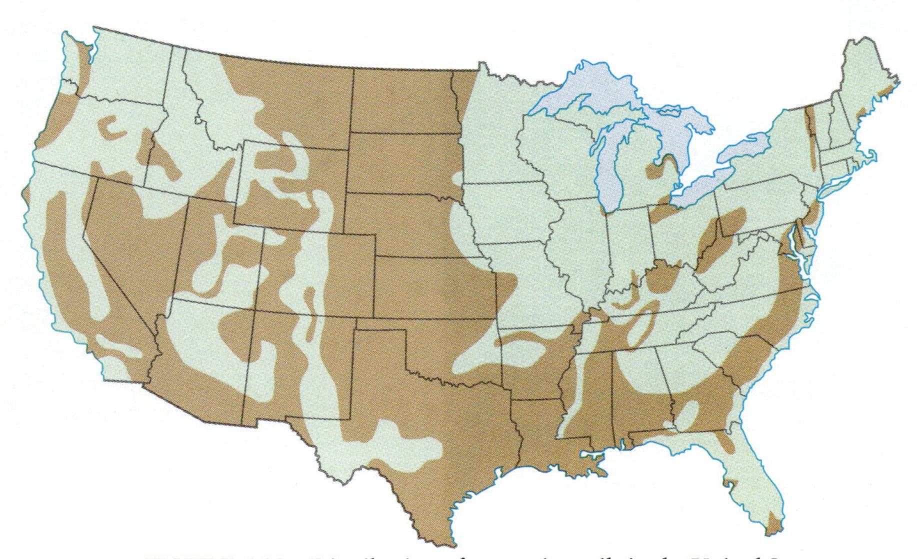

➢ **FIGURE 6.22 Distribution of expansive soils in the United States.**

Permafrost

The term **permafrost,** a contraction of *permanent* and *frozen,* was coined by Siemon Muller of the U.S. Geological Survey in 1943 to denote soil or other surficial deposits in which temperatures below freezing are maintained for several years. More than 20 percent of the earth's land surface is underlain by permafrost, and it is not surprising that most of our knowledge about frozen ground has been derived from construction problems encountered in Siberia, Alaska, and Canada. Permafrost becomes a problem for us humans when we change the surface environment by our activities in ways that thaw the near-surface soil ice, since this results in soil flows, landslides, subsidence, and related phenomena. Permafrost forms where the depth of freezing in the winter exceeds the depth of thawing in the summer (➢ Figure 6.23). If cold ground temperatures continue for many years, the frozen layer thickens until the penetration of surface cold is balanced by the flow of heat from the earth's interior. This equilibrium between cold and heat determines the thickness of the permafrost layer. Thicknesses of 1,500 meters and 600 meters (almost 5,000 and 2,000 ft) have been reported in Siberia and Alaska, respectively. Such thick layers are very old; they must have formed during the Pleistocene Epoch of geologic time, the time known as the Great Ice Age.

The top of the permanently frozen layer is called the **permafrost table** (➢ Figure 6.24). Above this is the **active layer,** which is subject to seasonal freezing and thawing and is always unstable during summer. When winter freezing does not penetrate entirely to the permafrost table, water may be trapped between the frozen active layer and the permafrost table. These unfrozen layers and lenses, called *talik* (a Russian word), are of environmental concern, because the unfrozen ground water in the talik may be under pressure. Occasionally, pressurized talik water bursts through to the surface, forming a domical mound 30–50 meters high called a *pingo.* The presence of ice-rich permafrost becomes evident when insulating vegetation, usually tundra, is stripped away for a road, runway, or building. Thawing of the underlying soil ice results in subsidence, soil flows, and other gravity-induced mass movements (➢ Figure 6.25). Even the casual crossing of tundra on foot or in a jeep can upset the thermal balance and cause thawing. Once melting starts, it is impossible to control, and a permanent scar is left on the landscape (➢ Figure 6.26).

Structures that are built directly on or into permafrost settle as the frozen layers thaw (➢ Figure 6.27). Both active and passive construction practices are used to control the problem. Active methods involve complete removal or thawing of the frozen ground before construction, whereas passive methods build in such manners that the

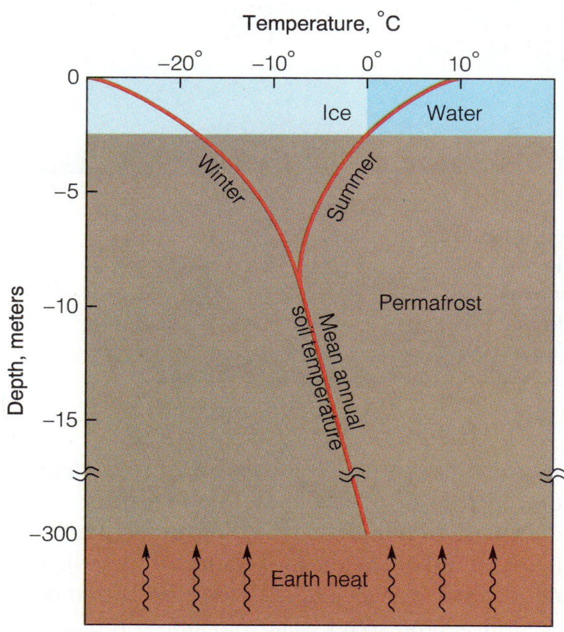

➢ **FIGURE 6.23 The interaction of soil temperature and depth establishes the lower and upper limits of permafrost. The upper limit is found where and when the summer soil temperature exceeds 0° C. The lower limit may be at great depth. It is the point where earth heat raises the soil temperature above the freezing temperature of water. Note the break in the depth scale.**

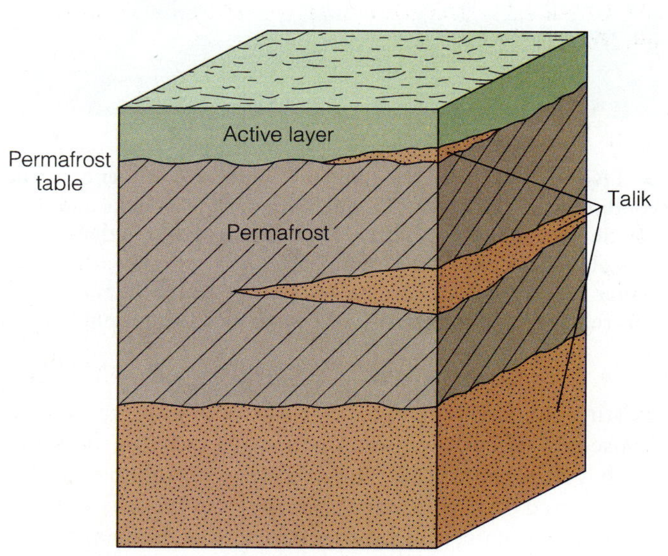

➢ **FIGURE 6.24 Permafrost zones. Unfrozen layers or lenses, called *talik,* may occur in the active layer and in the permanently frozen layer.**

➢ FIGURE 6.25 Caterpillar tractor mired in thawed perennially frozen lake clay; northern edge of Copper River Basin, Alaska.

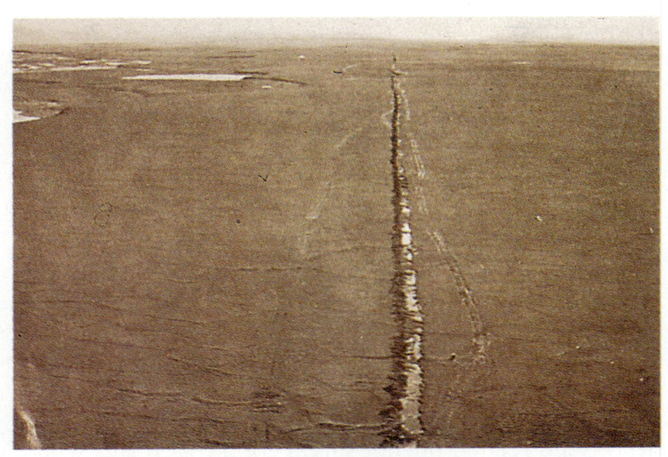

➢ FIGURE 6.26 Scars made by vehicles on tundra soil remain for many years; Alaska.

(a)

(b)

➢ FIGURE 6.27 Damage to houses due to improper construction for permafrost conditions in Alaska. (*a*) A modern home in Fairbanks and (*b*) Bert and Mary's Roadhouse on the Richardson Highway sinking into thawing ice-rich sediments. The log cabin was built in 1951 and the greatest subsidence is toward the front end of the center portion of the structure, since this is where the furnace is located. The structure was removed in 1965 because of distress due to continued sinking.

existing thermal regime is not disturbed. For example, a house can be built with an open space beneath the floor so that the ground can remain frozen, a building can be constructed on pipe foundations that use heat dissipators or actual coolant to keep them cold, and roads are constructed on top of packed coarse-gravel fill to allow cold air to penetrate beneath the roadway surface and keep the ground frozen (➢ Figure 6.28).

The most ambitious construction project on permafrost terrane is the 800-mile oil pipeline from Prudhoe Bay on Alaska's North Slope to Valdez on the Gulf of Alaska. About half the route is across frozen ground, and because the oil is hot when it is pumped from depth and also during transport, it was decided to place the pipe above ground to avoid the thawing of permafrost soil (➢ Figure 6.29). The 4-foot-diameter pipeline also crosses several active faults, three major mountain ranges, and a large river. The pipeline's above-ground placement allows for some fault displacement and for easier maintenance if it should be damaged by faulting. This placement has the added advantage of not blocking the migratory paths of herding animals such as caribou and elk.

(a)

(b)

➢ **FIGURE 6.28 Passive construction methods for mitigating damage due to permafrost. (*a*) Raised WW II–era Quonset hut; Nome, Alaska. (*b*) Heat-radiating foundation piers on a modern structure; Kotzebue, Alaska.**

(a)

(b)

➢ **FIGURE 6.29 The 800-mile Alaskan oil pipeline was constructed to accommodate permafrost and wildlife. (*a*) The meandering pattern allows for expansion and contraction of the pipe and for potential ground movement along the Denali fault, which it crosses. The pipeline supports feature heat-dissipating radiators. (*b*) Moose, elk, and caribou are able to migrate under the elevated pipeline.**

Settlement

Settlement occurs when a structure is placed upon a soil, rock, or other natural material that lacks sufficient strength to support it. Settlement is an engineering problem, not a geological one. It is treated here, however, because settlement and poor soil seem inextricably associated. All structures settle, but if settlement is uniform, say 2 or 3 centimeters over the entire foundation, there is usually no problem. When differential settlement occurs, however, cracks appear or the structure tilts. The Leaning Tower of Pisa is probably the best-known example of poor foundation design, and Galileo is said to have conducted gravity experiments at the 54.5-meter (163 ft) structure. Begun in 1174, the tower started to lean when the third of its eight stories was completed, sinking into a 2-meter-thick layer of soft clay just below the surface. To compensate for the lean, the engineer in charge made the stories taller on the leaning side. The added weight caused the structure to sink even further. It now leans about 5.2 meters (17 ft) from the vertical, about 5°, which gives one an ominous feeling when standing on the down-tilted side of the tower. The tower even has a slight bend like that of a banana, because the builders continued to construct vertically after tilting began (➢ Figure 6.30). It is interesting that one must descend to the main entrance, because the entire tower has settled 2 meters in its lifetime. Tilting has been aggravated by excessive withdrawals of underground water by the city of Pisa.

Attempts were made to stabilize the foundation by pumping cement grout (cement–water mixture, or "slurry") into the soft clays through 36 half-inch drill holes on the leaning side. This slowed the rate of tilting to about 0.6 millimeter per year. Some experts predicted that even at that rate of tilt, Pisa's tower would fall over in the not-too-distant future. Many schemes, mostly harebrained ideas, have been proposed over the years for halting the leaning. These include installing a giant wind machine to blow against the tower's leaning side and the stringing of cables to keep it from tipping farther. The most feasible solution developed recently is to install a huge concrete block under the tower to act as a "deadman," or counterweight, and then to tie, or "tension," the tower to the deadman with cables. The tower was closed to the public in 1990, and it is expected to reopen after repairs in late 1993 or 1994.

It should be noted that tilting towers are not exceptional in Italy. Few of the 170 campaniles (bell towers) still standing in Venice are vertical, for example, and many other towers throughout Italy and Europe tilt. These graceful towers are certainly testimony to past architects' ability and talent, but also to their lack of understanding that a structure can be truly beautiful only if local geology permits.

➢ FIGURE 6.30 Inadequate foundation investigation and design; Pisa, Italy.

Other Soil Problems

Laterite soils, the red soils of the tropics, are soft when they are moist but become brick-hard and durable when they dry out. The Buddhist temple complex at Angkor Wat in Cambodia is partly made of laterite bricks that have endured since the thirteenth century. Deforestation of areas underlain by laterites is an invitation to this hardening, as U.S. forces discovered in Vietnam when they attempted to grade roads and runways on lands that had been defoliated for military reasons. Deforested laterites make for swampy conditions during the rainy season, because they lack permeability; there is little infiltration of the water. This can raise havoc with wheeled vehicles and restrict mobility in such regions.

Although laterite soils support some of the densest rain forests in the world, they are relatively infertile because of extensive leaching of minerals. The nutrients that support the rain forest are in the vegetation, and they are recycled when the plants die and decompose into the soil. Deforested laterites used for agricultural purposes must be fertilized intensively after a few harvests in order to support productive crops.

In some areas of extensive irrigation agricultural soils form impervious clayey layers called **hardpans** in the B horizon. Hardpans inhibit waters from percolating and thus reduce crop yields. They must be broken up by deep plowing using large, heavy tractors pulling long, single-tooth rippers that penetrate the subsoil. Hardpans and claypans are the scourge of farmers, as they can render their farmland unproductive.

The presence or absence of beneficial elements in soils can result in various dietary benefits or deficiencies in humans. This is because the soil is the actual beginning of the food chain. Soil elements are taken up by plants (vegetables, fruits, and grains), which are then eaten by animals. Some examples of soil elements and health are discussed in Case Study 6.3.

CASE STUDY 6.3

Soils, Health, and Loco Weed

Humans need trace elements in their systems in order to function properly. **Trace elements** occur in small amounts in the body, usually a few parts per million, as opposed to the bulk elements, which are found in large quantities. The body obtains these elements naturally by this sequence: rocks to soils to plants and finally to animals. Humans get trace elements by eating plants or by eating animals that ate plants that grew on soils containing them or by using dietary supplements of them. Some water-soluble elements, such as fluoride, are sometimes added to municipal drinking-water supplies.

The best-known relationship between soils and human health is the one between soil iodine deficiency and the enlargement of the thyroid gland known as *goiter*. Besides being cosmetically demoralizing, goiter is debilitating in other ways. Iodine deficiency during pregnancy can lead to "cretinism" in newborns, a severe mental and physical disorder. Endemic goiter has been firmly established in the northern United States in the so-called goiter belt (see ➢ Figure 1). Commercial table salt is now iodized, and this disease has practically disappeared except among the poorly educated. Additionally, little of the U.S. diet is "home grown" anymore. This is another reason goiter is no longer an endemic problem. Soil-iodine thyroid diseases are also known in England, Thailand, Mexico, the Netherlands, and Switzerland.

Several trace elements—zinc, copper, iron, cobalt, manganese, and molybdenum—are associated with and stimulate enzymes, the catalysts of biochemical reactions. These enter the body mostly in plant and animal foods. Their importance might be better appreciated when we realize that silver, mercury, and lead are toxic because they are enzyme *inhibitors*. Some of the beneficial effects of trace elements are that zinc aids in healing wounds and burns, iron transports oxygen in the blood, and cobalt is a vital component of vitamin B_{12}. Deficiencies of these may result in anemia, slowed growth, vitamin-B_{12} deficiency, and other ailments.

A most interesting trace element is selenium. It is poisonous at high dosages, but deficiencies cause abnormalities in plants and animals. It is abundant in soils of the high plains and western United States, and certain plants there concentrate it in their tissue. Eating these plants, known as *locoweeds,* causes cattle to get the "blind staggers" and to appear drugged. Figure 1 shows areas where mineral-nutritional diseases are known to occur in animals. These are all localities where soils have deficiencies or excesses of vital or toxic trace elements.

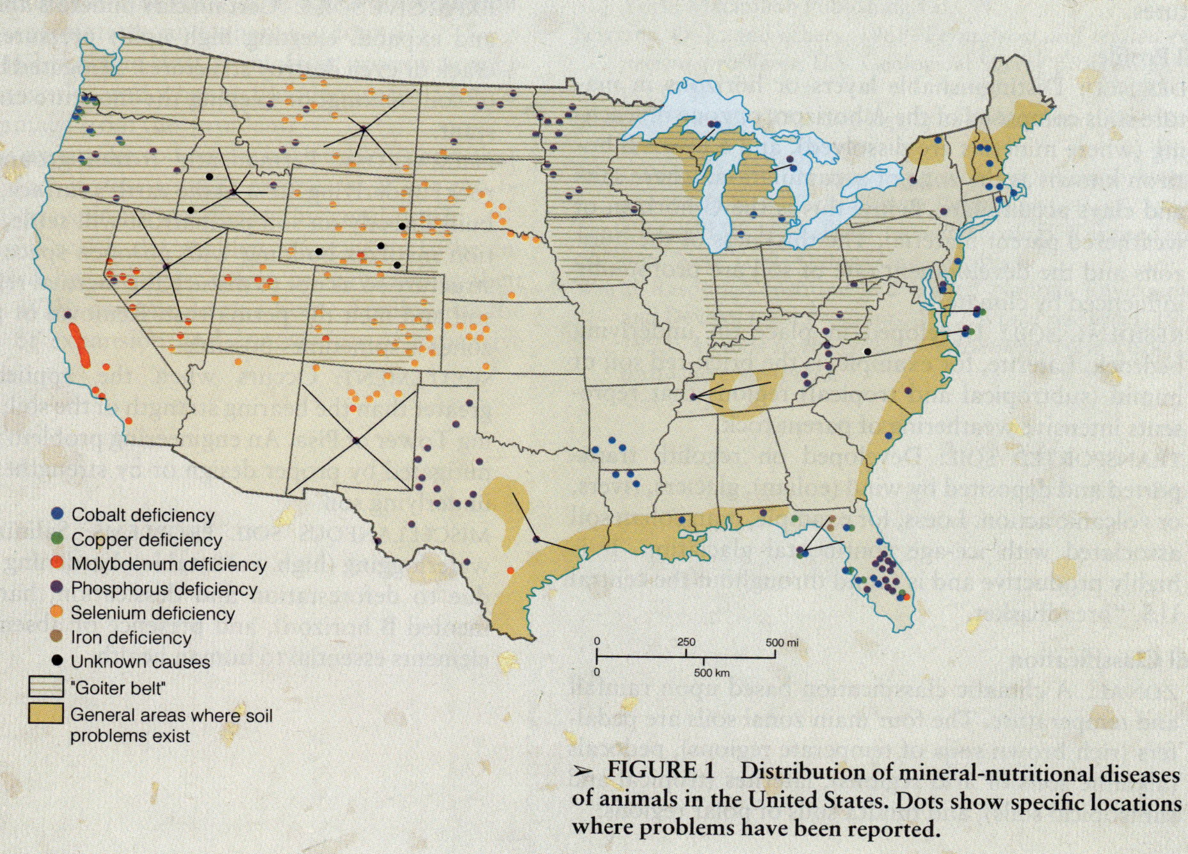

➢ **FIGURE 1 Distribution of mineral-nutritional diseases of animals in the United States. Dots show specific locations where problems have been reported.**

CHAPTER

7

LANDSLIDES AND MASS WASTING

Nature to be commanded,
must be obeyed.

FRANCIS BACON, PHILOSOPHER (1561–1626)

Mass wasting is the general term that denotes any downslope movement of soil and rock under the direct influence of gravity. Mass-wasting processes, which include landslides, rapidly moving debris flows, slow-moving soil creep, and rockfalls of all kinds, are a significant geologic hazard throughout North America. As a hazard to humans, mass wasting is more likely to occur than are volcanic eruptions, earthquakes, and floods, but it is much less spectacular and thus less newsworthy. Annual damage from landsliding alone in the United States is estimated at between one and two billion dollars. If we include other ground failures, such as subsidence, expansive soils, and construction-induced slides and flows, total losses are many times greater than the annual combined losses from earthquakes, volcanic eruptions, floods, hurricanes, and tornadoes.

Mass wasting is also a global environmental problem. Earthquake-triggered landslides in Kansu Province in China killed an estimated 200,000 people in 1920, and debris flows left 600 dead and destroyed 100,000 homes near Kobe, Japan, in 1938. The largest loss of life from a single landslide in U.S. history, 129 fatalities, occurred at Mameyes, Puerto Rico, in 1985.

The term **landslide** encompasses all moderately rapid falls, slides, and flows that have well-defined boundaries and that move downward and outward from a natural or

Crown scarp of Bluebird Canyon landslide, Orange County, California. The slide occurred on October 2, 1978 and resulted in the destruction of 24 homes but no loss of life or injury. Successful repair of the landslide required close cooperation between homeowners, city, county, and state officials—including the Office of the President of the United States.

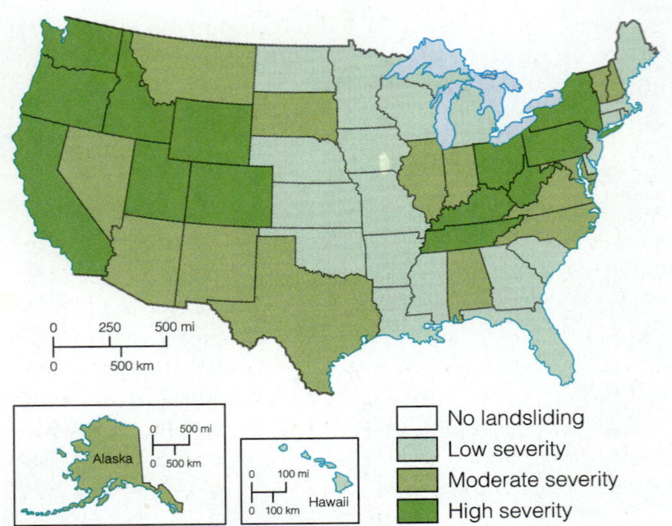

➢ **FIGURE 7.1 Severity of landsliding in the United States. Note that the most serious problems are found in the Appalachian Mountains, the Rocky Mountains, and the coastal mountain ranges of the Pacific Rim.**

artificial slope. They occur in all the states and are an economically significant factor in more than 25 of them (➢ Figure 7.1). Wherever unstable slopes exist, some form of mass wasting usually occurs. Areas with the greatest topographic relief are at highest risk, because gravity acts to smooth out topography by reducing the high areas through mass wasting and filling in the low areas with slide debris. For example, more than two million mappable slides exist in the Appalachian Mountains from New England to Alabama. Almost 10 percent of the land area of Colorado is landslide terrane, and landslides and debris flows in Utah caused $300 million worth of damage in 1983–84. In Southern California, eight wet years between 1950 and 1993 *averaged* $500 million in damage. Slope stability problems have such impact that a national landslide-loss-reduction program was implemented by the U.S. Geological Survey in the mid-1980s. A high priority of the program is to identify and map landslide-prone lands.

The classification of slope movements used in this book is similar to the one widely used by engineering geologists and soil engineers (➢ Figure 7.2). The bases of the classification are:

- the type of material involved, such as rock or earth (fine-grained soil);
- how the material moves, such as by sliding, flowing, or falling;
- the moisture content of the material; and
- the velocity at which the material moves.

For instance, soil creep occurs at imperceptibly slow rates and is relatively dry, debris flows are water-saturated and move swiftly, and slides are coherent masses of rock or soil that move along one or more discrete failure surfaces or **slide planes.** Failure occurs when the force that is pulling the slope downward (gravity) exceeds the strength of the earth materials that compose the slope. Because the gravity acting on a given slope is essentially constant, either the resistance of the earth materials to sliding must decrease or the gradient of the slope must increase in order to initiate sliding. In this chapter we restrict our discussion to the mechanics of slides and flows, because rockfalls are discussed in Chapters 6 and 11.

➢ **FIGURE 7.2 Classification of landslides by mechanism, material, and velocity.**

MECHANISM		MATERIAL: Rock	Fine Soil	Coarse Soil	VELOCITY ft/sec
SLIDE		Slump	Earth slump	Debris slump	Slow $< 10^{-6}$
		Block glide	Earth slide	Debris slide	Rapid 10^{-3}
FLOW		Rock avalanche	Earthflow, avalanche	Debris flow, avalanche	Very rapid 1–10
		Creep	Creep	Creep	Extremely slow
FALL		Rockfall	Earthfall	Debrisfall	Extremely rapid <10

FLOWS

Types of Flows

Creep is the slow (a few millimeters per year), essentially continuous downslope movement of soil and rock on steep slopes. It involves either freezing and thawing or alternate wetting and drying of a hill slope, which causes upward expansion of the ground surface perpendicular to the face of the slope. As the slope dries out or thaws, the soil surface drops vertically, resulting in a net downslope movement of the soil (➢ Figure 7.3). Burrowing animals and other biological processes that produce openings in the soil also contribute to soil creep. Bent trees, leaning fence posts and telephone poles, and bending of tilted rock layers downslope are all evidence of soil creep. Homes built with conventional foundations 18–24 inches deep may develop cracks due to soil creep. This is seldom catastrophic; typically this is a cosmetic and maintenance problem. The influence of soil creep can be overcome by placing foundations through the soil and weathered bedrock of the creep zone (Figure 7.3).

Debris flows are dense, fluid mixtures of rock, sand, mud, and water. They may be generated quickly during heavy rainfall or snowmelt where there is an abundant supply of loose soil and rock. Moving with the consistency of wet concrete, they are very destructive, with velocities up to many meters per second. Because of their high density, commonly 1.5–2.0 times the density of water, debris flows are capable of transporting large boulders, automobiles, and even houses in their mass. The house shown in ➢ Figure 7.4 is in Shields Canyon in the San Gabriel Mountains of southern California. Thirteen cars were packed around the house, and several others were deposited in the backyard swimming pool. The family was rescued from inside the house, floating on mud that reached nearly to the ceiling.

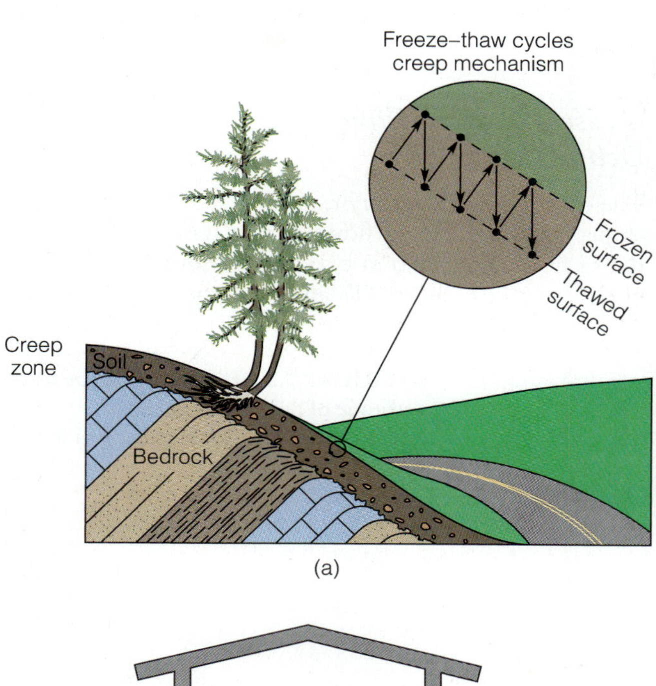

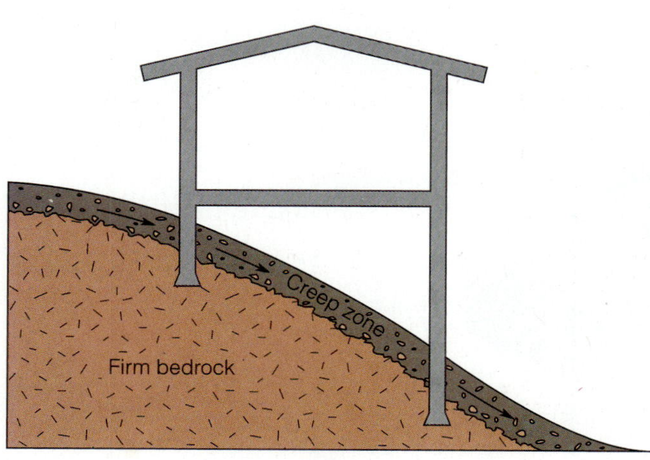

➢ **FIGURE 7.3** **(*a*) Diagram and (*b*) photograph of soil and rock creep; east slope of the Sierra Nevada, California. (*c*) Securing a structure on pilings that extend through the creep zone into firm bedrock can prevent damage due to soil motion.**

➢ **FIGURE 7.4 Debris flow in Shields Canyon, San Gabriel Mountains, 1977. Note that the car and the man are on the roof of the house.**

The areas most subject to debris flows are characterized either by sparse vegetation and intense seasonal rainfall, or are in regions that are subject to drenching rains associated with hurricanes. Debris flows are relatively common in Canada, the Andes (particularly Peru), and alpine and desert environments worldwide. Debris flows associated with volcanic eruptions (lahars) are potential hazards in all volcanic zones of the world. They pose a particular danger in populated areas near volcanoes in Washington, Oregon, and California. The eruption of Mount St. Helens in Washington displaced water in Spirit Lake at its base and produced a lahar that filled the north fork of the Toutle River with the largest (2.8 km^3; 0.67 mi^3) and most destructive debris flow of modern times (look back to Figure 5.27). These types of debris flows are the most significant threat to life of all the landslide hazards.

A **debris avalanche** is simply a fast-moving (> 4 m/sec or 15 km/h) debris flow (➢ Figure 7.5). A structure in the path of one of these flows can be severely damaged or completely flattened with attendant injury or loss of life. During the winter of 1982, debris flows and avalanches in the San Francisco Bay area killed 25 people and resulted in $66 million worth of damage (➢ Figure 7.6). In Mill Valley, a small town north of San Francisco, a debris flow ripped a house from its foundations and deposited it 45 meters downslope—without serious injury to the residents (passengers). The house effectively dammed the canyon and prevented debris-flow damage farther downslope.

Causes of Debris Flows and Debris Avalanches

The combination of heavy rainfall and loose soil on steep slopes promotes debris flows and avalanches. Studies of debris flows indicate that two rainfall conditions are necessary to cause them: (1) an initial period of rainfall,

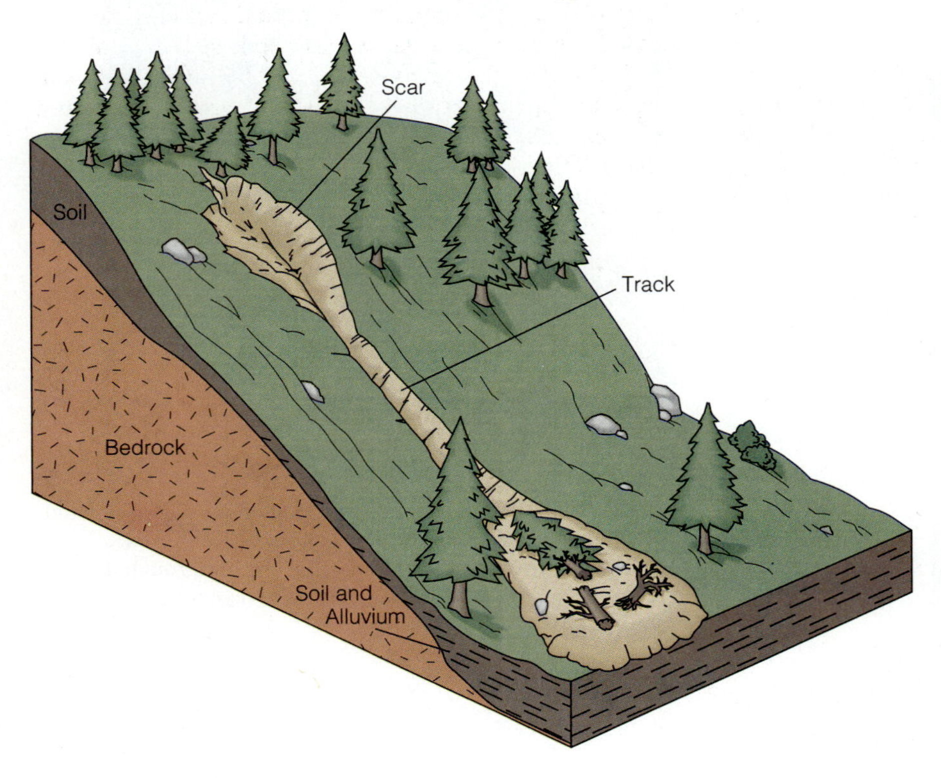

➢ **FIGURE 7.5 Debris avalanche track and zone of deposition. Such an avalanche may travel thousands of feet beyond the base of the slope.**

➢ **FIGURE 7.6 Debris avalanche tracks on a hillside near San Rafael, California.**

known as *antecedent rainfall,* that saturates the soil and (2) a subsequent period of intense rainfall that puts even more water into the soil and initiates flowage. Thus the onset of debris flows may be estimated during the rainy season if we have data on the amount of rainfall that can be expected to initiate flowage. Such data have been compiled for flows and slides in the San Francisco Bay area of northern California (➢ Figure 7.7). They indicate that short periods of intense rainfall or less-intense rainfall over a longer duration can produce debris flows. Specifically, the curves of Figure 7.7 show that for a rainfall intensity of 0.5 inch per hour, the threshold time for the onset of debris flows in Marin County is 8 hours, and that for Contra Costa County it is 14 hours. The differing rainfall thresholds in these relatively close areas are due to the variability of geologic materials and topography. The lower curve of Figure 7.7 shows that less-intense rainfall will produce debris flows in semiarid and arid areas of California. This is because these dry areas have little vegetation with root systems that retain soils and abundant loose surface debris. Although these curves are only preliminary, they will serve to alert residents in critical areas once the threshold conditions for flows have been attained, and they will be refined as more data become available.

Mountain slopes burned by range and forest fires are also susceptible to debris flows during the wet season. The loss of active root systems to bind soil particles can result in an extremely dangerous condition. In addition, debris flows from burned slopes have been found to have longer runout distances into foothill areas than those from vegetated slopes.

Not all debris flows and avalanches are triggered by intense rainfall. Slide Mountain, about halfway between

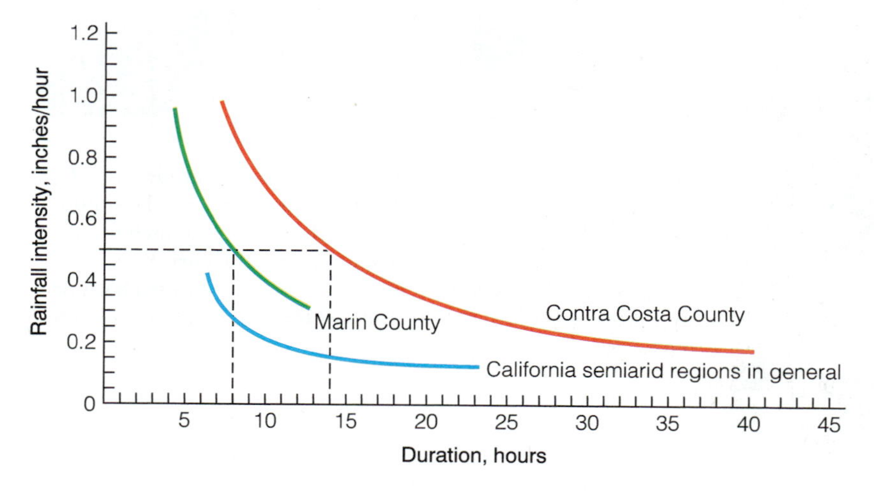

➢ **FIGURE 7.7 Rainfall thresholds for debris avalanches and landslides for two central California counties and California semiarid regions in general. For example, a rainfall intensity of 0.5 inch/hour for 8 hours is sufficient to initiate flows in Marin County. The same intensity for 14 hours is the threshold for Contra Costa County.**

Reno and Carson City, Nevada (and between Washoe Lake and Lake Tahoe), is not named for the ski area on its slopes; it is the site of many flows and slides. On 30 May 1983, a debris flow was generated on Slide Mountain that killed one person, injured many more, and destroyed a number of homes and vehicles, all in less than 15 minutes (➢ Figure 7.8). About 720,000 cubic meters of weathered granite gave way from the mountain's steep flank and slid into Upper Price Lake. This mass displaced the water in the lake, causing it to overflow into a lower lake, which in turn overflowed into Ophir Creek gorge as a water flood. Picking up sediment as it went, it became a debris flow that emerged from the gorge, spread out, destroyed homes, and covered a major highway. Because this is a popular recreational area and debris flows can be sudden and hazardous to one's health, geologic hazard warning signs are posted on trails throughout the area. There is a history of debris flows in this part of Nevada, as documented by the 8 July 1890 edition of the *Carson Appeal:* ". . . On Sunday afternoon about quarter of 5

(a)

(b)

(c)

(d)

➢ FIGURE 7.8 Slide Mountain in western Nevada, site of the 1983 disaster. (*a*) Looking up Ophir Creek and the runout area of the debris flow. The scar of the debris avalanche source area is the treeless area on the mountain. (*b*) Debris flow in lower Ophir Creek and a damaged home. (*c*) Abraded bark on tree serves to record the thickness of the flow. (*d*) A warning to be taken seriously.

o'clock, Price's Reservoir at the foot of Slide Mountain burst, and the water, rushing down the canyon, submerged the V&T Railroad track at Franktown. . . ." The 1983 Slide Mountain debris flow and the one associated with an earthquake at Yungay, Peru, that killed 18,000 people (discussed in Chapter 4) were both initiated by landslides into water masses.

LANDSLIDES

The Mechanics of Slides

Landslides do not just happen. They are explainable as a change in the balance between **driving forces,** the gravity forces that pull a slope downward, and **resisting forces,** the cohesive and frictional forces that hold a slope in place. A change in the balance may result from water seepage into slope material, an oversteepening of the slope by natural erosion or artificial cutting, or the addition of weight at the top of a slope. If any of these things happen, the driving forces may eventually exceed the resisting forces, which will cause a landslide to occur.

This can be demonstrated using a simple model of a sliding block on an inclined plane (➢ Figure 7.9). The driving force, *d,* is the component of gravity acting parallel to the inclined plane at an angle α with the horizontal. The coefficient of friction between the block and the plane is *f*. Static friction between the block and slide plane increases as the normal force, *n,* acting across the slide plane increases. Friction between the block and the plane resists sliding. Thus,

$$\text{resisting forces} = \text{friction coefficient} \times \text{normal force}$$

or

$$r = f \times n$$

A delicate balance exists, and sliding is imminent when:

$$d \text{ (driving force)} = f \text{ (friction coefficient)} \times n \text{ (normal force)} = r \text{ (resisting force)}$$

or

$$d = f \times n = r$$

The Factor of Safety (F.S.) relates resisting force to driving force. When the two forces are equal, the F.S. equals 1.0. When resisting forces are greater than driving forces, the F.S. is greater than 1.0; and when resisting forces are less than driving forces, the F.S. is less than 1.0. Most modern building codes require that manufactured or natural slopes have a Factor of Safety of 1.5 or greater. In practice, one simply sums up all the driving forces and resisting forces acting along a potential slide plane. From this it can be calculated whether the F.S. of the selected

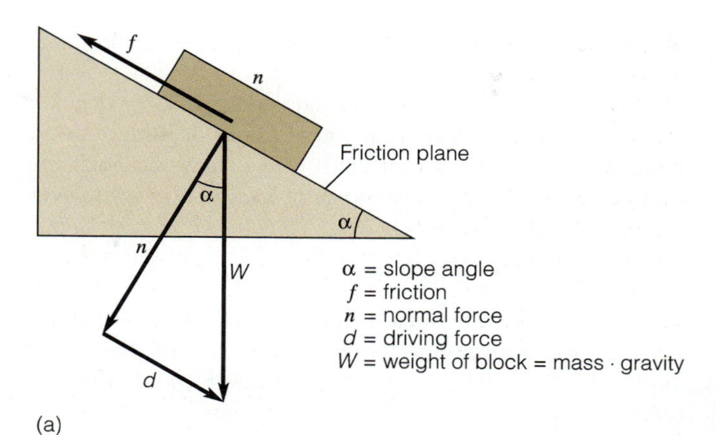

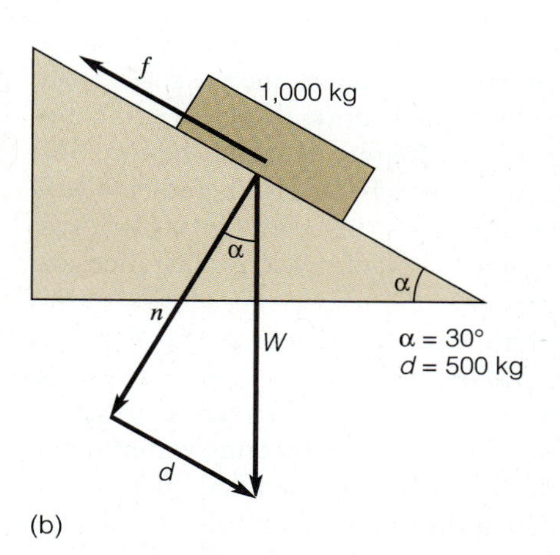

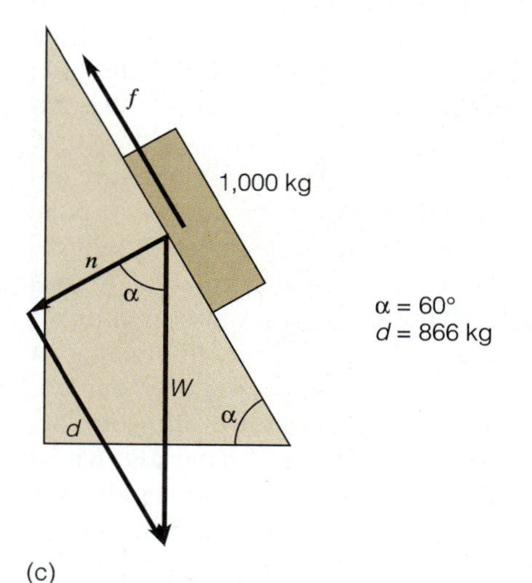

➢ **FIGURE 7.9 (*a*) Resolution of driving forces (*d*) and resisting forces (*f* × *n*) acting on a slide plane inclined at an angle α. Note the increase in driving force with an increase in slope angle from (*b*) 30° to (*c*) 60°.**

slope is greater than 1.0 (relatively stable) or less than 1.0 (relatively unstable).

This concept is simplified considerably if we use **vectors,** which are lines with both magnitude (length proportional to the forces exerted) and direction. Thus in Figure 7.9 we can see that tangent α (side opposite over side adjacent) is

$$\tan \alpha = d/n, \quad \text{or} \quad d = n \tan \alpha$$

where α equals the slope angle. The effect of slope on driving force is apparent in the figure; increasing the slope angle increases the driving force. At a slope of 30° the driving force is 500 kg; at 60° it increases to 866 kilograms.

The shear strength of a soil ($f \times n$ in our example) is derived by laboratory tests that determine the soil's **angle of internal friction.** Internal friction is the friction between soil grains and is analogous to static friction. **Soil cohesion** is the attractive forces between silt and clay particles and is a factor of a soil's shear strength. Clays are cohesive, but certain other soils, dry sand for instance, are cohesionless.

If water pressure builds up in the pore spaces between sediment grains, it pushes them apart and reduces the soil's effective grain-to-grain friction—referred to as its **effective stress.** This is analogous to injecting water under pressure beneath the block on an inclined plane. The water pressure will support a portion of the block's weight, thereby reducing the normal force n and the sliding resistance of the block. Water in soil or sediment pore space supports part of the grain-to-grain pressure, thereby reducing the frictional resistance across the grain contacts. The effective stress increases as the normal force increases and decreases as the pore-water pressure increases. A "quick" condition exists when the pore pressure is equal to intergranular friction; in other words, when the net effective stress is reduced to zero. At this point, sand becomes a dense sand–water mixture known as *quicksand.* It is often said that water "lubricates" a failure plane, making it easier to slip. It has been demonstrated, however, that a certain amount of moisture actually increases the strength of some materials. Moist sand, for example, builds a better castle and stands on a steeper slope than does dry sand. Water itself has no lubricating qualities. Rather, it is pore-water pressure that reduces grain-to-grain friction and therefore soil strength.

Types of Landslides

Two kinds of slides are recognized according to the shape of the slide surface. They are **slumps** or **rotational slides,** which move on curved, concave-upward slide surfaces and are self-stabilizing, and **block glides** or **translational slides,** which move on inclined slide planes. A block glide moves like a glacier until it meets an obstacle or until the slope of the slide plane changes.

SLUMPS. Slumps are the most common kind of landslide and range in size from small features a few meters wide to huge failures that can damage structures and transportation systems (➢ Figure 7.10). They are spoon-shaped, having a slide surface that is curved concave-upward and exhibiting a backward rotation (➢ Figure 7.11). They occur in geologic material that is fairly homogeneous, such as soil or badly weathered or fractured bedrock, and the slide surface cuts across geologic boundaries. Slumps move about a center of rotation, that is, as the toe of the slide rotates upward, its mass eventually counterbalances the downward force, causing the slide to stop (Figure 7.11b). Slumping produces depressions and flat areas on otherwise regularly sloping ground surfaces. Slumps grow headward (upslope) because as one slump mass forms, it removes support from the slope above it. Thus a stair-step surface is common in landslide terrain (Figure 7.11c). Slump-type landslides are the scourge of highway builders, and repairing or removing them costs millions of dollars every year.

BLOCK GLIDES. Block glides are coherent masses of rock or soil that move along relatively planar sliding surfaces (failure planes), which may be sedimentary bedding planes, metamorphic foliation planes, faults, or fracture surfaces. For a block glide to occur, it is necessary that the failure plane be inclined *less steeply* than the inclination of the natural or manufactured hill slope. Slopes may be stable with respect to block glides until they are steepened for subdivisions or for roads, thus leading to landsliding (➢ Figure 7.12).

A classic block glide of about 10 acres is seen along a sea cliff undercut by wave action at Point Fermin in San Pedro, California. Movement was first detected there in January 1929, and by 1930 the landslide had moved 2

➢ FIGURE 7.10 Rotational landslide along the Pacific Coast Highway near Malibu, California. The highway had to be relocated around the toe of the slide and a street around the scarp at the top.

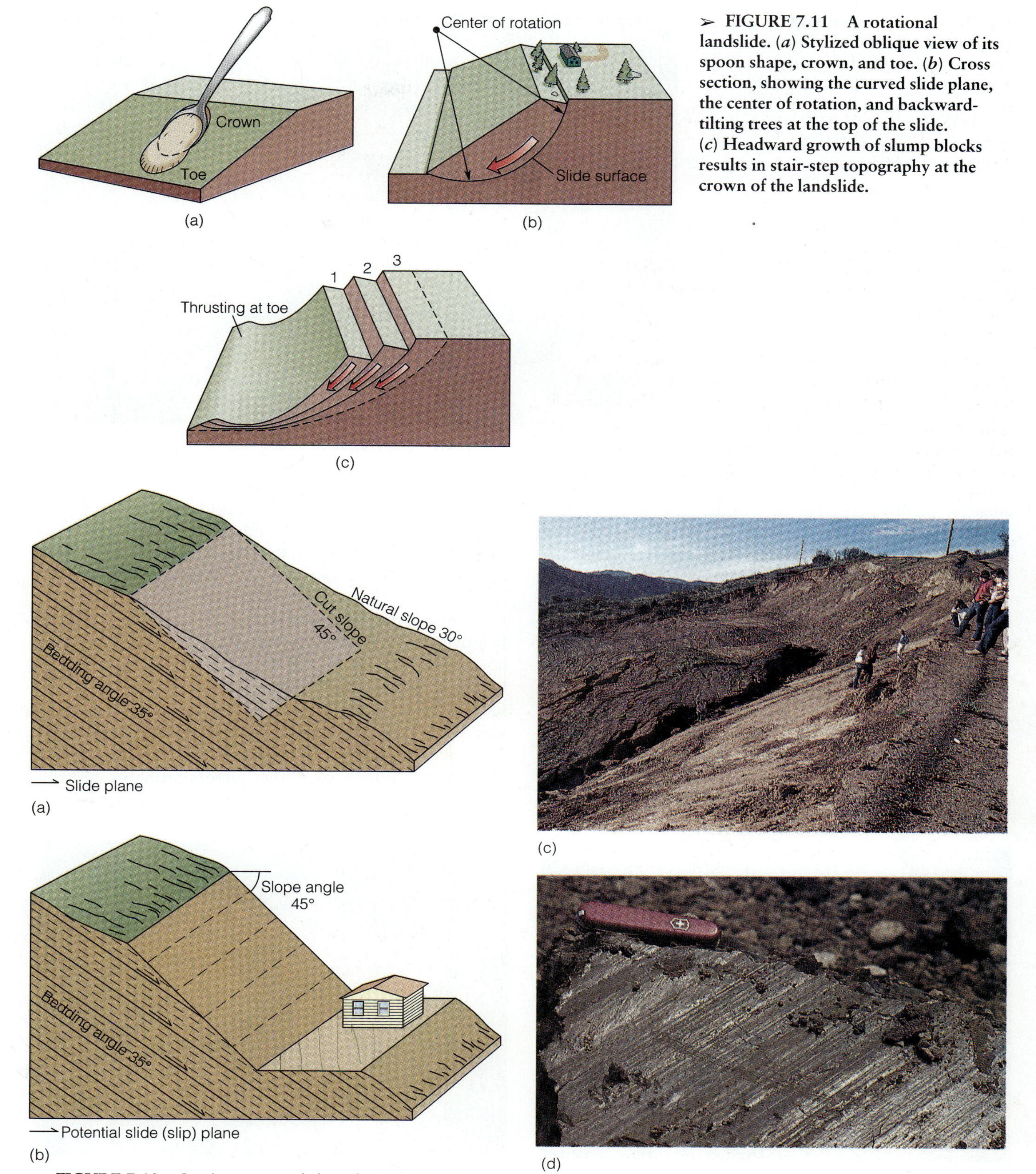

➢ **FIGURE 7.11 A rotational landslide. (*a*) Stylized oblique view of its spoon shape, crown, and toe. (*b*) Cross section, showing the curved slide plane, the center of rotation, and backward-tilting trees at the top of the slide. (*c*) Headward growth of slump blocks results in stair-step topography at the crown of the landslide.**

➢ **FIGURE 7.12 Cutting a natural slope for homesites leads to potential block slides on bedding or foliation planes. (*a*) Natural slope and proposed cut. (*b*) After cutting, the slope exhibits unstable conditions. (*c*) A slide plane in clay shale shortly after the overlying slide mass began moving; Santa Monica Mountains, California. (*d*) Striations gouged in wet clay by the landslide mass shown in (*c*).**

➢ FIGURE 7.13 Point Fermin landslide in San Pedro, California, is a rock block glide that has moved intermittently since 1929. Damaged homes have been removed from the slide area in the foreground. The slide plane emerges just in front of the houses in the background, which are on relatively stable ground.

meters seaward. It has been intermittently active ever since (➢ Figure 7.13). The rock of the slide mass is coarse sandstone (not the type of earth material usually involved in block glides), but a thin layer of **bentonite** dipping 15° seaward forms the slide plane. Bentonite is volcanic ash that has chemically weathered to clay minerals, which become plastic and slippery when wet. Bentonite is very commonly involved in slope failures; addition of water is all that is needed to initiate a landslide where dips may be as slight as 5° (see Case Study 7.1).

Some of these translational slides are large and move with devastating speed and tragic results. In 1985, a tropical storm dumped a near-record 24-hour-rainfall total of 560 millimeters (22 in) on a mountainous region near the city of Ponce on the south coast of Puerto Rico. At 3:30 A.M. on Monday, October 7, much of the Mameyes residential district of the city was destroyed by a rock block glide that materialized during the most intense period of rainfall. This resulted in the worst loss of life from a landslide in U.S. history—129 deaths. The landslide was in a sandstone whose stratification parallels the natural slope of the slide mass, a condition called a *dip slope.* It moved at least 50 meters (165 ft), probably on a clay layer in the sandstone, before breaking up into the large blocks that destroyed 100 homes (➢ Figure 7.14). The scarp at the top of the slide is 10 meters high, and the maximum thickness observed at the toe of the landslide is 15 meters. In addition to the heavy rainfall and adverse geologic conditions, two other factors probably contributed to the disaster: (1) Mamayes was a densely popu-

(a)

(b)

➢ FIGURE 7.14 The 1985 Mameyes landslide near Ponce, Puerto Rico. (*a*) Disrupted surface and destroyed homes. (*b*) Two phases of movement are visible from the left side of the photograph to the right side.

lated district with no sewer system; sewage was discharged directly into the ground. (2) A water main at the top of the slide was reported to have been leaking for some time. Thus the subsurface rocks were saturated with water, and the heavy rains of Tropical Storm Isabel were all that was needed to trigger the slide. The slide mass closely followed the boundaries of the Mamayes residential district, testimony to the impact of urban development on the landscape.

In August 1959 near Hebgen Lake in southern Montana, a major earthquake triggered a block glide in schist with foliation dipping into the canyon of the Madison River. The slide moved at more than 150 kilometers per hour (90 mph), burying campers, cars, and trailers (➢ Figure 7.15). Even people nearby who were outside the immediate slide area did not avoid the slide's effects. Many survivors were initially knocked down by a violent air blast and then moments later were engulfed by a surge of water displaced from the river.

A similar landslide occurred in the Alps of northeastern Italy in 1963 at the site of Vaiont Dam, the highest thin-arch dam in the world at the time (275 m). The geology at the reservoir consists of a sedimentary-rock structure that is bowed downward into a concave-upward **syncline** with its axis parallel to the Vaiont River canyon. Limestones containing clay layers dip toward the river and reservoir from both sides of the canyon due to this downfolded structure (➢ Figure 7.16, page 173). Slope movements and slippage along clayey bedding planes above the reservoir had been observed before the dam was constructed. This condition gave engineers and geologists sufficient concern that they placed survey monuments on the slope above the dam for monitoring such movement. Heavy rains fell for two weeks before the disaster, and slope movements as large as 80 centimeters per day were recorded. On the night of October 9th, without warning, a huge mass of limestone slid into the reservoir sufficiently fast to generate a wave 100 meters (330 ft) high. The wave burst over the top of the dam and flowed into the Piave River valley, destroying villages in its path and leaving 2,500 people dead. The landslide velocity into the reservoir was very high; so speedy was it that the

➢ FIGURE 7.15 Catastrophic rock block glide near Hebgen Dam, Montana, that was triggered by a major earthquake in 1959. The slide plane was along foliation in schist. The Madison River was dammed by the slide, forming the so-called quake lake at the left.

CASE STUDY 7.1

The Portuguese Bend Landslide—Tuff Business

Portuguese Bend is a hillside community in the Palos Verdes Hills overlooking the Pacific Ocean about 40 kilometers (25 mi) south of downtown Los Angeles. It was identified as an ancient (Pleistocene, ice-age) landslide on a geologic map of the hills published in 1941. Nevertheless, the area was developed in the 1940s and 50s into about 200 homesites. The land beneath the development began moving in 1956, eventually encompassing 250 acres and destroying 160 ranch-style homes (➢ Figure 1). The landslide first revealed itself in cracked driveways and floor slabs, doors that would not close, and broken water and gas pipes. Damage ultimately reached the point that few of the homes were safe for occupancy, and then only with constant maintenance (➢ Figure 2).

The Portuguese Bend slide is a classic block glide complicated by large slump blocks. The rocks of the slide mass are

➢ FIGURE 1 The Portuguese Bend landslide as seen from the air. Two basalt intrusions buttress the stable points around which the slide is moving.

➢ FIGURE 2 An early sign of landsliding at Portuguese Bend, 1956.

the Monterey Formation (siltstone and shale), and the Portuguese Tuff, containing layers of altered volcanic ash known as bentonite. Bentonite is composed mostly of clay minerals that are capable of absorbing huge amounts of water, which causes the clay to become plastic. Although the sedimentary layers dip seaward at an angle of only 6°, the bentonitic slide plane is so plastic that the rock mass moves on this gentle ramp toward a 30-meter-high sea cliff (➢ Figure 3). Thus, both the rocks and the structure of Portuguese Bend are very conducive to landsliding.

The reactivation of the ancient landslide occurred when water was introduced underground from several hundred cesspools. Adding to the problem and to the driving forces was the placement of earth material excavated for a new road on the head of the slide. These two factors, excess water and the addition of weight to the slide mass, are basic contributors to geologic instability.

Twenty-three large-diameter (4 feet) reinforced-concrete columns were placed vertically across the bentonite slide plane in 1957 as a means of stopping the movement. Although this effort slowed the landslide, the columns eventually either broke or simply rotated as the mass moved seaward. In the early 1980s the huge columns began emerging from the sea-cliff face, hundreds of feet from where they were emplaced (➢ Figure 4).

Between 1956 and 1986, the landslide moved 3–15 millimeters per day on average, with extremes ranging from no movement at all during a summer period to as much as 35 millimeters/day during wet winters. Total movement in that period was 190 meters (620 ft) along the eastern boundary, which caused extensive damage to transportation and utility systems. The damage to road systems has been so expensive to repair that a Redevelopment District was formed and the state allocated $2 million in aid to stabilize the slide mass.

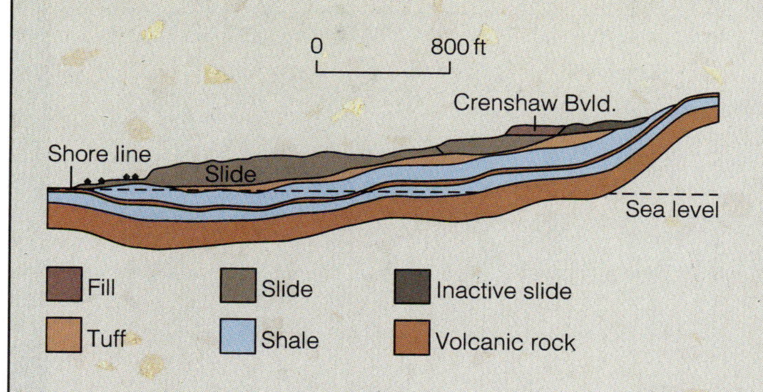

➢ FIGURE 3 North-south cross section of the landslide, showing the rocks and the bentonite slide plane.

➢ FIGURE 4 Reinforced concrete piles that were installed across the slide plane in an early attempt to stop movement are now emerging at the toe of the landslide, 700 feet from where they were inserted. The columns are 4 feet in diameter and 20 feet long.

CASE STUDY 7.1 *Continued*

Upon the advice of a local geology professor, Dr. Perry Ehlig, 13 wells were installed to pump out underground water and thus lower the water table, and the surface of the slide was regraded in 1986–1987. The regrading was an effort to seal the surface from water infiltration and to achieve a better balance between driving and resisting forces (➢ Figure 5). The slide had slowed significantly and some parts had stabilized by 1993. When the entire slide mass is sufficiently stabilized (Safety Factor 1.2), the slide area will be reclaimed for recreational uses.

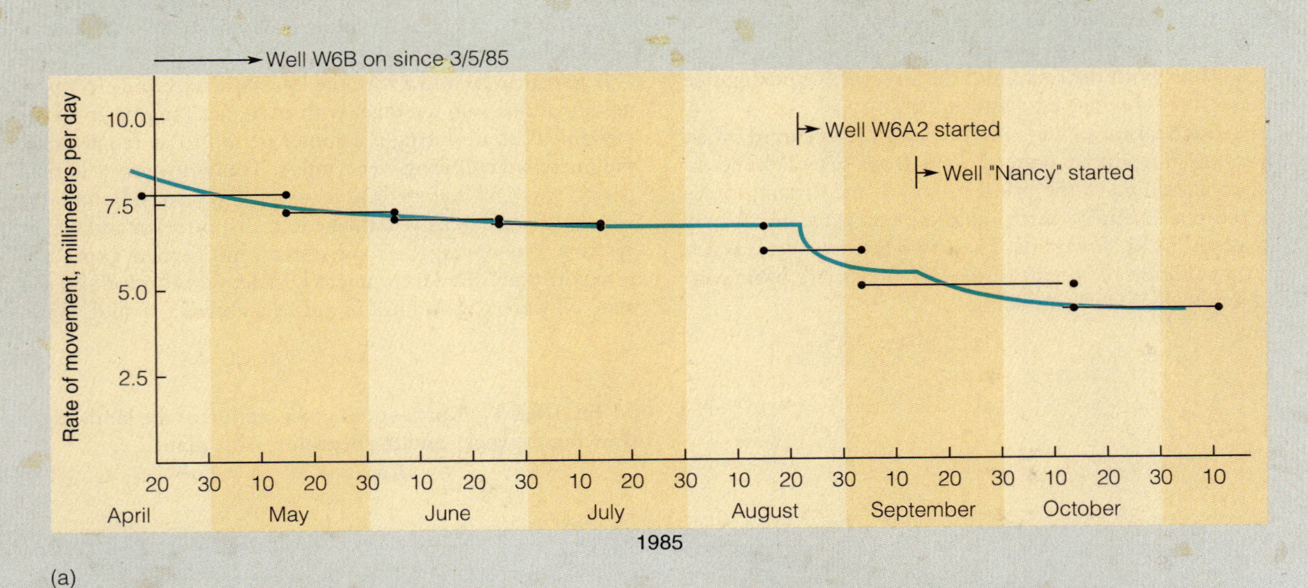

(a)

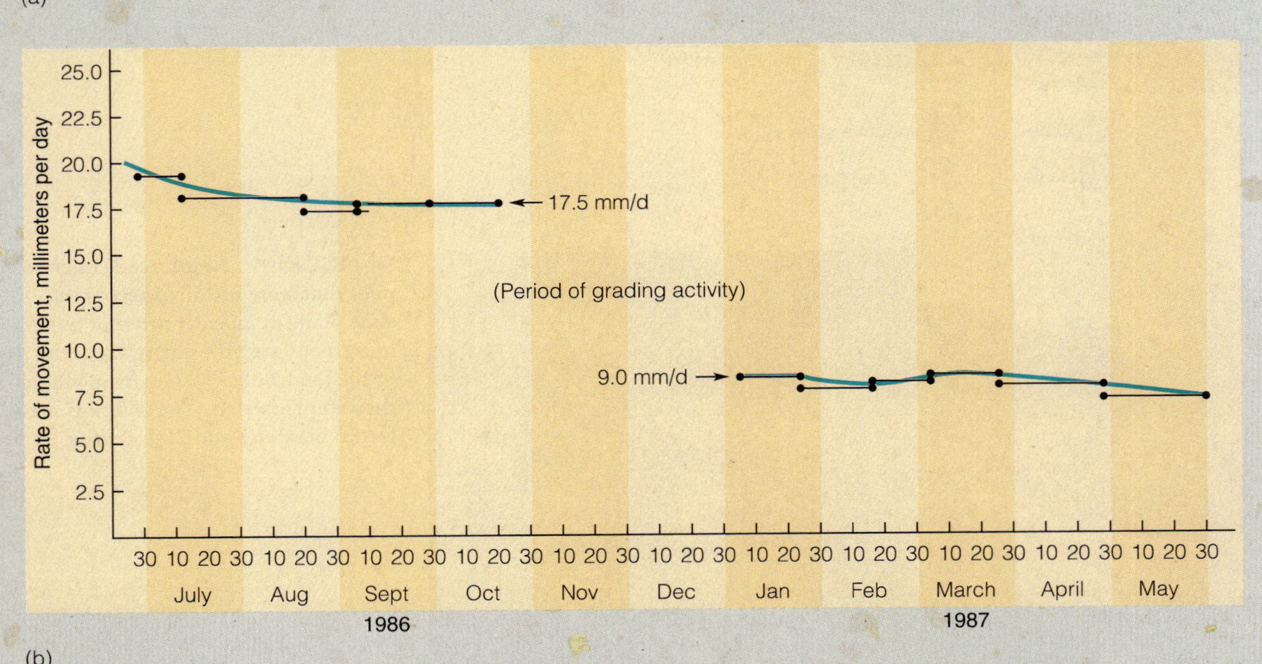

(b)

➢ **FIGURE 5 Slowing of the Portuguese Bend landslide attributed to (*a*) installing dewatering wells to lower the water table in 1985 and to (*b*) regrading and removing material from the top of the slide and compacting it at the toe, October 1986 to January 1987.**

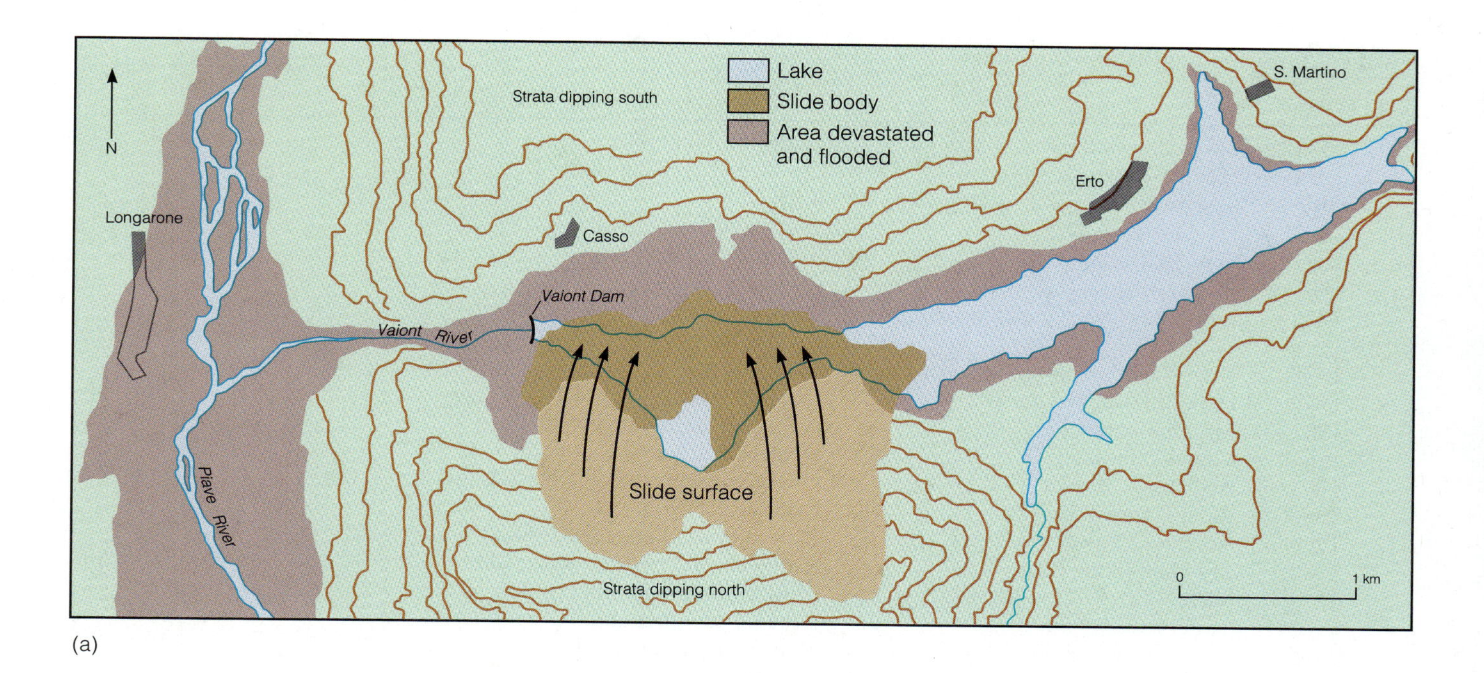

(a)

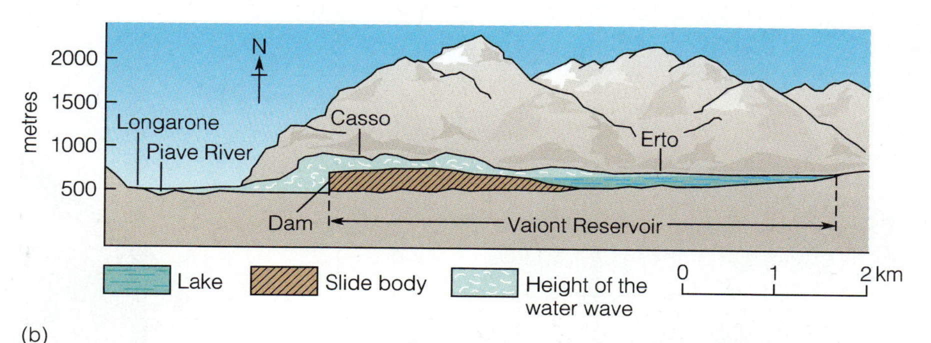

(b)

➢ FIGURE 7.16 The 1963 Vaiont Dam landslide. (*a*) Map of the slide mass that hurtled into the reservoir and the area that was impacted by the giant wave and flooding. Longarone and several other villages along the Piave River below the dam were devastated. (*b*) Cross section of the area showing the elevation of the water displaced over the dam and down the Vaiont River to Longarone.

slide mass almost emptied the lake. The dam did not fail and it is still standing today—a monument to excellent engineering but poor site selection.

Selected significant historic landslides are listed chronologically in Table 7.1. All of these slides were considered major disasters, and some of them were triggered by earthquakes, which allowed little if any advance warning. Most of them moved with high velocity, and many were accompanied by shock waves or by walls of water from displaced lakes or rivers (see Case Study 7.2, page 176). Some of these disastrous slides' masses deformed and disintegrated as they moved, especially those that were transformed into debris flows as velocity and water content increased.

➢ Figure 7.17 portrays the kinds of failures that might be experienced in regions of abundant rainfall, steep slopes, and weak rock or soil. The diagram illustrates the principal slide and flow types: slump, block glide, rock avalanche, soil creep, and rockfall.

Lateral Spreading

Horizontal movement of a mass of soil overlying a liquefied or plastic layer characterizes a form of mass wasting called **lateral spreading.** Such slides are complex, as they involve elements of translation, rotation, and flow. Typically triggered by earthquakes, lateral spreading may result in the spontaneous liquefaction of water-saturated sand layers or in the collapse of **sensitive clays—**also known as **quick clays** (see Chapter 4). Spreading failures in the United States occur mostly in sand layers; however,

TABLE 7.1 Significant Historic Landslides

YEAR	LOCATION	TYPE	DEATHS
1512	Biasca, Switzerland	Landslide dam broke	>600
1556	Hsian, China	Quake-triggered landslides	~1,000,000
1806	Goldau, Switzerland	Rock glide	457
1843	Mt. Ida, Troy, N.Y.	Slump and flow	15
1881	Elm, Switzerland	Rockfall	115
1903	Frank, Alberta, Canada	Rock glide	70
1920	Kansu Province, China	Quake-triggered landslides and cave collapse	~200,000
1938	Kobe, Japan	Debris flows	600
1959	Hebgen Dam, Montana	Quake-triggered landslide	~26
1962	Mt. Huascarán, Peru	Ice avalanche and debris flow	~4,000
1963	Vaiont Dam, Italy	Landslide into reservoir, flood	~2,000
1964	Anchorage, Alaska	Quake-triggered quick-clay landslide	114*
1966	Rio de Janeiro, Brazil	Landslides	279
1966	Aberfans, Wales	Mine-dump collapse/debris flow	144
1970	Mt. Huascarán, Peru	Avalanche/debris flow after quake	25,000
1972	Buffalo Creek, West Virginia	Mine-dump collapse/debris flow	400
1982	San Francisco Bay area	Debris avalanches and flows	25
1985	Mameyes, Puerto Rico**	Hurricane-triggered debris flows	129
1992	Mt. Pinatubo, Philippines	Typhoon-triggered lahars	~350

* Combined toll from quake and slide
** Largest in U.S. history
SOURCES: Various.

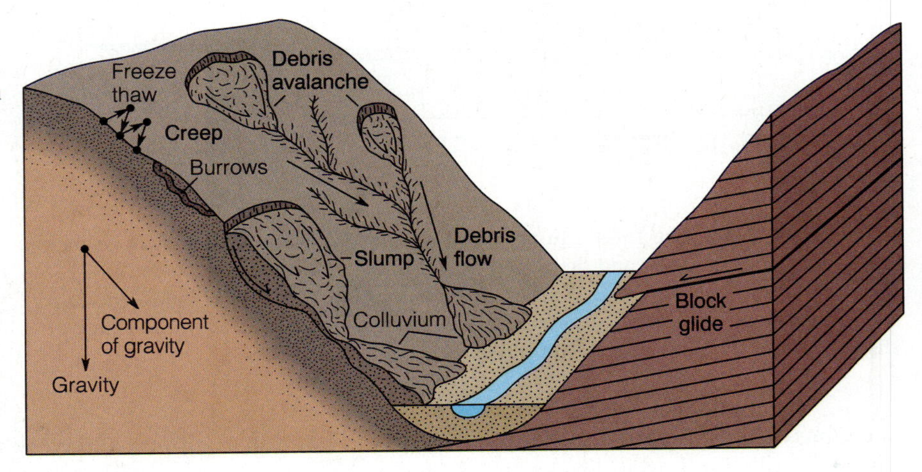

➢ FIGURE 7.17 Kinds of mass movement commonly found on steep-sided stream valleys in humid regions. Colluvium is soil moved downslope by gravity.

areas underlain by glacial sediments may spread because of quick clay layers within the deposits.

Much of the damage in the Marina district of San Francisco in the 1906 and 1989 earthquakes was due to lateral spreading caused by liquefaction (see Figure 4.1). Most of the damage in Anchorage, Alaska, during the 1964 quake was due to quick-clay-induced lateral spreads in the Turnagain Heights residential district and the downtown area. The spreading was so extensive that two houses that had been more than 200 meters apart collided within the slide mass. The Turnagain Heights area is now a tourist attraction known as "Earthquake Park" (➢ Figure 7.18).

Probably no lateral spread in history has received more attention in geology and soils textbooks than the one at Nicolet, Quebec, Canada, in 1958. Here, water-saturated quick clays collapsed because of vibrations generated by artillery fire at a nearby military installation. The result was that a large section of the town and a historic monastery slid into the Nicolet River (➢ Figure 7.19). It was discovered later that the old monastery's wooden sewer system had leaked water into the subsurface glacial-marine clays, causing them to liquefy.

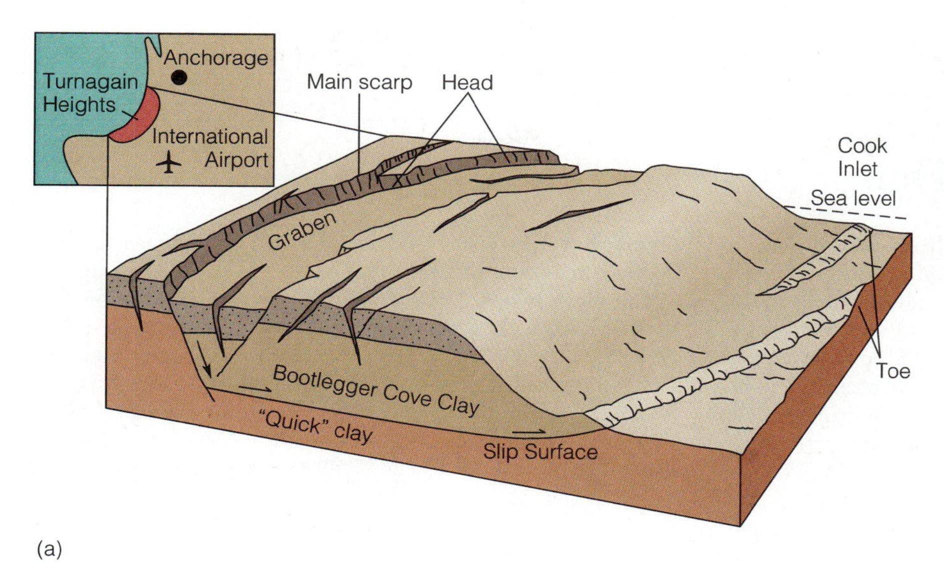

➢ **FIGURE 7.18 The Turnagain Heights landslide following the 1964 magnitude-8.6 Alaska earthquake. (*a*) Simplified geology of the slide. The earthquake caused failure of the sensitive (quick) Bootlegger Cove Clay and liquefaction of the sand and silt layers. (*b*) Turnagain Heights after the landslide.**

➢ **FIGURE 7.19 Quick-clay slide at Nicolet, Quebec, Canada. The house on the slide (to the right of the bridge) traveled several hundred feet with relatively little damage.**

CASE STUDY 7.2

A Failed Landslide Dam

The term *geologic hazard* had not yet been coined when a disastrous dam failure and ensuing flood completely wiped out the village of Kelly, Wyoming, on the morning of 18 May 1927. The dam was not human-made; rather, it was a result of a great landslide into the valley of the Gros Ventre River (pronounced "grow vänt") that had occurred some two years earlier. A lake formed behind the landslide rubble some 60 kilometers (35 mi) south of Yellowstone National Park, and the rising lake waters eventually overtopped the dam. The landslide was easily eroded, and when the downward-cutting water reached the critical point, the impounded waters were suddenly released in a deluge that flooded areas downstream.

The Gros Ventre landslide of 1925 was a translational slide involving: (1) permeable sandstone strata overlying impermeable clay layers (the slide plane); (2) sedimentary layers inclined toward the valley; and (3) the Gros Ventre River undercutting the base (toe) of the slope and exposing the slide plane. Saturated by heavy rains and snowmelt, an enormous rock mass suddenly slid on the clay layers into the river valley below (see ➢ Figure 1). The landslide dammed the river, and a lake formed behind it (➢ Figure 2).

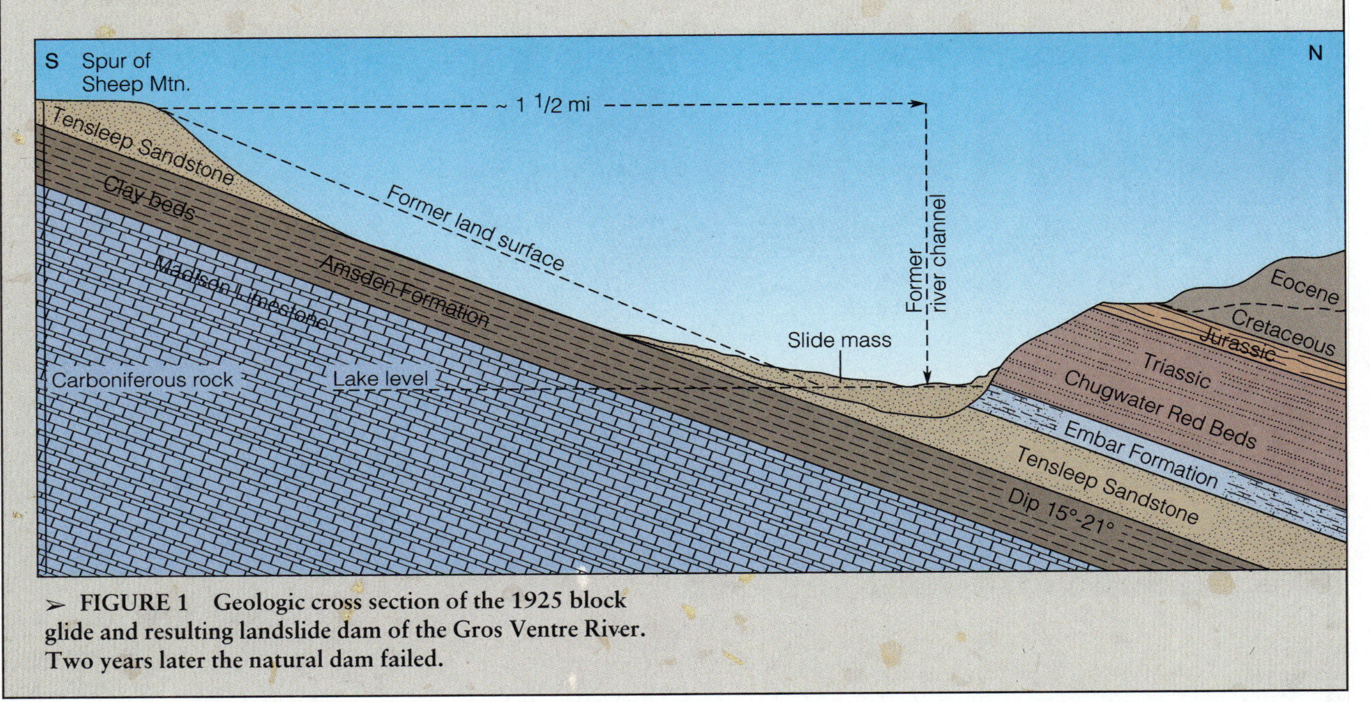

➢ **FIGURE 1 Geologic cross section of the 1925 block glide and resulting landslide dam of the Gros Ventre River. Two years later the natural dam failed.**

Factors that Lead to Landslides

Seldom can a landslide be attributed to a single cause; rather, landslides result from series of events that lead to failure. Nonetheless, the weakening of slope materials due to the addition of water is the most important causative factor of all slides and flows. Thus heavy rainfall, rapid snowmelt, leaking water mains, private sewage-disposal (cesspool) inflow, and poor building-pad drainage can all lead to landsliding. Excess water causes a buildup of pore-water pressure, which weakens the materials supporting the slope. This is why mass wasting is closely correlated with series of heavy-rain years in semiarid Southern California and with torrential flooding in the Eastern United States.

Recall that when driving forces exceed resisting forces, a failure is imminent. Factors that increase driving force are

- an increase in the slope angle,
- removal of lateral support at the toe of a slope, and
- added weight at the top of a slope.

The effect of increasing slope angle is best illustrated by the natural **angle of repose** of granular material such as dry sand or gravel. The angle of repose of such material

The lake waters overtopped the dam on May 17, but no alert to the potential flood hazard was issued. Early on the 18th, someone noticed kitchen utensils floating down the swelling river at Kelly and, going up the valley, found a washed-out ranch house about a mile below the eroding landslide dam. At 11:00 A.M. a wave of water 15-feet high came rushing down the river gorge and swept the plain where Kelly had stood. All of the buildings in Kelly except the schoolhouse and a church were washed away. Miraculously, only 7 of Kelly's 70 inhabitants drowned, most of them while trying to retrieve personal possessions from their inundated homes. Bridges downstream were washed out, and the flood affected communities as far downriver as Idaho Falls, 190 kilometers (115 mi) away.

The overtopping waters had eroded a 100-foot-deep channel through the landslide dam in a matter of hours and released an estimated 13 billion gallons (43,000 acre-feet) of water. The flood hazard could have been mitigated by constructing a rock-covered spillway through the slide mass to allow rising lake waters to be released gradually and prevent rapid downward erosion of the landslide dam.

➢ **FIGURE 2 Sheep Mountain and Gros Ventre landslide and debris flow. Note the lake that formed behind the slide mass, located four miles above Kelly, Wyoming.**

is the maximum slope angle at which it can remain stable. For example, no matter how steeply you try to pile dry sand, it will always form a slope of about 32°–34°, its angle of repose (➢ Figure 7.20). Moist sand has a greater angle of repose because of the temporary cohesion (added strength) imparted to it by moisture. This is why a castle built of moist sand collapses when the sand dries and loses its temporary cohesion. The angle of repose represents the point at which the frictional resistance between sand grains (shear strength) and the downslope component of gravity are in balance.

Theoretically, mass movements will not occur as long as the angle of repose of the particular slope material is not exceeded. Because undercutting the toe of a slope is equivalent to increasing the slope angle, it causes instability. Adding weight to the slope, as with fill material or an unusually heavy structure, increases the pore pressure and can thus induce failure. Factors that reduce a slope's strength and, therefore, its resistance to sliding are

- infiltration of water underground,
- weathering and breakdown of minerals, especially when clays are formed, and
- burrowing animals.

In addition, it should be noted that sliding is also facilitated when a slope is cut in such a way that its bedding planes are exposed or inclined (dipping) out of the slope face.

A slope with a Factor of Safety less than 1 is ready to slide, but it might not do so unless it is *triggered* by an earthquake, heavy traffic, sonic boom, detonation of

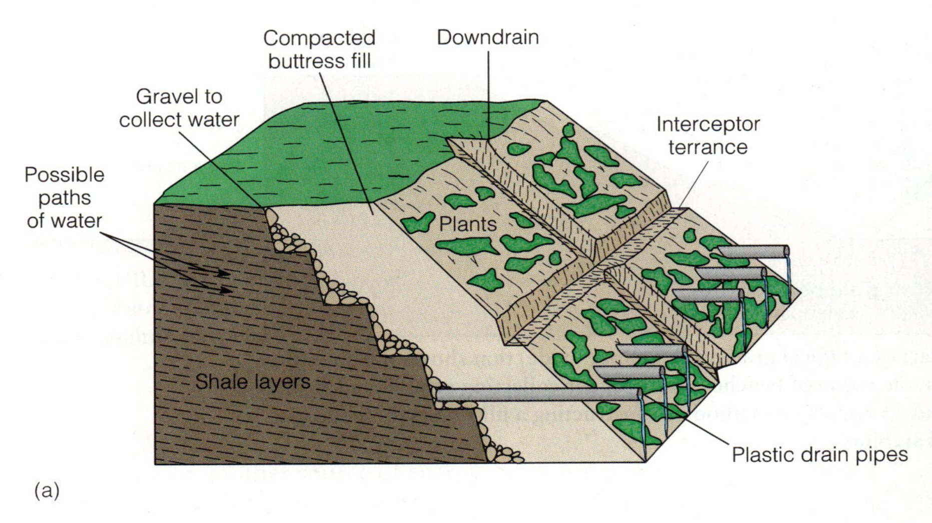

➢ FIGURE 7.23 **(*a*) Buttress fill with horizontal drains for removing water and surface drains for preventing erosion. The fill acts as a retaining wall to hold up unstable slopes such as those that result from excavation. (*b*) A slope with surface drains built according to accepted standards.**

instability remains. The method is cost-effective only for small landslides, however.

Retaining Devices

Many slopes are oversteepened by cutting back at the toe, usually for the purpose of obtaining more flat building area. The vertical cut can be supported by constructing steel-reinforced concrete-block retaining walls containing drain (weep) holes for alleviating water-pressure buildup behind the structure (➢ Figure 7.24). Retaining devices are constructed of a variety of materials, including rock, timber, metal, wire-mesh fencing, and a sprayed concrete known as *shotcrete* (➢ Figure 7.25).

Steep or vertical rock slopes that are jointed or very seamy are often strengthened by inserting 5–7-meters-long rock bolts into holes drilled perpendicular to the planes of weakness. This binds the planes together much as a beam is bound. Rock bolts are used extensively to support tunnel and mine openings. They add considerably to the safety of these operations by preventing sudden rock "popouts." They are also installed on steep roadcuts to prevent rockfalls onto highways (➢ Figure 7.26). Retaining devices are also effective for mitigating rock and debris falls.

Diversion Techniques

Debris flows present a different challenge, because they may originate some distance away from the site of interest. It is simply not good practice to build at the bottom or the mouth of a steep ravine or gully, especially if loose soils are present on the higher slopes. This is particularly true in areas of high mean annual rainfall, areas that are subject to sudden cloudbursts, and areas that have been burned, such as in the tragic October, 1993 firestorms in southern California. As long as people continue to build houses in the "barrels" of canyons, as is done in the foothills of southern California, debris flows will continue to take their toll. Debris flow insurance is expensive—about $1,500 per $100,000 of house value per year in 1993—but the cost can be less if *deflecting walls* are

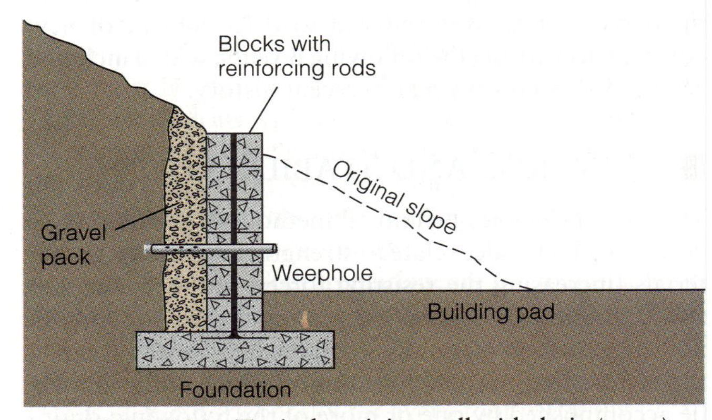

➢ FIGURE 7.24 **Typical retaining wall with drain (weep) holes to prevent water retention behind the wall and the buildup of hydrostatic pressure. The wall is constructed of concrete blocks with steel reinforcing rods and concrete in their hollow centers.**

➢ **FIGURE 7.25** "Shotcrete" stabilization of a roadcut in fractured igneous rock; Kobe, Japan.

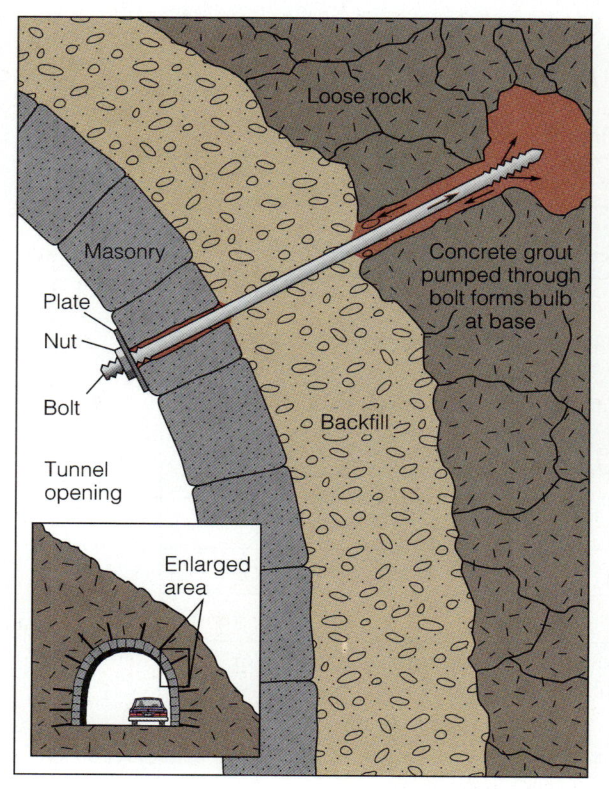

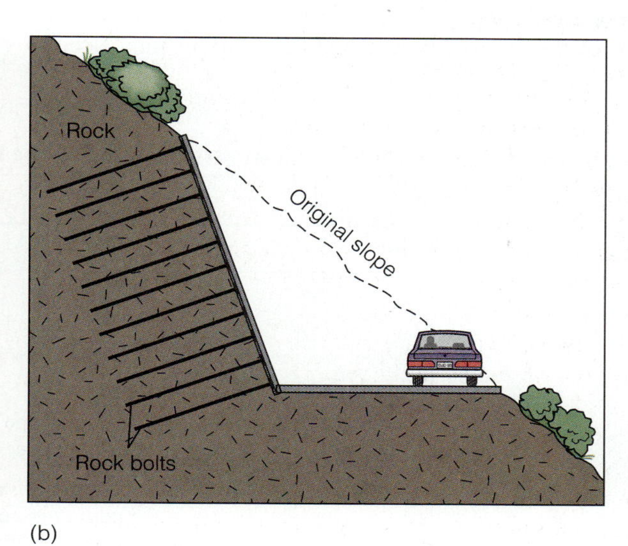

➢ **FIGURE 7.26** (*a*) Typical rock bolt and array of bolts as they are installed in a tunnel or mine opening. (*b*) Rock bolts support jointed rocks above a highway.

(a)

(b)

➢ **FIGURE 7.27 (*a*) The outlet of a debris-flow diversion structure; San Fernando Valley, California. (*b*) One effect of the May 1968 Heath Canyon mudflow due to rapid snowmelt; Wrightwood, California. A diversion wall or dike might have saved this mountain cabin in the San Gabriel Mountains. Heath Canyon is now leveed to contain debris flows within the river channel.**

designed into the house plans (➢ Figure 7.27). These walls have proven to be effective in protecting dwellings, but they could have adverse consequences if they divert debris onto neighbors' structures, in which case neither an architect nor a geologist would be as helpful as a lawyer.

Snow avalanches like landslides require a plane or planes of weakness along which the sliding may occur. The mechanics of avalanching is similar to landsliding, the difference being mainly in the material (snow and ice) and the rapidity of sliding, which is much greater in avalanches (see Case Study 7.3).

Table 7.2 summarizes mitigating measures for the various mass-wasting processes.

TABLE 7.2 Summary of Mass-Wasting Processes and their Mitigation

LANDSLIDES		
Cause	*Effect*	*Mitigation*
Excess water	Decreased friction (effective stress)	Horizontal drains, surface sealing
Added weight at top	Increased driving force	Buttress fill, retaining walls, decrease slope angle
Undercut toe of slope	Increased slope angle and driving force	Retaining walls, buttress fill
"Daylighted" bedding	Exposure of unsupported bedding in cut or natural slope	Buttress fill, retaining walls, decrease slope angle
OTHER MASS-WASTING PROCESSES		
Process	*Mitigation*	
Rockfalls	Rock bolts and wire mesh on the slope, concrete or wooden cribbing at bottom of the slope, cover with "shotcrete"	
Debris flows	Diversion walls or fences	
Soil creep	Deep foundations	
Lateral spreading	Dewater, buttress, retain (difficult to mitigate)	

CASE STUDY 7.3

Look Out Below

Although snow avalanches have been around as long as there have been mountains and snow, the relatively recent emergence of the recreational skiing industry as big business has transformed avalanche control from an art to a science. Unfortunately, it's too late for Hannibal. His crossing of the Alps in 218 B.C. was plagued by avalanches. Purportedly 18,000 of his troops and who knows how many of his elephants were killed by them. It's also too late for others. In World War I 6,000 troops were buried in one day by avalanches in the Dolomite Mountains of northern Italy. Avalanches in high mountain regions have wiped out villages, disturbed railroad track alignments, blocked roads, and otherwise made life difficult if not impossible for millenia.

Avalanches are caused by heavy new snowfall or a weakening of older snow. A **slab avalanche** consists of a coherent mass of snow and ice that hurtles down a slope of more than 30° as an entity—much like a magazine sliding off a tilted coffee table. It may be a thick, wet snow accumulation sliding on ice, or a coherent slab of snow that detached along a *depth hoar* layer. Depth hoar begins to form at the surface when snow particles evaporate in very cold, calm weather and form a loose, open array of snow crystals. As the layer of crystals is buried, evaporation continues, resulting in a very weak porous layer at depth. This hoar layer may collapse when it is disturbed, forming a compressed-air layer upon which the overlying snow can move with great velocity. Slab or "wet" avalanches associated with depth hoar are common on the east side of the Rocky Mountains because of the cold temperatures and relatively light snowfall there. These avalanches account for the majority of avalanche deaths because of their sheer weight and their tendency to solidify when they stop. About 25 avalanche fatalities occur each year in North America, mostly in backcountry areas, and the victims tend to be cross-country skiers, snowmobilers, climbers, and hikers.

Powder-snow or "dry" avalanches are more common in the Sierra Nevada and Coast Ranges, where depth hoar is rare and abundant moisture results in deep snow packs. Dry avalanches are dangerous because they entrain a great amount of air and behave like a fluid. Furthermore, the strong shock wave that precedes them can uproot trees and flatten structures. They are most common in areas of heavy new snowfall on top of older snow with weak layers at depth. Because they usually occur during storms, they are difficult to predict. Victims of dry avalanches tend to be buried in light snow close to the surface, which makes their chances of being rescued alive better than those of wet-avalanche victims. Regardless of whether an avalanche is wet or dry, it results from a weak boundary layer between potentially unstable snow above and a stable mass below.

Avalanches can be triggered by skiers, rockfalls, loud noises, or explosives. Avalanche control is thus achieved by triggering an avalanche with explosives or cannon fire when dangerous snow conditions exist in an area frequented by humans. The explosions collapse the depth hoar or detach powder snow, causing a "controlled" avalanche.

Various means have been developed for mitigating the destructive might of avalanches. Fences are built to divert them away from structures. In some areas in the Alps, the uphill sides of chalets are built similar to a ship's prow so that they will divert avalanching snow. Railroads and highways can be protected by building snowsheds along stretches adjacent to steep slopes (➢ Figure 1). In the event of an avalanche, the snow slides *over* the road or track, rather than across it. Much of the transcontinental railroad across the Donner Pass of the Sierra Nevada is protected by snowsheds.

➢ FIGURE 1 A snowshed protects the tracks of the White Pass and Yukon Railroad in Alaska.

GALLERY: SLIP AND SLIDE

Undeniably, the results of mass-wasting processes can be tragic and devastating, as explained throughout this chapter. In some cases, however, the results are whimsical. For example, a block glide occurred along Highway 89 in Wyoming when the toe of a dormant ancient slide mass was removed to make room for the road (➢ Figure 7.28). This caused the landslide to reactivate, endangering a telephone line across the moving mass. Over time the solution to this problem became increasingly obvious: the telephone pole supporting the line should be cut off near its base. The pole now hangs from the telephone wires, and the base of the pole has moved 70 feet downslope. Clay-rich shale inclined toward the river in the background of the photograph formed the slide plane.

A 1978 debris flow in Glendale, California, uncovered 35 burial vaults at Verdugo Hills Cemetery (➢ Figure 7.29). Some of the vaults were moved more than a half-mile downslope into a residential area. Even the deceased are not free from geologic hazards. Explanations for a few other "light" landslide examples (➢ Figures 7.30 to 7.33) are given in their captions. All of the areas shown had been developed before landsliding occurred.

➢ FIGURE 7.28 Active block glide on U.S. 89 in Montana. The telephone pole, hanging from the lines, was cut from its base, which has moved 23 meters (70 ft) downslope in three years. The person standing downslope is near the "stump" of the pole.

➢ FIGURE 7.29 Slope with exposed burial vaults after surface material was removed by a debris flow; Verdugo Hills Cemetery, Glendale, California.

➢ FIGURE 7.30 "Pedestal pavement" formed by landsliding and erosion; Point Fermin, California. This former street was lined with houses that were destroyed by the landslide.

➢ FIGURE 7.31 Watch that centerline! Offset along the east side of the Portuguese Bend landslide looking toward stable ground. The slide moves continuously, and the road requires periodic repair and centerline repainting. (See Case Study 7.1.)

➢ FIGURE 7.32 Sliding tennis court; Abalone Cove, California. Note the consistent right-slip movement on shear planes in this part of the landslide.

➢ FIGURE 7.33 Very appropriate use for landslide terrain. A mobile-home park is tucked onto an inactive landslide near Malibu, California. Should the landslide reactivate, residents can quickly remove their homes. Note spoon shape of this large slump failure.

Summary

Mass Wasting

DEFINED: Downslope movement of rock and soil under the direct influence of gravity.

CLASSIFICATION BY:

1. Type of material—rock or earth.
2. Type of movement—flow, slide, or fall.
3. Moisture content.
4. Velocity.

Flows

DEFINED: Mass movement of unconsolidated material in the plastic or semifluid state.

TYPES:

1. Creep—the slow, imperceptible downslope movement of rock and soil particles.
2. Debris flows—dense fluid mixtures of rock, sand, mud, and water that are fast-moving and destructive.
3. Debris avalanche—fast-moving (15 km/h) debris flow.

CAUSES: Heavy rainfall on steep slopes with loose soil and sparse vegetation. Most common in arid, semiarid, and alpine climates and where dry-season forest fires expose bare soil.

Slides

DEFINED: Movement of rock or soil (landslide) on a discrete failure surface or slide plane.

MECHANICS: Driving force (gravity) exceeds resisting forces (friction and force normal to the slide surface).

TYPES:

1. Rotational slide or slump on a curved, concave-upward slide surface.
2. Translational slide or block glide on an inclined plane surface.
3. Lateral spreads of water-saturated ground (liquefaction of sand and quick clays).

CAUSES:

1. Weakening of the slope material by saturation with water, weathering of rock or soil minerals, and plants and burrowing animals.
2. Steepening of the slope by artificial or natural undercutting at toe.
3. Added weight at the top of the slope such as by a massive earth fill.
4. A trigger such as an earthquake, vibrations from explosives, heavy traffic, or sonic booms.

Reduction of Losses

GEOLOGIC MAPPING: To identify active and ancient landslide areas.

BUILDING CODES: To limit the steepness of manufactured slopes and specify minimum soil and fill conditions and surface-water drainage from the site.

Control and Stabilization

LATERALLY INSERTED PIPES (hydraugers): To drain subsurface water.

EXCAVATION: To redistribute soils and rock from the head of the potential slide to the toe. This decreases driving force and increases resisting force.

BUTTRESS FILLS AND RETAINING DEVICES: To retain active slides and potentially unstable slopes.

ROCK BOLTS: To stabilize slopes or jointed rocks.

DEFLECTION DEVICES: To divert debris flows around existing structures.

Key Terms

angle of internal friction
angle of repose
bentonite
block glide
creep
debris avalanche
debris flow
driving force
effective stress
landslide
lateral spreading
mass wasting
resisting force
rotational landslide
quick clay
sensitive clay
slab avalanche
slide plane
slump
soil cohesion
syncline
translational landslide
vectors

Study Questions

1. What grading and land-use practices can reduce the likelihood of landslides?
2. What factors caused activation of the long-dormant Portuguese Bend landslide?
3. How can vegetation, topography, and sediment characteristics help to identify old or inactive slumps, slides, and debris flows?
4. Name the processes that move earth materials downslope (gravity is the *cause,* not the *process*). On what factors does each depend?
5. Why are block glides more common in schist, as in the case of the Hebgen Dam landslide, than in marble or gneiss?
6. Modern building codes require a Factor of Safety of 1.5 or greater in order to build on or at the top of a cut slope. What is the Factor of Safety, and why should it be 1.5?
7. Suppose that you wish to build a home at the top of a steep slope that is 3.1-meters high (10 feet) and supported by poorly cemented sand. How far back from the edge of the slope should you locate your house to be perfectly safe? (*Hint:* Think about the angle of repose.)
8. What is effective stress, and how is it related to the formation of quicksand?

9. How do building codes and regulations reduce the risk of landslides? Are they effective?
10. What are some common, cost-effective methods of landslide control and prevention?
11. What caused the disaster at Vaiont Dam?

Further Reading

Alger, C., and Earl Brabb. 1985. *Bibliography of U.S. landslide maps and reports.* U.S. Geological Survey open file report 85-585.

Coates, Donald, ed. 1977. *Landslides. Reviews in engineering geology,* no. 8. Boulder, Colorado: Geological Society of America.

Dolan, R., and H. Grant Goodell. 1976. Sinking cities. *American scientist,* no. 1.

Fleming, R. W., and T. A. Taylor. 1980. *Estimating the costs of landslides in the United States.* U.S. Geological Survey circular 832.

Krohn, J. P., and James Slosson. 1976. Landslide potential in the United States. *California geology,* October.

McPhee, John. 1989. Los Angeles against the mountains. *The control of nature,* ch. 3. New York: Farrar, Straus & Giroux.

National Academy of Sciences. 1978. *Landslides: Analysis and control.* Transportation Research Board special report 176. Washington, DC: National Academy Press.

Nilsen, Tor H., and others. 1979. *Relative slope stability and land-use planning in the San Francisco Bay region, California.* U.S. Geological Survey professional paper 944. Washington, DC: U.S. Government Printing Office.

Schultz, A., and R. W. Jibson. 1989. *Landslide processes of the eastern United States and Puerto Rico.* Geological Society of America special paper 236.

Tremper, Bruce. 1993. Life and death in snow country. *Earth: The science of our planet,* no. 2.

U.S. Geological Survey. 1982. *Goals and tasks of the landslide part of a ground-failure hazards reduction program.* U.S. Geological Survey circular 880. Washington, DC: U.S. Government Printing Office.

Varnes, David. 1984. *Landslide hazard zonation: A review of principles and practice.* Paris: UNESCO.

CHAPTER

8

SUBSIDENCE AND COLLAPSE

When [the geologist's] final report is written, the longest and most important chapter will be upon the latest and shortest of the geological chapters.

G. K. GILBERT, GEOLOGIST

Two million people live below the high-tide level in Tokyo. The canals of Venice overflow periodically and flood its beloved tourist attractions. The Houston suburb of Baytown has subsided almost 3 meters since 1900, and 80 square kilometers (31 mi^2) of this coastal region are permanently under water. Earthen dikes are needed in some places in California to prevent the sea from flooding the land, and Mexico City's famous Palace of Fine Arts rests in a large depression in the middle of the city (Table 8.1). These areas suffer from natural and human-induced **subsidence,** a sinking or downward settling of the earth's surface. Loss of life due to subsidence is rare, and it is not usually catastrophic, but each year subsidence causes damage estimated at tens of millions of dollars in the United States. Planners and decision makers need to be aware of the causes and impacts of subsidence in order to assess the risks and reduce material losses.

High tide and storm surge frequently flood the beautiful city of Venice, Italy. The frequency of these high waters, known as the "aqua alta" to Venetians, has increased dramatically since 1960 because Venice is sinking into the Adriatic Sea. Subsidence of the city is due to poor foundation conditions and withdrawal of underground water.

CASE STUDY 8.1

It's a Dam Disaster—I

About 11:15 A.M. on Saturday, 14 December 1963, the caretaker on duty at the Baldwin Hills Reservoir in Los Angeles heard unusual sounds coming from a gallery of pipes beneath the reservoir. Upon inspection, he found the area to be full of mud and water and "just blowing right out and making a terrific racket." He alerted his supervisor, who immediately ordered the lowering of the water in the reservoir. Neither man realized he was witnessing a catastrophe in the making. At 3:38 that afternoon, the earth-filled dam holding back 65,000,000 gallons of water would rupture. Flood waters would devastate the residential area below the dam, destroying scores of homes and causing 5 deaths. Only early warning and prompt action by the water department and the police prevented a greater tragedy.

The Baldwin Hills Dam and Reservoir were completed in 1950. The project was built as a backup facility for serving water needs during peak-use hours. Some redesign was required during construction, because the project geologist found that two normal faults ran through the reservoir and earthen dam (➢ Figure 1). Because the geologist believed that one of the faults was active, special subsurface drains were placed below the reservoir to collect any water that might percolate downward along the faults and then drain out below the dam. Percolating underground water in sandy materials can create conduits by an erosion process called **piping**, which can seriously weaken a dam foundation. The Newport–Inglewood fault, the active fault of most concern in the greater Los Angeles area, is only a few hundred yards

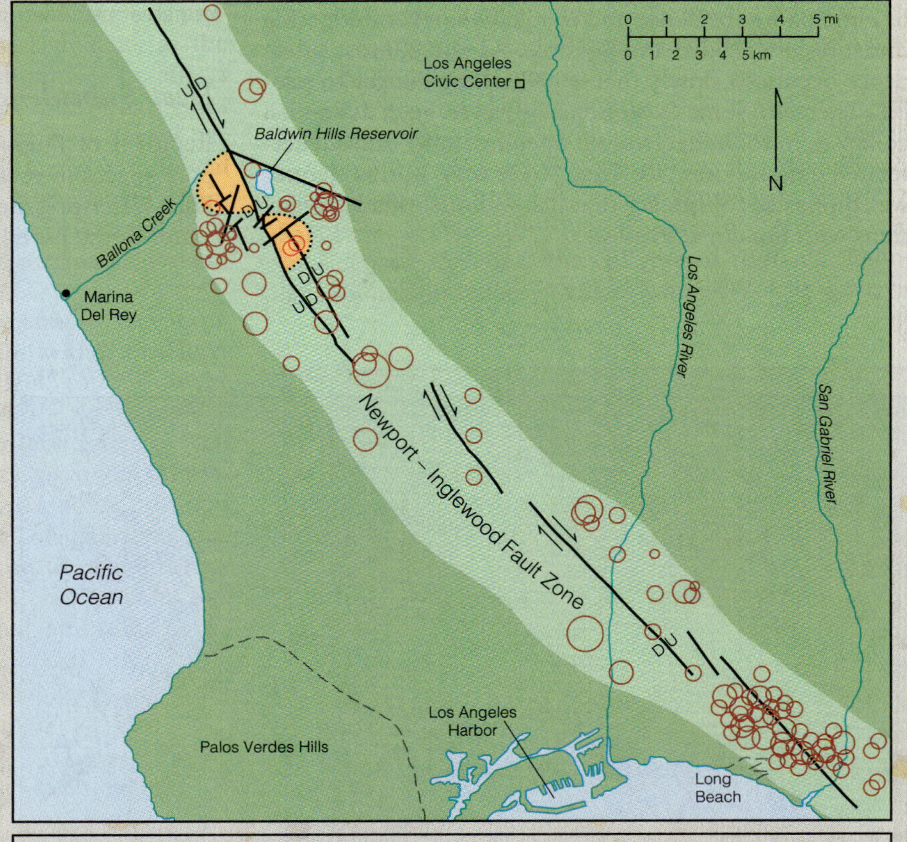

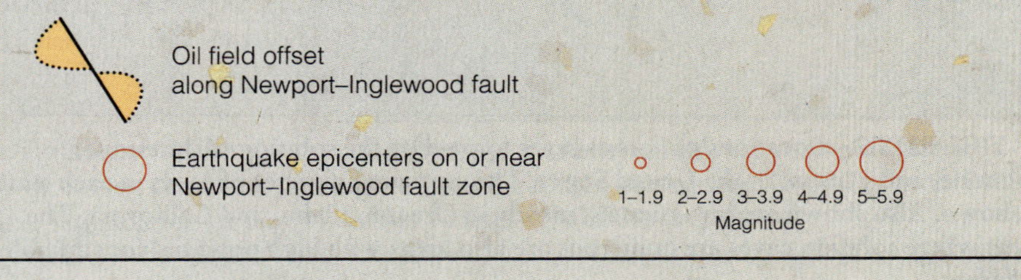

➢ FIGURE 1 Location of Baldwin Hills Reservoir and the oil field relative to the active Newport–Inglewood fault zone. Note the offset of the oil field on the right-lateral strike-slip Newport–Inglewood fault. Only earthquake epicenters on or near the main fault are shown.

from the dam, and the surrounding Inglewood Oil Field is an area of known subsidence. The design of the reservoir (➢ Figure 2) incorporated an impervious 10-foot-thick earth seal lined with asphalt and containing gravel-packed tile drains below the seal. This design was intended to capture any water that might leak through the dam if one of the faults should move and cause the lining to crack.

Believing they had designed around all potential geological problems, the engineers boldly built a dam (*a*) near a major active fault, (*b*) in an area of known subsidence, and (*c*) with two other, inactive faults running through the site. On the day of the failure, the reservoir cracked, the tile drains were clogged, and full-reservoir water pressure piped through the foundation beneath the dam.

By 1:00 P.M. the reservoir was being drained as quickly as possible, but water began appearing on the downslope toe of the dam. Leakage was so severe by 1:30 that a decision was made to evacuate the residential area below the dam. By 2:00 the leak in the dam face was described as "a sheet of water shooting through that little crack." The streets below the dam were still dry at the time, however, and some residents were not aware of the problem. It was not until 8 minutes before the dam breached that the retention basin at the toe of the dam filled and overflowed onto the home-lined street below it. The photo sequence of ➢ Figure 3 was taken over a period of about 15 minutes during the final failure of the dam. Flooding was extensive and 5 people who were unaware of the failure drowned or were swept away.

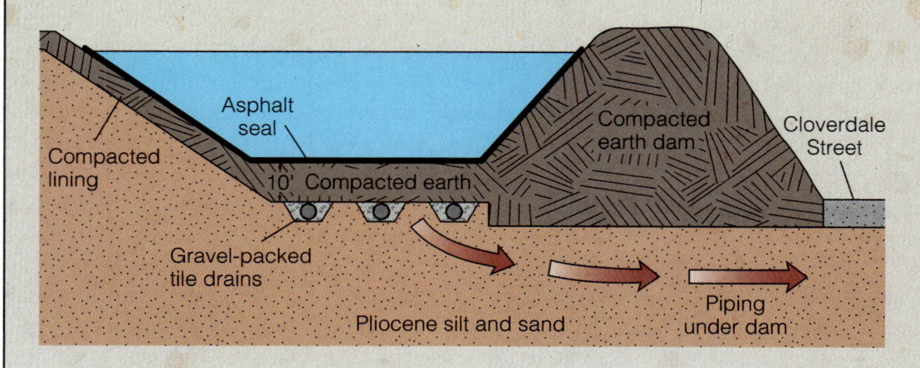

➢ FIGURE 2 Crack-proof lining and tile drains designed to protect against leaking and piping under the Baldwin Hills Dam. The lining did crack, and the drains failed to function as designed.

(a)

(b)

(c)

(d)

(e)

➢ FIGURE 3 (*a–d*) Sequence of photos taken during the last 15 minutes in the life of the Baldwin Hills Dam, 14 December 1963. (*e*) Breach in the dam, with the city in the background. The crack in the foreground resulted from subsidence on the fault running below the dam, the location of which is directly below the V-shaped notch in the middle of the photo. Piping took place along the fault zone.

CASE STUDY 8.1 *Continued*

An investigation by the mayor's "blue-ribbon" committee of geologists and engineers determined that subsidence caused by oil extraction from the Inglewood Oil Field ultimately led to the collapse of the Baldwin Hills Dam. On the day of the failure, slippage occurred along the two inactive faults that ran through the reservoir—slippage induced by oil-field subsidence. Although the slippage was only a matter of inches, it was enough to rupture the reservoir lining, forcing water under full-reservoir pressure along the faults into the underlying sands. This percolating water eventually "piped" an opening under the earthen dam embankment, which led to the dam failure and flooding below.

Indeed, this is a dam that never should have "happened." It was built in an area of known subsidence, close to a major active fault, and with inactive faults running through the reservoir. Any location with these geological problems would receive little consideration today as a potential dam site. The dam sits empty today, its floor covered by vegetation growing through cracks in the lining, mute testimony to faulty site selection. We learn much from tragedies like this one, and the education is very expensive.

7)—helps to support the overlying material. As pressure is reduced by extraction, the weight of the overburden gradually transfers to mineral-and-rock-grain boundaries. If the sediment was originally deposited with an open structure, the grains will reorient into a closer-packed arrangement, thus occupying less space, and subsidence will ensue (➢ Figure 8.3). Because clays are more compressible than are sands, most compaction takes place in clay strata.

Twenty-two oil fields in California have subsidence problems, as do many fields in Texas, Louisiana, and other oil-producing states. A near world record for subsidence is held by the Wilmington Oil Field in the city of Long Beach, California, where the ground has dropped 9 meters (about 30 ft). This field is the largest producer in the state, with more than 2,000 wells tapping oil in an upward-arched geologic structure called an *anticline* (see Chapter 13). The arch of the Wilmington field's anticline gradually sagged as oil and water were removed, causing the land surface to sink below sea level (➢ Figure 8.4). Dikes were constructed to prevent the ocean from flooding the adjacent facilities of the Port of Los Angeles and the naval shipyard at Terminal Island. People who have their boats moored on Terminal Island must actually walk *up* to board them at sea level (➢ Figure 8.5)! When it rains it is necessary to pump water out of low spots upward to the sea, and oil-well casings and underground pipes have "risen" out of the ground as the land has sunk. To remedy the situation, the city of Long Beach initiated in 1958 a program of injecting water back into the ground to replace the oil being withdrawn (➢ Figure 8.6). Subsidence has now stopped, and the surface has even rebounded very slightly in places. However, fluid injection cannot attain the necessary pressures to bring the surface back to its original level.

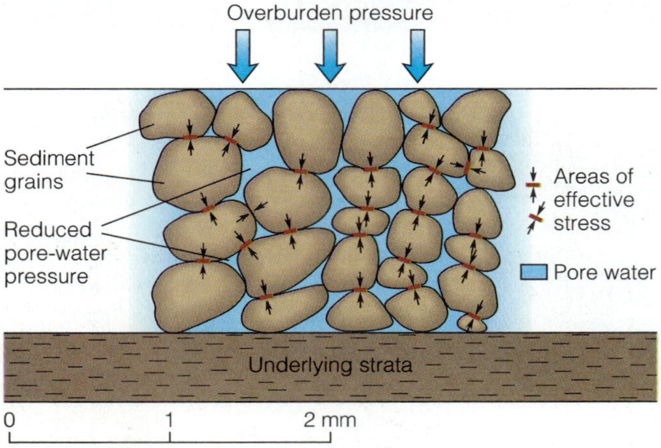

➢ **FIGURE 8.3 A reduction of pore pressure causes increased effective stress and rearrangement and compaction of sediment particles.**

Visible effects related to ground-water withdrawals are generally restricted to water-well damage or to cracking of long structures such as canals. As water levels are lowered, however, unconsolidated sediments compact, and tension cracks and fissures can develop. In the Antelope Valley, California, home of Edwards Air Force Base, the amount of water pumped from underground storage has exceeded the amount replacing it since 1930, and water levels beneath the dry lake have been lowered by more than 60 meters (200 ft). The dry lakebed at Edwards is the landing strip for the space shuttle and other high-performance aircraft. A ground fissure appeared in the lakebed that, initially only a few centimeters wide, became greatly enlarged by surface-water runoff and erosion (➢ Figure 8.7). This halted operations on the lakebed until repairs could be made. Similar fissuring has occurred in Arizona, New Mexico, Texas, and Las Vegas, Nevada. Coincidently, the magnitude of subsidence due to ground-water withdrawals is similar to oil-field subsidence. In the San Joaquin Valley of California, for

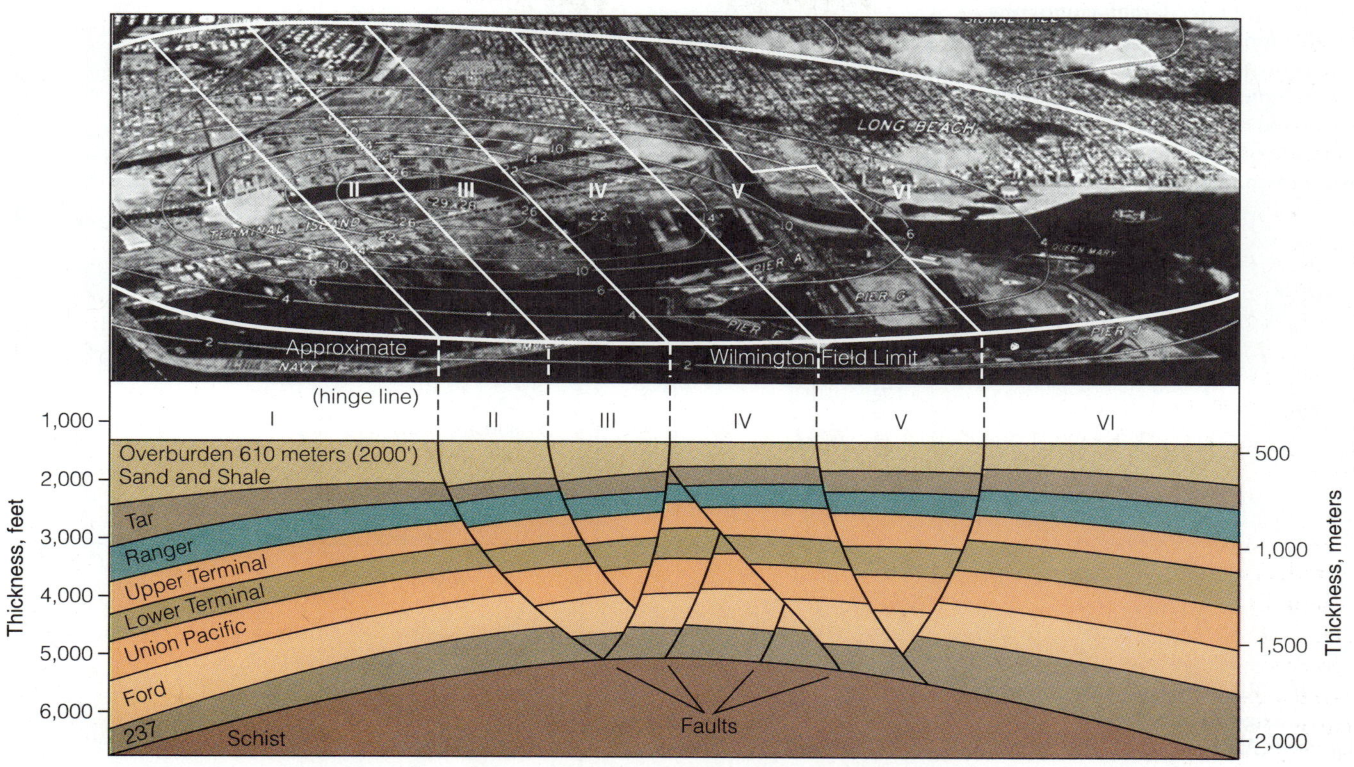

➢ **FIGURE 8.4 Contours of subsidence due to oil withdrawal at the Wilmington Oil Field, Long Beach, California, 1928–1974. The oil is geologically trapped in the faulted anticline shown in cross section below the photograph.**

(a)

(b)

➢ **FIGURE 8.5 The effects of subsidence on the coastal communities of Wilmington and Long Beach, California. (*a*) The power plant (*left*) is at the point of maximum subsidence in the Wilmington Oil Field, about 9 meters. Well below sea level, the building is protected from the sea by dikes. The yachts are moored at sea level. (*b*) This fire hydrant extruded as the ground around it sank; Long Beach.**

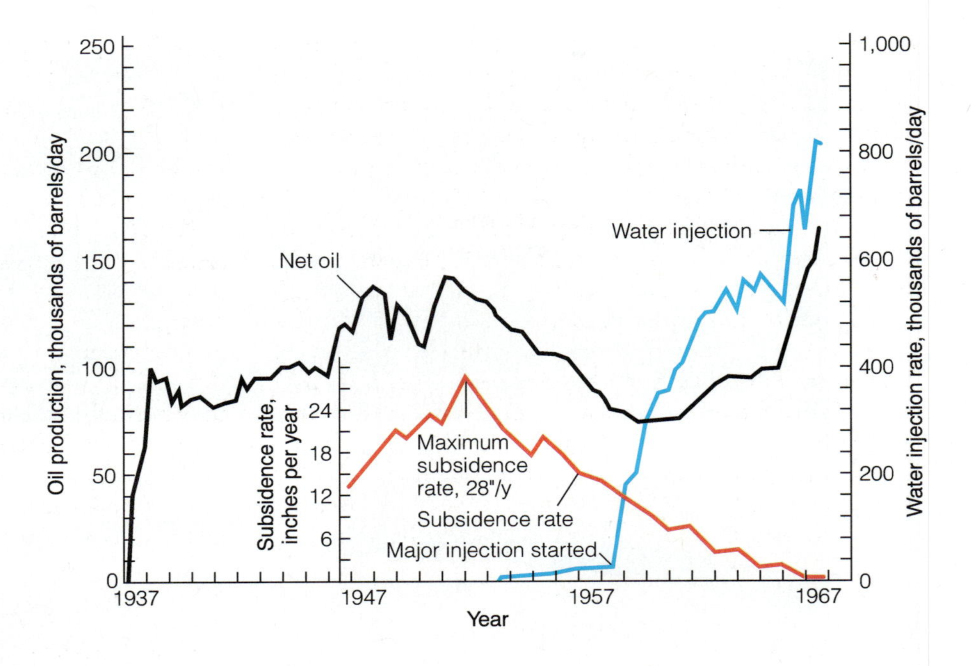

➢ **FIGURE 8.6 Graph comparing production, subsidence, and water injection at Wilmington Oil Field, 1937–1967. Note that the rate of subsidence decreased rapidly with the onset of the injection of water to replace the oil withdrawn from underground.**

➢ **FIGURE 8.7 Giant ground fissure across Rogers Lake, the dry lakebed of Edwards Air Force Base, due to groundwater withdrawals in the Antelope Valley. This dry lake is one of the two landing strips for the space shuttle.**

example, 8.9 meters (29 ft) of subsidence resulted from 50 years' water extraction for agricultural irrigation (➢ Figure 8.8). How much the land subsides depends mostly on how significantly the underground-water levels are lowered and the thicknesses of the compactable dewatered sediment layers.

The "sinking" of Venice, Italy, is of special interest because of the city's historical importance, priceless antiquities, and beauty (➢ Figure 8.9). The central square, Piazza San Marco, is awash several times a year, and it has been raised in attempts to compensate for subsidence. Venice is sinking at a rate of about 25 centimeters (1 ft) per century, in part due to settlement into soft sediments, and also from water withdrawal from 20,000 wells in the region. Sitting on a low mudbank in a lagoon sheltered by barrier islands, the city has dropped 3 meters relative to sea level since it was founded. During periods of extremely high water, locally called the *aqua alta* (Italian, "high water"), the avenues and squares of Venice become flooded, and gondola traffic comes to a halt. The *aqua alta* occurs when an onshore flow of water from the Adriatic Sea, known as a *storm surge,* coincides with exceptionally high tides. There were fifty-eight high waters between 1867 and 1967, thirty of them in the last decade of that period. Although Venice has subsided only 22 centimeters since 1910, the frequency of flooding has increased dramatically since 1950 (➢ Figure 8.10). This demonstrates the sensitivity of coastal areas to subsidence. Movable barriers to prevent flooding have been under study by the Italian government since 1989; most if not all of Venice's water wells have now been shut down, and water is imported by aqueduct.

Mexico City sits in a fault valley that has accumulated nearly 2,000 meters (6,600 ft) of sediments, mostly of

➢ FIGURE 8.8 Subsidence due to excessive ground-water withdrawal for agriculture in the San Joaquin Valley, California. The numbers on the pole indicate the ground level for the years indicated. The area is still subsiding, only at a much slower rate.

➢ FIGURE 8.9 The "aqua alta" (high water) floods St. Mark's Square in Venice, Italy, requiring tourists and other pedestrians to walk on raised boardwalks.

CASE STUDY 8.2

New Orleans—A City with the Blues

As the Mississippi River meanders through New Orleans it is not readily apparent that much of the city lies below the level of the great river (➢ Figure 1). In its early history the city was much closer to the Gulf of Mexico, but it is now 175 kilometers (108 mi) north of the mouth of the river because the delta has built seaward at a rapid pace (see Chapter 9). About half the city area was low wetlands at one time. These lowlands have now subsided below sea level and are surrounded by dikes and levees built to contain the river and prevent flooding. Tectonic downwarp due to the sheer annual load of 400–500 million tons of deltaic and floodplain sediment dumped by the river and the subsequent compaction of the sediments has resulted in 12–24 centimeters (5–10 in) of subsidence in the past 100 years. In addition to this natural subsidence, 3 meters of subsidence was caused by the intentional withdrawal of underground water to lower the water tables and improve surface-water drainage. The net result is that 45 percent of New Orleans is now below sea level, and therefore below river level (➢ Figure 2). The lowest point in the city is 2 meters below sea level in the famous French Quarter, noted for its jazz and dixieland music, restaurants, and the Mardi Gras. To protect the city from river floods and from surges of water from hurricanes in the Gulf of Mexico and Lake Pontchartrain, a system of 5-meter-high levees completely surrounds New Orleans. In terms of geologic hazards, the parts of the city that are built upon the levees are safer than the adjacent lowlands of the floodplain.

Because the river sediments underlying New Orleans are highly compactable, deep foundations are required for providing solid footings for high-rise and massive low-rise structures. The Superdome stadium, for example, rests on 2,266 piles driven to a depth of more than 50 meters (165 ft). The area is so low-lying and the ground-water table so near the land surface that flood waters from tropical storms are prevented from draining. This causes large areas of the city to become inundated. Flood waters here, as in many other areas of subsidence, must be pumped out of the city and up to the river.

➢ FIGURE 1 High-altitude false-color image of New Orleans, a city almost completely surrounded by water. The light spot in the lower center is the city. Note that the Mississippi River meanders through it, Lake Pontchartrain is to the north, and smaller Lake Borgne is to the northeast. Low, swampy delta lands appear brownish to the east and south. The area in the photo is approximately 200 kilometers (125 mi) across.

➢ FIGURE 2 New Orleans from the Mississippi River. The higher areas adjacent to the river are levees, and the lowlands in the distance are on the floodplain of the river. Much of the city visible in the photo is below the river level and below sea level.

(a)

(b)

(c)

➢ **FIGURE 8.14 Sinkhole in Winter Park, Florida. (*a*) Shortly after initial collapse. Note the sports car going down the sink. (*b*) Air view that shows its almost perfect symmetry. (*c*) Repaired and landscaped, the sinkhole area is now an urban park.**

sinkhole in May 1981 realized the extent of the sinkhole problem in the southeastern United States.

Sinkholes are most commonly found in carbonate terranes (limestone, dolomite, and marble), but they are also known to occur in rock salt and gypsum. They form when surface soil and rock collapse into shallow caverns created by natural acids attacking and dissolving limestones and dolostones (see Chapter 6). A landscape that is pitted with sinkholes, disappearing streams, lakes, and caverns is described as **karst terrane,** after the German name for the Kars limestone plateau in northwest Yugoslavia where such solution features are very common. In the United States the areas most affected by sinkhole collapse are in Alabama, Florida, Georgia, Tennessee, Missouri, and Pennsylvania. Total U.S. property damage from sinkhole collapse is estimated at several hundred million dollars (➢ Figure 8.15).

Thousands of naturally formed sinkholes exist in the United States, but their collapse may be accelerated by human activities. For example, excessive water withdrawal that lowers underground-water levels reduces or removes the buoyant support of shallow caverns' roofs (➢ Figure 8.16). In fact, the correlation between water-table lowering and sinkhole formation is so close that "sinkhole seasons" are designated in the Southeast. These occur whenever ground-water levels drop naturally because of decreased rainfall and in the summer, when there is heavy demand for water. Vibrations from construction activity or explosive blasting also may trigger sinkhole collapse. Other mechanisms that have been proposed for sinkhole formation are fluctuating water tables, which cause alternate wetting and drying of cover material reducing its strength, and high water tables that erode the covers of underground openings.

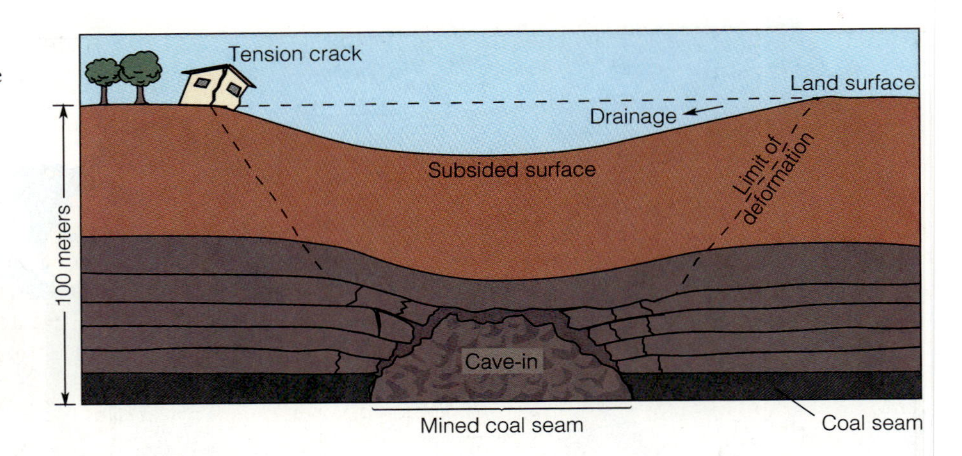

➢ FIGURE 8.18 Pattern of deformation in rocks and ground surface over an extracted coal seam.

➢ FIGURE 8.19 Surface effects of underground mining along the Tongue River in Wyoming. The pits, troughs, depressions, and cracks have formed since the mine was abandoned in 1914. (The road in the upper right corner of the photo is approximately 6 meters (20 ft) wide.)

alternatives: physically removing the burning strata or evacuating the town.

There is a conflict between maximum recovery of coal resources and mining methods that minimize subsidence. In the past, wide pillars of coal were left in a mine to support the roof during mining activities, and after the level was mined out, the pillars were removed. The roof of the shaft might then collapse into the opening, causing surface subsidence equal to about two-thirds the thickness of the coal seam. Environmental considerations now dictate leaving the pillars in place and backfilling the mine with sand or finer-grained mine waste. With this procedure, as much as 50 percent of the resource may be left underground.

Mitigation

The best strategy for minimizing losses is to restrict human activities in areas that are most susceptible to subsidence and collapse. Unfortunately, most instances of land subsidence due to fluid withdrawals have not been anticipated up to now. Our knowledge of the causes, mechanics, and treatment of deep subsidence has increased with every damaging occurrence, so that we can now implement measures to reduce the incidence and impact of these losses.

Some of the measures now taken to minimize subsidence are controlling regional water-table levels by monitoring and limiting water extraction, replacing fluids by water-injection, and importing surface water for domestic use in order to avoid drawing on underground supplies. The State of Texas established the Harrison–Galveston Subsidence District in 1979 to regulate the withdrawal of water and thereby control subsidence and prevent coastal flooding in the Galveston area. Underground mining operators are now required to leave pillars to support mined-out coal beds, and mine openings are being backfilled with compacted mine waste to support the overburden. This latter procedure, however, may introduce chemicals that could contaminate underground water unless the backfill is first stripped of pollutants. Pennsylvania's Mine Subsidence Insurance Act of 1962 allows some property owners to buy subsidence insurance after inspection of their property. Another Pennsylvania bill allows homeowners to buy the bituminous coal beneath their property at a fair price and therefore exercise control over mining operations.

Collapsible soils along canals are now routinely precollapsed by soaking, thereby anticipating in advance any adjustments of the canal gradient. Relatively inexpensive geophysical techniques are available for locating shallow limestone caverns beneath potential building sites. Subsidence problems will continue, but we should employ every means possible to foresee the problem and minimize impact.

Property owners' ability to control subsidence on their own land is hampered by legal problems. The legal recovery theories (tort law) that establish one's right to sue for damages due to subsidence seem to contradict laws that establish others' rights to recover resources from underground. Although it is known that pumping fluids out of the ground can cause subsidence, existing laws may not enable one to force a halt to pumping. It may be time to change the law and work toward a balance between the need to control subsidence and the need to extract underground water, coal, and petroleum products.

■ THE FUTURE

Subsidence rates have decreased in most areas of the world where they have been a serious problem, mainly because of the growing awareness of the causes of subsidence. Unfortunately, the ground surface does not rebound once subsidence has been stopped, and many of the world's major cities remain at risk of flooding. If sea level rises in response to global warming, as most models predict it will (see Chapters 11 and 12), then coastal flooding will become more frequent and severe. Sea level has risen 15–20 centimeters (6–8 in) per century over the past 300 years, and recent estimates predict a sea-level rise of between 1 and 3 meters from the present level by the end of the next century. If the actual rise is even close to the upper estimate, many heavily populated areas will be inundated. Impressive bulwarks and dikes will be needed for holding back the ocean unless we choose to relocate cities, which is unlikely. Living behind such barriers will be a challenge, and it presents significant risks. It has been said that if sea level does rise significantly, construction of new dikes or the raising of the old ones will be the most expensive undertaking yet faced by humans!

■ GALLERY: WONDERS OF KARST TERRANE

Karst terrane exists in many parts of the United States, but it is particularly prevalent in parts of Florida and Kentucky (see Figures 8.2 and 8.15). Perhaps the most spectacular karst area in the world is along the Li River in the predominantly limestone region of Kwangsi in southern China near Canton (Guangzhou). Solution of limestone has reduced the landscape there but left enormous pinnacles of resistant limestone, called "tower karst," that present an almost extraterrestrial appearance (➢ Figure 8.20). The towers and other karst features draw tourists from all over the world.

➢ FIGURE 8.20 Residual karst towers; Kwangsi region, southern China. The former limestone plateau has been lowered by solution except for resistant-limestone pinnacles.

In many places, water containing dissolved calcium carbonate percolates downward through soil and rock into caverns, where evaporation and precipitation form needlelike **stalactites**—extending down from the ceiling—and the more irregular **stalagmites**—rising up from the floor (➢ Figure 8.21). These fresh-water limestone features in caves are called **dripstone,** because they build up drip-by-drip. Carlsbad Caverns in New Mexico and Mammoth Cave in Kentucky offer spectacular examples, but literally hundreds of other limestone caves also exhibit these features and are open to the public (see Figure 8.2). **Speleology** is the study of caves, and "caving" is an exciting recreational pastime for many people. Cavers explore and map natural underground openings. Much of what we know about caves and caverns is attributable to these hobbyists' efforts and talents.

(a)

(b)

➢ FIGURE 8.21 **(*a*) Dripstone stalactites, stalagmites, and pillars in a limestone cavern. (*b*) A "caver" descends into a large room that is accessible only by rope.**

SUMMARY

Subsidence

DEFINED: A sinking of the earth's surface caused by natural geologic processes or by human activity such as mining or the pumping of oil or water from underground.

TYPES:

1. Deep subsidence caused by the removal of water, oil, or gas from depth.
2. Shallow subsidence of ground surface caused by underground mining or by hydrocompaction, the consolidation of rapidly deposited sediment by soaking with water.
3. Collapse of ground surface into shallow underground mines or into natural limestone caverns, forming sinkholes.
4. Settlement occurs when the applied load of a structure is greater than the bearing capacity of its geologic foundation.

Causes

DEEP SUBSIDENCE: Removal of fluids confined in pore spaces at depth transfers the weight of the overburden to the grain boundaries, causing the grains to reorient into a closer-packed arrangement. Twenty-two oil fields in California have subsidence problems, and parts of the Texas Gulf Coast, San Joaquin Valley of California, Venice (Italy), and Mexico City have subsided because of withdrawal of underground water.

SHALLOW SUBSIDENCE AND COLLAPSE: The ground surface may collapse into shallow mines (usually coal). Sinkhole collapse is due to a lowering of the water table, which reduces the buoyant forces of underground water. Some sinkholes result from the collapse of soil and sediment cover into an underground limestone or dolomite cavern. Heavy application of irrigation water causes some loose, low-density sediments and soils to consolidate. This is known as hydrocompaction.

Problems

Subsidence may cause ground cracking, deranged drainage, flooding of coastal areas, and damage to structures, pipelines, sewer systems, and canals. Whole buildings have disappeared into sinkholes.

Mitigation

Underground water can be monitored and managed. Extracted crude oil can be replaced with other fluids. Mine operators are now required to leave pillars to support the mine roof, and rock spoil can be returned to the mine to provide support. Collapsible soils along canals can be precollapsed by soaking. Relatively inexpensive geophysical techniques can be used to locate limestone caverns with roofs that might collapse. Subsidence insurance is available in Pennsylvania.

Features

Land that is underlain by limestone or dolomite and characterized by caves, caverns, lakes, and disappearing streams is called *karst* topography.

KEY TERMS

dripstone
hydrocompaction
karst terrane
piping
settlement
sinkhole
speleology
stalactites
stalagmites
subsidence

STUDY QUESTIONS

1. What are the causes of subsidence and of collapse?
2. What causes sinkholes to form, and how have humans contributed to this problem?
3. What is the difference between settlement and subsidence? What should be done in advance to prevent settlement?
4. Distinguish between subsidence and collapse due to natural causes and due to human causes.
5. What can be done to slow or stop subsidence caused by withdrawal of water or petroleum?
6. Name the common rocks that develop into karst topography. Where in the United States would you go to see karst terrane?
7. What engineering problems make constructing canals and aqueducts in areas of hydrocompaction such an insidious geologic hazard?
8. What surface manifestation of subsidence due to ground-water withdrawal makes it a serious problem, particularly in California and Arizona?
9. What hazard does subsidence present in coastal areas?
10. What is speleology, and what might be the value of this study?

FURTHER READING

Amandes, C. B. 1984. Controlling land surface subsidence: A proposal for a market-based regulatory scheme. *U.C.L.A. law review* 31, no. 6.

Dolan, R., and H. Grant Goodell. 1986. Sinking cities. *American scientist* 74, no. 1: 38–47.

Halliday, William R., M.D. 1976. *Depths of the earth*. New York: Harper and Row.

Holzer, T. L., ed. 1984. Man-induced land subsidence. Geological Society of America, *Reviews in engineering geology*-VI.

Menard, H. W. 1974. *Geology, resources, and society*. New York: W. H. Freeman and Co.

State of California, Resources Agency. 1965. *Landslides and subsidence*. Geologic Hazards Conference, Los Angeles.

CHAPTER

9

WATER ON LAND

Rain added to a river that is rank
perforce will force it overflow the bank.

WILLIAM SHAKESPEARE (1564–1616)

If extraterrestrials were to approach earth from afar, they couldn't help but be struck by our planet's heavy cloud cover and its expanse of ocean, clear indications that earth is truly "the water planet." Water is indeed what distinguishes our planet from other bodies in our solar system. Three quarters of the earth's surface is covered by water, and even the human body is 65 percent water. Water provides humans with means of transportation, and much human recreation is in water—and *on* it where it exists in solid form as ice and snow. Since it is not equally distributed on the earth's surface, we have droughts, famine, and catastrophic floods. The absence or abundance of water results in such marvelous and diverse features as deserts, rain forests, picturesque canyons, and glaciers. In this chapter we investigate the reasons for this uneven distribution and its consequences. Water is involved in every process of human life; it is truly our most valuable resource.

WATER AS A RESOURCE

Water is the only common substance that occurs as a solid, a liquid, and a gas over the normal temperature range found at the earth's surface. About 97 percent of the earth's water is in the oceans, and 2 percent is in ice caps and glaciers; less than 1 percent is readily available to humans as underground or surface fresh water (Table 9.1). Fortunately, water is a renewable resource as illustrated by the **hydrologic cycle,** the earth's most important natural cycle (➢ Figure 9.1). Water that is evaporated

The Mississippi River engulfs a farm south of Des Moines, Iowa, during the great flood of 1993.

CASE STUDY 9.1

The Colorado—A River in Peril

The thought grew in my mind
that the canyons of this region
would be a Book of Revelations
in the rock-leaved Bible of geology.

JOHN WESLEY POWELL, GEOLOGIST AND EXPLORER

The Colorado River begins as a mere rivulet on the west side of the drainage boundary called the **Continental Divide.** All water that falls on the continent west of the divide flows to the west, and only two large rivers in the contiguous 48 states flow to the Pacific Ocean: the Colorado and the Columbia (➢ Figure 1). The Colorado's flow, 15 million acre-feet per year (af/y), is puny compared to those of the Mississippi and the Columbia Rivers at 400 million and 192 million af/y, respectively. An acre-foot of water is the amount that would cover an acre of land 1-foot deep; it equals 325,000 gallons, or enough to supply the water needs of a family of 5 for a year. However, the Colorado is one of the saltiest and siltiest rivers in the United States. The river and its two major tributaries, the Green River of Utah and the Gunnison River of Colorado (➢ Figure 2), flow over sedimentary rock outcrops containing salt deposits, which contributes to the Colorado's saltiness. Its steep gradient and predominantly sedimentary geology account for the large mass of sediment it transports, up to a half-million tons a day.

The Colorado and Green Rivers figured prominently in the exploration and settlement of the West. The Colorado's most celebrated explorer, John Wesley Powell, who explored the area in 1869 and 1871–72, noted prophetically that "all waters of . . . arid lands will eventually be taken from their natural channels." Today southern Utah's Lake Powell, impounded by the Glen Canyon Dam, bears his name. The Glen Canyon is one of 10 major dams that control the Colorado's flow and meter out its precious water to more than 20 million people. Because of drought, its 1991 level was almost 70 feet below its design level. Reservoirs of the Colorado River basin upstream from the Boulder (Hoover) Dam with storage capacities greater than 100,000 acre-feet are

RESERVOIR	STORAGE CAPACITY, MILLION ACRE-FEET
Flaming Gorge Reservoir	0.490
Strawberry Reservoir	0.160
Blue Mesa Reservoir	0.110
Navajo Reservoir	0.240
Lake Powell	3.400
Lake Mead	3.600
Total	8.000

The Colorado River has been called "a river drained dry" (see the Carrier entry in the Further Reading list). One of the most regulated streams in the world, legal battles over its water brew continuously among the river states and between the United States and Mexico because the water has been overappropriated. Problems began in 1922 when an agreement divided the river's catchment area into two water-resource basins. This agreement totally ignored Native American and Mexican claims dating back to the mid-

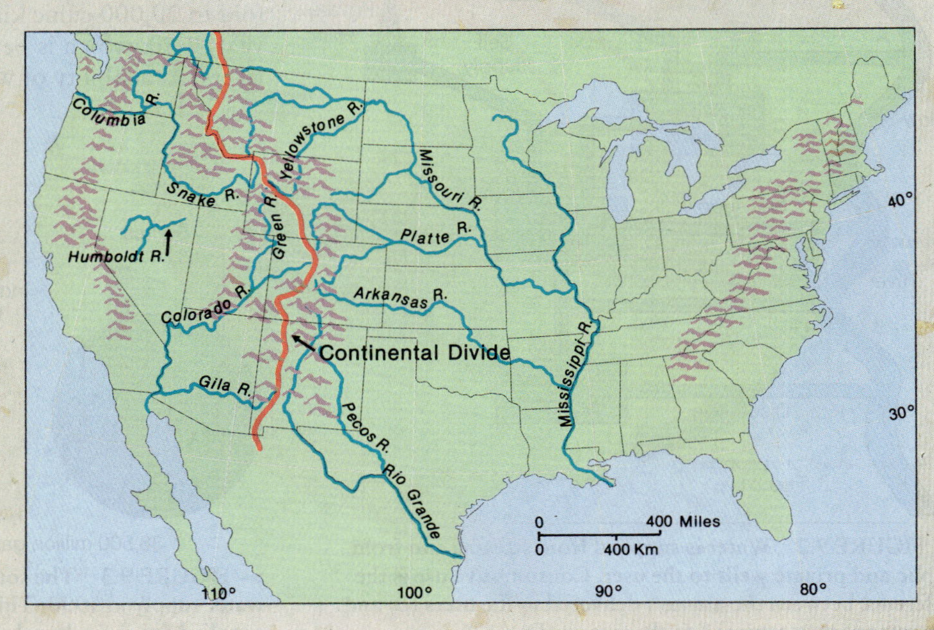

➢ **FIGURE 1 West of the Continental Divide streams flow to the Pacific Ocean. East of it, they flow to the Gulf of Mexico.**

1800s, which under the "first in time, first in right" doctrine, may establish their legal claim to significant amounts of river water. Under the 1922 agreement, Utah, Wyoming, Colorado, and New Mexico were to share the water of the upper basin, and Arizona, California, and Nevada were to share the lower-basin water. The users of each basin were allocated 7.5 million af/y. In addition, Mexico, by a 1944 treaty, was allocated a minimum of 1.5 million af/y.

Unfortunately, the allocations exceed the Colorado's total sustained yield, which fell to 9 million af/y during the drought of the late 1980s and early 1990s. The variability in river flow before construction of the Glen Canyon Dam in the mid-1960s and extensive offstream use is shown here.

PERIOD	AVERAGE ANNUAL FLOW, million acre-feet	REMARKS
1896–1968	14.8	measured and estimated flow before construction of Glen Canyon Dam (74 years)
1896–1929	16.8	34-year wet period
1914–1923	18.8	wettest 10-year period
1930–1968	13.0	38-year dry period
1931–1940	11.8	driest 10-year period

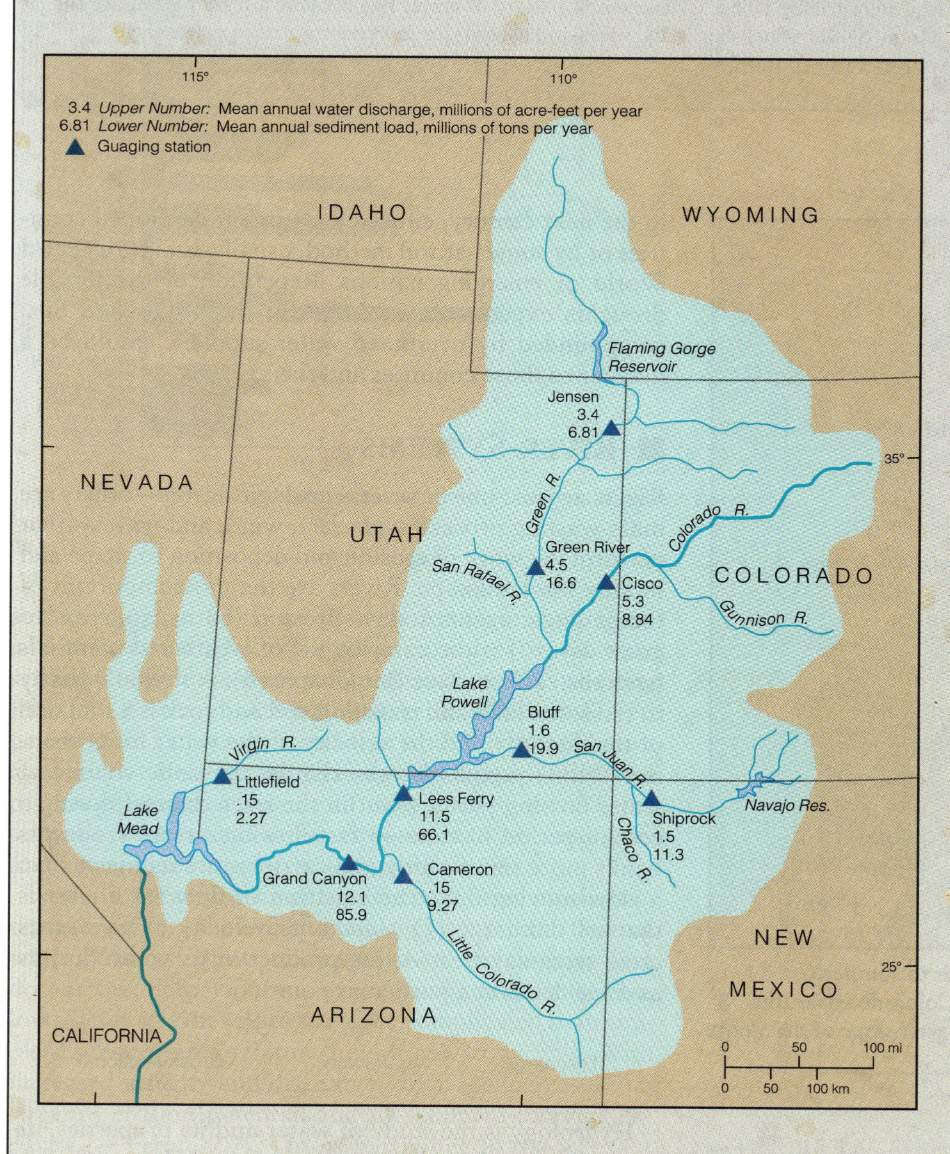

➢ **FIGURE 2 Colorado River drainage basin with its major tributaries. The mean annual water discharge and sediment load (1941–1957) at selected locations are shown.**

continued

causes the particles to settle. The saddle-shaped lower boundary of the "erosion" field in ➢ Figure 9.8 shows

CASE STUDY 9.2

River Velocity and Friction

Hydraulic engineers have found that a stream's velocity depends on the **channel shape,** and is a function of the stream's cross-sectional area and its wetted perimeter. The **wetted perimeter** is the amount of contact the stream has with its channel in cross section—in other words, how much streambed area provides frictional resistance to the flow of water. Specifically, the relationship is stated as velocity (V) is proportional to the stream's cross-sectional area (A) divided by its wetted perimeter (p), or

$$V \approx A/p$$

➢ Figure 1 shows three streams of identical cross-sectional area but of significantly different wetted perimeters. The stream of shape 1 has the smallest wetted perimeter and would be the swiftest, whereas the stream of shape 3 would be the slowest.

The effect of wetted perimeter on velocity is illustrated in ➢ Figure 2, which shows three small tributaries joining to form a single stream. The cross-sectional area of the main stream is equal to the sum of the areas of the three tributaries, but the main stream's wetted perimeter is less than the sum of the wetted perimeters of the three tributaries. Thus A/p is greater for the main stream, and the extra water is accommodated by increased velocity.

This example assumes there is no drastic change in the roughness (frictional drag) of the streambed under consideration. All other things being equal, the smaller the wetted perimeter, the greater is the velocity. Obviously, if the slope (gradient) increases, the velocity also will increase. Thus, the velocity of a stream really depends on four variables:

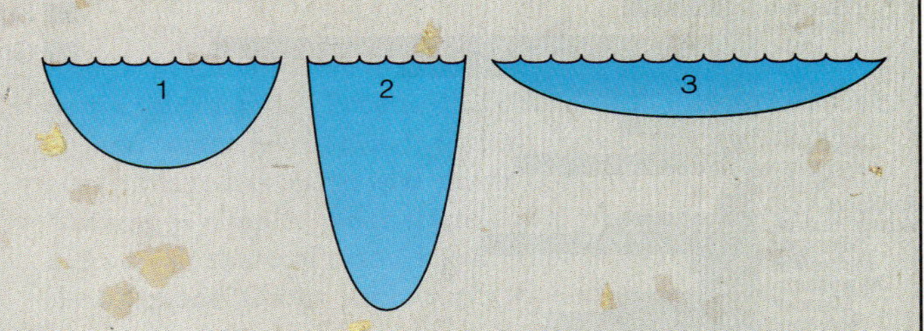

➢ FIGURE 1 Three stream cross sections of equal areas. Section 1 has the smallest wetted perimeter, and, assuming equal gradients, this stream's velocity would be greatest.

- gradient (the slope),
- channel shape (A/p),
- discharge, which influences both area and wetted perimeter, and
- roughness (the frictional drag of boulders, sand, silt, clay, or whatever composes the bed).

On alluvial fans in arid regions (see Figure 9.10), just the opposite happens. The wetted perimeter increases because of changes in the channel shape and also because the fan channels divert or spread out into many *distributary channels,* rather than *coalescing* into a single main stream (see Figure 9.32). When this happens, the water slows and eventually sinks into the ground.

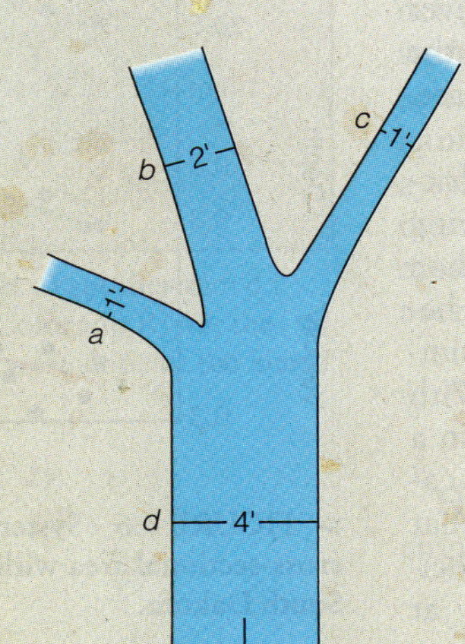

	Area, ft² A	Wetted perimeter, ft P	A/P
1' a 1', 1'	1	3	1/3
1' b 1', 2'	2	4	1/2
1' c 1', 1'	1	3	1/3
1' d 1', 4'	4	6	2/3

➢ FIGURE 2 Cross sections of three tributaries (*a, b,* and *c*) and the main stream (*d*). The cross-sectional area of the main stream equals the sum of the three tributaries' cross-sectional areas, but the ratio of the main stream's area to its wetted perimeter, *A/p*, is more than the *A/p* of any tributary. Because the main stream has less bed friction (wetted perimeter), it speeds up to accommodate increased discharge.

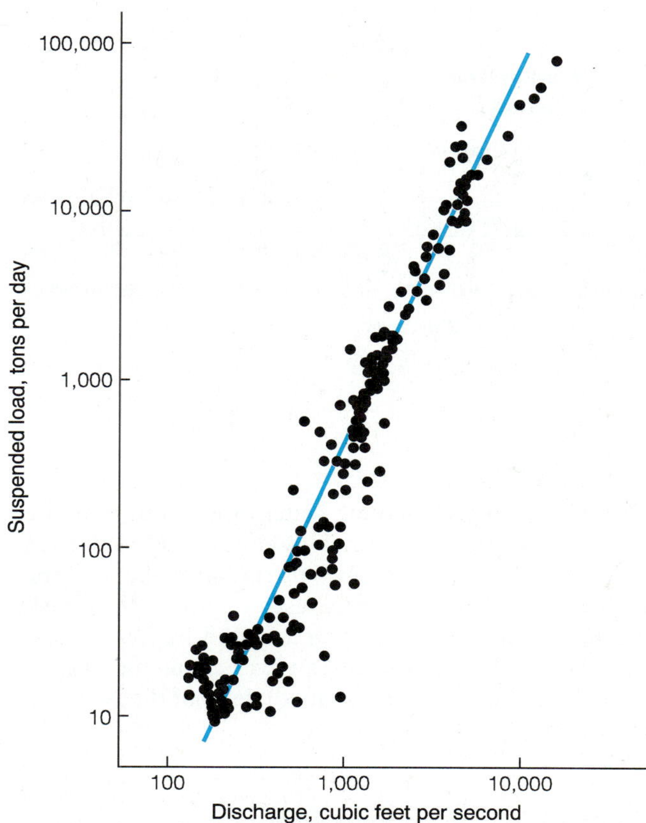

➢ **FIGURE 9.7 The relationship between water discharge (Q) and the suspended load for a typical stream. The points denote separate measurements. A tenfold increase in water discharge results in almost a 100-fold increase in sediment load.**

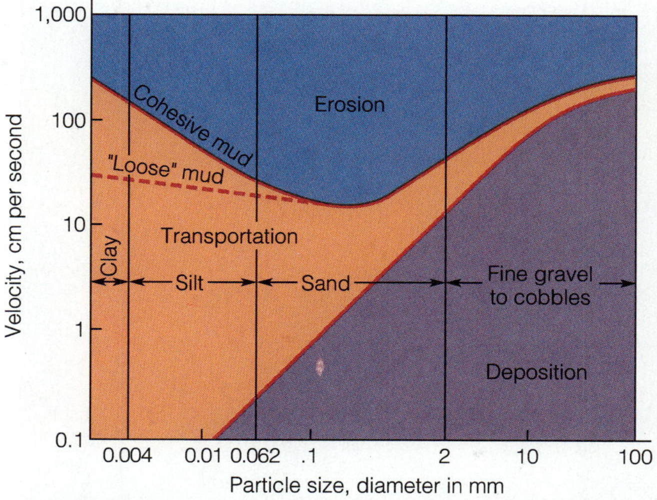

➢ **FIGURE 9.8 So-called Hjulstrom diagram of threshold stream velocities for erosion, transportation, and deposition of varying particle sizes. Note that a higher water velocity is required to erode clay and silt than to move fine sand.**

the velocities required to lift or suspend clay, silt, sand, gravel, and cobbles; the middle, "transportation" field shows the velocity range in which those particles remain in suspension. At flow velocities below the lower "transportation" curve, particles settle to the bottom of the stream (are deposited). Note that although higher velocities are required to erode clay and silt particles than to erode sand grains, once they are suspended in the stream, they can be transported at much lower velocities than can sand. This is perhaps surprising, but clay and silt particles are more cohesive than sand grains and, therefore, require a higher velocity to be picked up (suspended) into the stream flow. You may have experienced this phenomenon when walking on dry fine-grained sediment on a windy day. As your foot breaks the cohesive sediment surface and creates air turbulence, the sediment becomes airborne as choking dust.

The longitudinal profile, or side view, of a stream channel has a slope known as its **gradient,** which can be expressed as the ratio of its vertical drop to the horizontal distance it travels (in meters per kilometer or feet per mile). The gradient of the Colorado River in the Grand Canyon is about 3.2 meters per kilometer, a moderately steep gradient, whereas in its lower reaches, where it flows into the Gulf of California, its gradient is only a few centimeters per kilometer. A mountain stream may have a gradient of several tens-of-meters per kilometer and an irregular profile with rapids and waterfalls. As the same stream approaches the lowest level to which it can erode, an elevation known as **base level,** its gradient may be almost flat. A stream's ultimate base level is sea level, of course, but it may have intermediate, temporary base levels such as a lake surface or a rock formation highly resistant to erosion. A stream whose gradient has adjusted such that an equilibrium exists between its discharge and its sediment supply is known as a **graded stream.** Longitudinal profiles of graded streams are generally concave-upward from source to end. ➢ Figure 9.9 shows the profiles of the Arkansas River, which drains the wide east slope of the Rocky Mountains, and the Sacramento River, which drains the much narrower Pacific Slope.

Gradients decrease in the lower reaches of a river, and sedimentation may occur as flooding streams overflow their banks onto adjacent **floodplains.** Decreased water velocity on the floodplain results in the deposition of fine-grained silt and clay adjacent to the main stream channel. This productive soil, commonly known as *bottom land,* is much desired in agriculture.

STREAM FEATURES

All sediment deposited by running water is called **alluvium** (Latin, a flooded place or flood deposit). In the arid West, where streams lose velocity as they flow from mountainous terrain onto flat valley floors, we find conspicuous depositional features known as **alluvial fans**

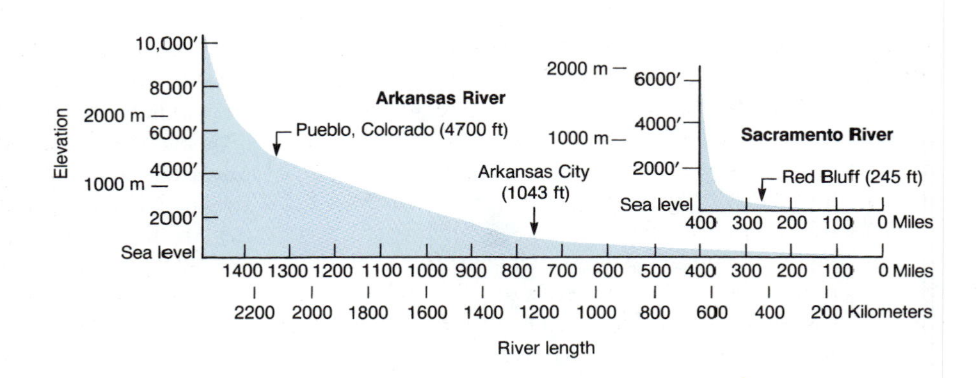

➢ **FIGURE 9.9 Measured longitudinal profiles for the Arkansas and Sacramento Rivers. The vertical to horizontal scale ratio is 275:1.**

(➢ Figure 9.10). They are in fact fan-shaped and they build up as the main stream branches, or splays, outward in a series of *distributaries,* each "seeking" the steepest gradient. Exemplified by Figure 9.10, alluvial fans can provide beautiful, interesting views of surrounding terrain, but can also be subject to flash floods and debris flows from intense seasonal storms in nearby mountains.

Deltas form where running water moves into standing water. The Greek historian Herodotus (484–420 B.C.) applied the term to the somewhat triangular shape of the Nile delta in recognition of its resemblance to the Greek letter delta (Δ). Deltas are characterized by low stream gradients, swamps, lakes, and an ever-changing channel system. The "arcuate" (arc-shaped) delta of the Nile and

➢ **FIGURE 9.10 Alluvial fan at the base of Day Canyon, San Gabriel Mountains, Claremont, California. The lighter-colored, straight feature trending across the middle of the fan is the scarp of the active Cucamonga fault. This area is becoming highly urbanized.**

the "bird's foot" delta of the Mississippi also are described by their appearance on maps (➢ Figure 9.11). Deltas also form at the mouths of small streams, as illustrated by the arcuate delta at Malibu Creek—location of the famous Malibu Colony of the rich and famous and the early beach-boy-type surfing movies (➢ Figure 9.12).

Three main types of deltas are recognized according to the relative importance of streams, wave action, and tides in their formation and maintenance. The Mississippi delta exemplifies *stream-dominated* deltas. It consists of long, fingerlike distributaries that have built the delta out into the Gulf of Mexico and imparted to it the appearance of a bird's foot. The Nile delta is categorized as *wave-dominated,* because its seaward margin consists of a series of sand-barrier islands formed by currents and wave action. The tremendous sand output of the Nile has created the huge coastal dunes east of Alexandria that formed the background for the famous World War II battle at El Alamein (see Case Study 12.2). The Ganges–Brahmaputra delta of Bangladesh is *tide-dominated,* consisting of tidal flats and many low-lying sand islands offshore (Figure 9.11). Because deltas are at or near sea level, they are subject to flooding by tropical storms. Coastal flooding is the most deadly of natural disasters, even

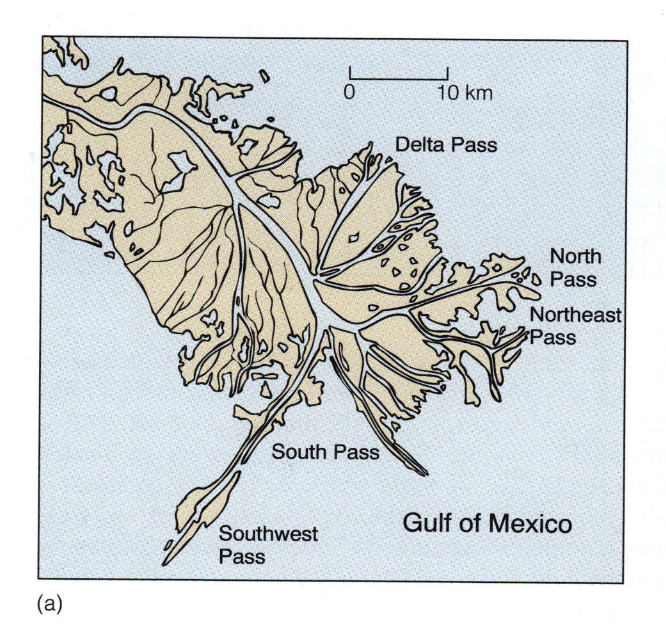

(a)

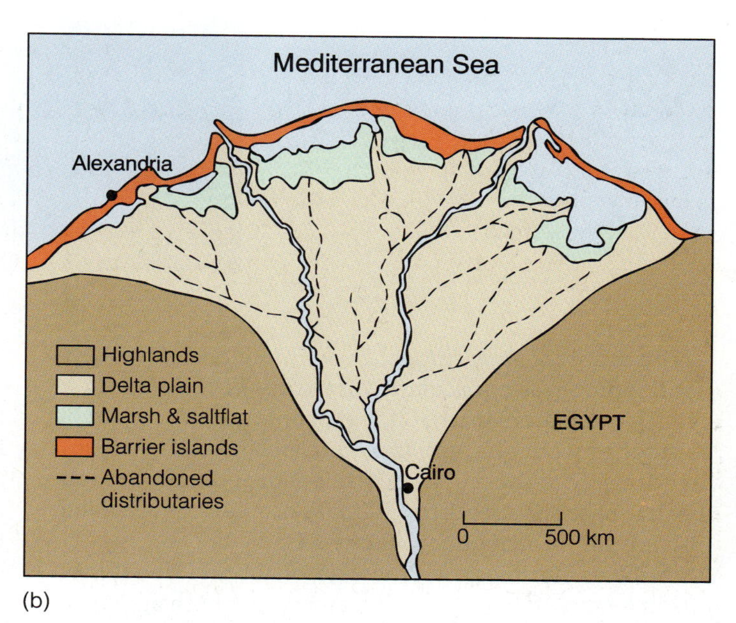

(b)

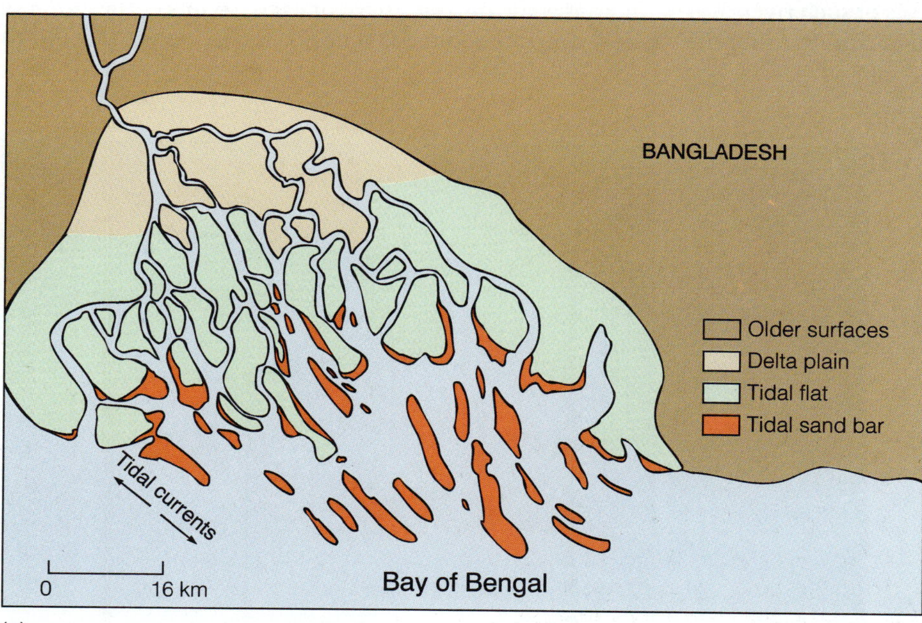

(c)

➢ FIGURE 9.11 (*a*) River-dominated "bird's-foot" delta of the Mississippi River. The main river channels are separated by swampy land. (*b*) Wave-dominated arc-shaped delta of the Nile River. (*c*) Tide-dominated Ganges–Brahmaputra delta of Bangladesh. Note the tidal flats and the islands which are subject to inundation.

➢ **FIGURE 9.12 The arcuate wave-dominated delta of Malibu Creek, Malibu, California, after a record rainfall in 1969. Note the suspended sediment in the ocean and the concentration of homes along the delta north (*left*) of the creek. Compare this shape with that of the Nile delta, Figure 9.11b.**

more deadly than earthquakes or river floods. This type of flooding was particularly disastrous in Bangladesh in 1970, when more than 500,000 people were drowned by a surge of water onshore from a tropical cyclone. A similar tragedy occurred in 1991, when another onshore water surge left a reported 100,000 dead on the low-lying islands off the delta (see Chapter 11).

Streams and rivers with low gradients and floodplains form a number of important features that both aggravate and alleviate flood damage. The courses of such rivers follow a series of S-shaped curves called **meanders** (after the ancient name of the winding Menderes River in Turkey; ➢ Figure 9.13). Straight rivers are the exception, rather than the rule, and it appears that running water of all kinds will meander—ocean currents, rivers on glaciers, and even trickles of water on glass. The key to these regularly spaced curves is the river's ability to erode, transport, and deposit sediments. The meandering process begins when the gradient is reduced to a slope at which a stream stops deepening its valley. As a stream meanders, it erodes the outside curve of each meander, the *cutbank*, and deposits sediment on each inside curve, the *slip-off*

➢ **FIGURE 9.13 Meanders on the Owens River in north-central California.**

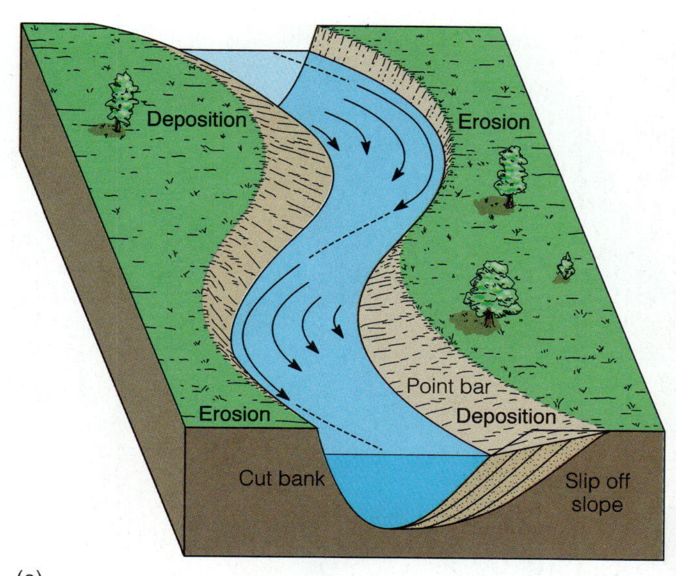

(a)

(b)

➢ **FIGURE 9.14 Erosion and deposition patterns on a meandering stream. (*a*) Erosion of the cutbank and deposition of a point bar on the slip-off slope. Arrow length is proportional to stream velocity. (*b*) Rillito Creek rampaging through Tucson, Arizona, during the flood of 1983. Damage occurred when the cutbank of the meander migrated west (*left*) into vacant property immediately upstream from the townhouses (*center left*). This allowed erosion to occur behind a concrete bank that was protecting the townhouse property. Note the prominent point bar that developed (*bottom center*) as the meander migrated westward.**

slope, producing a deposit known as a **point bar.** ➢ Figure 9.14 shows the velocity distribution on a stream meander, which explains the erosion–deposition pattern. Meanders also widen stream valleys by lateral cutting of valley side slopes, causing mass wasting and eventually creating an alluvium-filled floodplain adjacent to the river. It is across this surface that the river meanders, sometimes becoming so tightly curved that it short-cuts a loop to form a **cutoff** and **oxbow lake** (➢ Figure 9.15).

By the time a river has developed a wide floodplain with meanders, cutoffs, and oxbow lakes, a situation of poor drainage exists that increases the impact of overbank flooding. **Natural levees** build up on both banks of a river when it overflows and deposits sand and silt adjacent to the channel (➢ Figure 9.16). Levees, natural or manufactured, keep flood waters channelized so that the water level in the stream may be above the level of the adjacent floodplain. This is a tenuous situation, because a breach in the levee could cause flooding of the floodplain, which could force the river into a new course. New Orleans on the Mississippi River is an example of a city on a floodplain that is almost perenially below the level of the river that flows through it (see Chapter 8). Although levees may prevent flooding of adjacent floodplains, high waters restrained by levees can cause flooding downstream where there are no levees.

During the 1993 floods, artificial levees saved some cities in the Mississippi River drainage basin, notably St. Louis, where the flood crested at 49.4 feet and the levee height is 52 feet. Quincy, Illinois, was without artificial levees and was inundated, whereas the levees at Hannibal, Missouri, on the opposite side of the river held and the town was spared. Some levees had to be intentionally breached to reduce flood crest elevation downstream where there was extensive urbanization. Areas adjacent to the intentional breaches were mostly agricultural and were of course inundated. More than 800 of the 1,400 levees in the nine-state disaster area were breached or overtopped. Most of these were simple dirt berms built by local communities, which explains why so many small towns were almost completely submerged by the floodwaters. However, more than 30 levees built by the Corps of Engineers also failed during this record flood (➢ Figure 9.17).

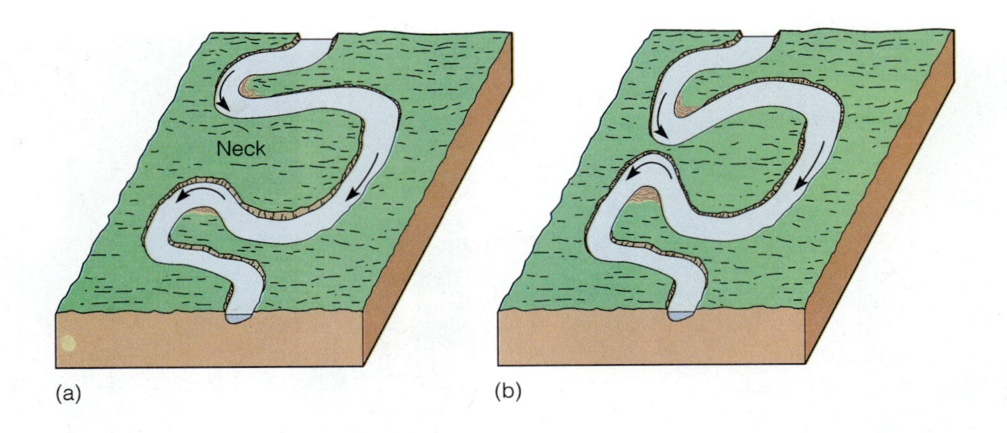

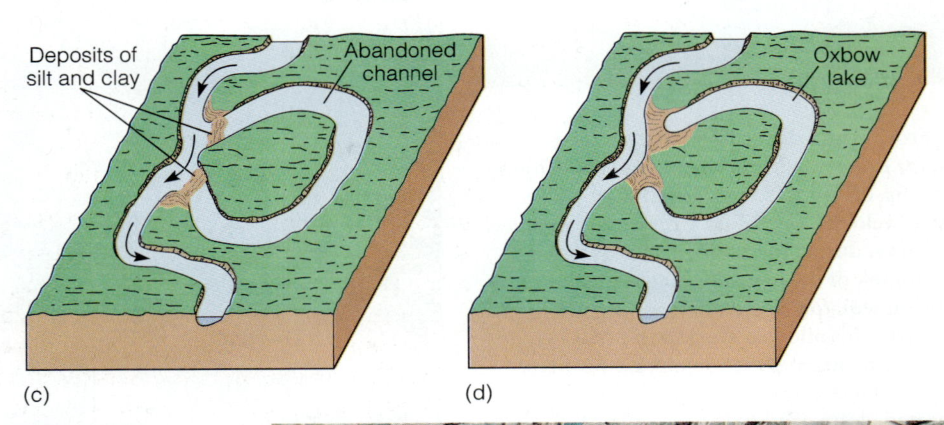

➢ **FIGURE 9.15** (*a–d*) **Meander pattern sequence of erosion leading to a cutoff and an oxbow lake. A cutoff forms in (*c*) and completely isolates the oxbow lake in (*d*). (*e*) Oxbow lake on the Tallahatchie River in Mississippi. Note the old scars from the migration of meanders.**

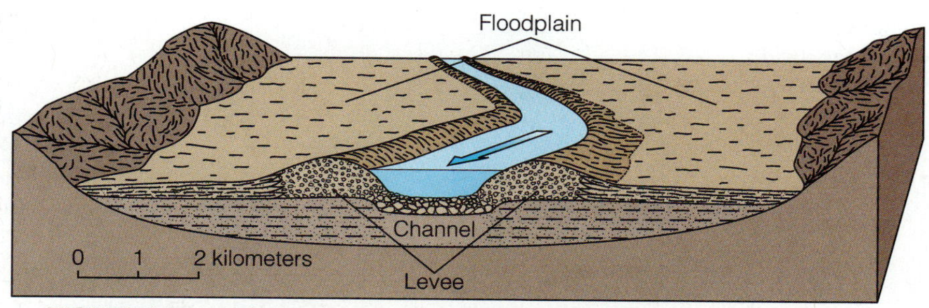

➢ **FIGURE 9.16 Relationship of natural levees and floodplain to a hypothetical stream. The coarsest sediment is deposited in the stream channel, where the stream's velocity is greatest, and finer silt and clay are deposited on levees and the floodplain as velocity diminishes during overflows.**

➢ **FIGURE 9.17 One of the 800 levees that failed or were overtopped along the Mississippi, Missouri, and Illinois rivers during the 1993 floods. This levee ruptured near Davenport, Iowa, on the Mississippi River.**

When There Is Too Much Water

Excess water, whether in rivers, lakes, or along a coast, leads to flooding. The basic hydrologic unit in fluvial (river) systems is the **drainage basin,** the land area that contributes water to a particular stream or stream system. (The map in ➢ Figure 9.18 provides an example.) Individual drainage basins are separated by **drainage divides.** A drainage basin may have only one stream, or it may encompass a large number of streams and all their tributaries. Most of the rain that falls within a particular drainage basin after a dry period infiltrates (seeps) into the uppermost layers of soil and rock; little water runs off the land. As the surface materials become saturated and the rain continues, however, rainfall exceeds infiltration, and runoff begins. Runoff begins as a general sheet flow, then forms tiny rills and gullies, then flows into small tributary streams, and finally is consolidated into a well-defined channel of the main stream in the watershed. Whether or not a flood occurs is determined by several factors: the intensity of the rainfall, the amount of antecedent (prior) rainfall, the amount of snowmelt if any, the topography, and the vegetation.

Upland floods occur in watershed areas of moderate or high topographic relief. They may occur with little or no warning and they are generally of short duration; thus the name *flash floods.* Such floods result in extensive damage due to the sheer energy of the flowing water. Lowland, riverine flooding, on the other hand, occurs when broad floodplains adjacent to the stream channel are inundated, and damage occurs because things get wet or silt-covered. The difference in these types of floods is due to watershed morphology (shape). Mountain streams occupy the bottoms of V-shaped valleys and cover most of the valley floor; the banks are narrow, and there is no adjacent

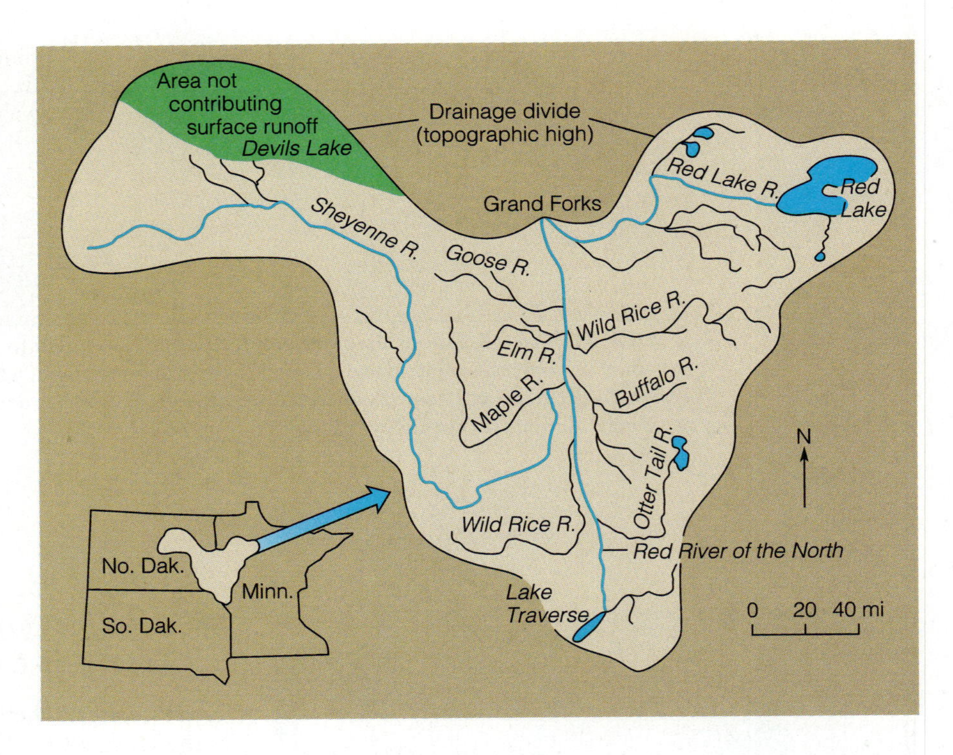

➢ **FIGURE 9.18 Drainage basin of the Red River, showing the drainage divide and the principal tributary streams.**

floodplain. Heavy rainfall at higher elevations in the watershed can produce a wall of fast-moving water that descends through the canyon, destroying structures and endangering lives. Lowland streams, on the other hand, have low gradients, adjacent wide floodplains, and meandering paths.

A **hydrograph** is a graph of discharge, velocity, stage (height), or some other characteristic of a water body over time, and it is the basic tool of hydrologists (➢ Figure 9.19). The "synthetic" hydrograph in ➢ Figure 9.20 graphically illustrates the basic difference between upland (flash) floods of short duration and lowland floods that last longer. The timing of flood crests on the main stream—that is, whether the crests arrive from the tributaries simultaneously or in sequence—in large part determines the severity of lowland flooding.

Floods are normal and, to a degree, predictable natural events. Generally, a stream will overflow its banks every 2 to 3 years, gently inundating a portion of the adjacent floodplain. Over long periods of time, catastrophic high

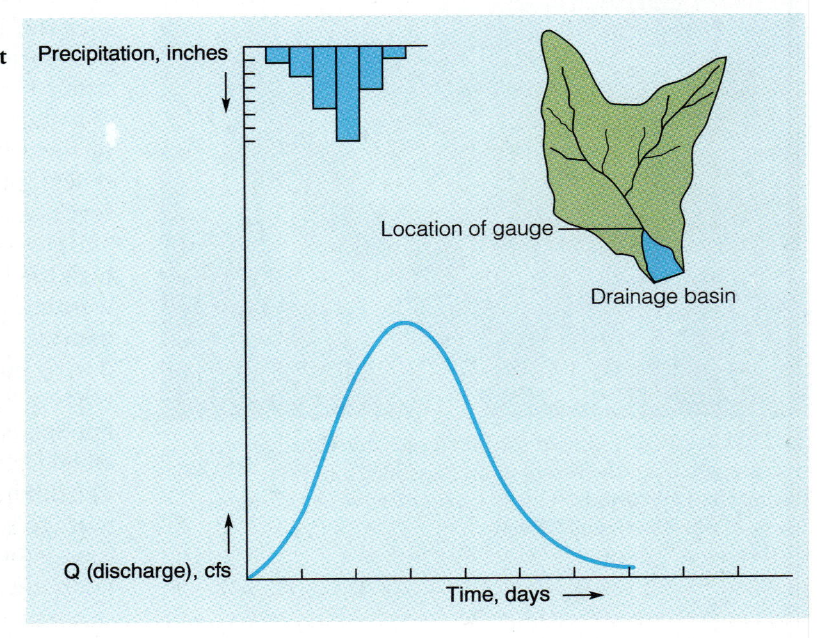

➢ **FIGURE 9.19 A hypothetical hydrograph that relates precipitation and discharge to time. Note that peak discharge occurs after the peak rainfall intensity.**

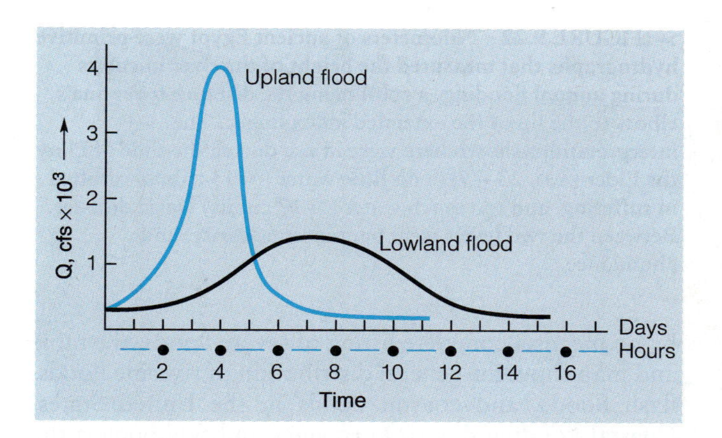

➢ **FIGURE 9.20 Hydrograph of a hypothetical upland flash flood lasting a few hours and of a hypothetical lowland riverine inundation lasting a number of days.**

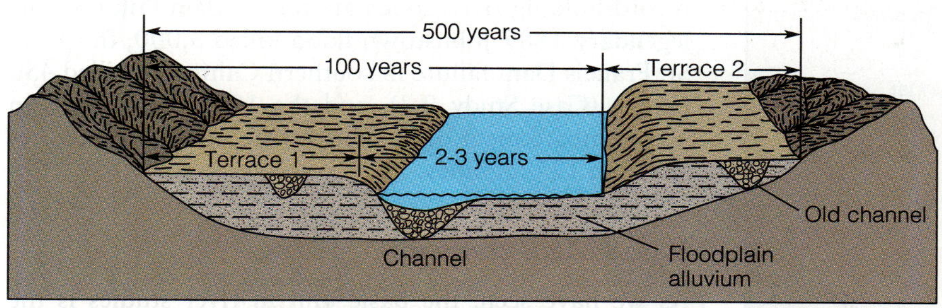

➢ **FIGURE 9.21 The probable extent of flooding over long time intervals on a hypothetical floodplain. The 2–3-year flood inundates the existing floodplain. The 100-year flood inundates both the existing floodplain and a higher one (Terrace 1), which formed when the river stood at a higher level. The 500-year event inundates an even higher terrace and all lower terraces and floodplains.**

waters can be expected. Depending upon their severity, these events are referred to as fifty-, one-hundred-, or five-hundred-year floods. The more infrequent the event, the more widespread and catastrophic the inundation (➢ Figure 9.21). The Midwest flood of 1993 that inundated 421 counties in 9 states is now known as the 500-year flood, with a peak flow of 420,000 cfs (3.1 million gallons per second) at Keokuk, Iowa (see Flood Frequency below).

Geologists, hydrologists, and farmers have long known that flooding can be beneficial as well as harmful, and the annual flooding on the Nile River has long served to illustrate this. These floods once replenished soil moisture and deposited rich silt for productive farming and the good life along its banks (➢ Figure 9.22). The water of these floods traveled hundreds of miles to Egypt from the Ethiopian mountains. As early as 457 B.C., Greek historian and naturalist Herodotus recognized the benefits of this annual flooding and called Egypt "the gift of the Nile."

In 1971 the Aswan High Dam upriver on the Upper Nile in southern Egypt was completed. Its purpose was to increase arable lands, generate electricity, and reduce the flooding on the Lower Nile. Severe environmental consequences followed, including the depletion of soil nutrients beneficial to farming and the proliferation of parasite-bearing snails in irrigation ditches and canals. Schistosomiasis is a human disease found in communities where people bathe or work in irrigation canals containing snails with parasitic worms (schistosomes). These revolting creatures, commonly known as *blood flukes,* invade people's skin when they step on the snails while wading in the canals. The flukes live in their human hosts' blood and release eggs that cause extensive cell damage. Next to malaria, schistosomiasis is the most serious parasitic infection in humans. Outbreaks of it in Egypt are directly connected to dam construction, because there are no more floods to flush the snails from the canals each year. Another negative consequence is that a large shrimp fishery off the Nile delta was ruined, because the dam trapped the nutrients that support plant plankton upon which the shrimp larvae feed. In addition, Lake Nasser behind the dam is experiencing severe siltation.

Floods from all causes are the number-one natural disaster in the world in terms of loss of life. Twenty-two disastrous U.S. floods are listed in Table 9.2, and their locations are shown on the map of ➢ Figure 9.23. Although

CASE STUDY 9.3

It's a Dam Disaster—II

St. Francis Dam is just one of many dams that have been constructed on faults and that later failed or presented problems because of it. This dam was to be the southern terminus of the 225-mile-long Los Angeles Aqueduct system to bring water from the east slope of the Sierra Nevada and the Owens Valley to arid southern California. Much has been written about the motives and the consequences of building the aqueduct; the award-winning movie *Chinatown*, for example, is a thinly disguised portrayal of the project's principals. The dam was completed in 1926, and the 38,000-acre-foot reservoir was full by 5 March 1928 (➢ Figure 1). Just 1 week later it failed, and 450 persons lost their lives.

The dam was built in San Francisquito Canyon, a short distance north of the now-heavily populated San Fernando Valley. The plan was to store aqueduct water here and distribute it by gravity to local storage reservoirs and eventually to users in the San Fernando Valley and the Los Angeles basin. The St. Francis Dam was a massive gravity dam; that is, it held back the water in its reservoir by its sheer weight alone, rather than with the support of an arch to distribute stresses into the foundation rock (➢ Figure 2). The right abutment (foundation) of the dam was in mica schist, which is landslide-prone, and the left abutment was in red sandstone and conglomerate. The contact between the two rock types was an inactive fault zone of crushed rock about five feet thick. Later it was found that the fault zone and sandstone contain considerable gypsum, which is soluble in water. Indeed, these were not desirable foundation conditions for a dam. No geologic examination was performed before the dam was built, and therefore the fault was not identified, but the rocks were tested dry and found to have adequate strength.

Shortly after the reservoir had filled, a leak was noticed at the left (sandstone) abutment. On March 10th damkeeper Tony Harnischfeger reportedly told friends who were arranging a fishing party that "if the dam is still here Wednesday, I will come down and we will go out (fishing)." On March 12th Harnischfeger noticed a new leak at the west (left) abutment and alerted William Mulholland and Harvey Van Norman, the men who were responsible for the dam's construction. They believed the seepage was not serious and had a photographer take their pictures on top of the dam.

Just before midnight that night, water burst through the foundation along the fault contact between the schist and sandstone, and the dam collapsed at this point. The swirling waters undermined the schist at the opposite abutment, carrying away the damkeeper's shelter and leaving only the dam's center section. A wall of water 185 feet high exploded down the canyon, wiping out powerhouses, farms, construction camps, and an Indian trading post owned by silent-movie actor Harry Carey. Huge pieces of the dam, some weighing 10,000 tons, were washed more than a half-mile downstream. Water submerged a construction-camp tent city so quickly that many of the sealed canvas tents acted as

➢ FIGURE 1 The St. Francis Dam in San Francisquito Canyon, March 1928.

air pockets, bringing floors, tents, and men rapidly to the surface, where they rode the wave and luckily survived the flood. A rancher shotgunned a hole in the roof of his house to provide an escape hatch for his family as the waters rose. As it turned out, this was unnecessary. The entire house was lifted and floated downstream. There it was gently deposited without harm to the house or any of the family.

In all, about 450 people drowned in the flood. William Mulholland died shortly afterward and was regarded as the dam's 451st victim. The dam's center section remained standing, like a monolithic gravestone (➢ Figure 3). Eventually it was blasted into rubble so that it would cease being a reminder of the tragedy and a tourist attraction.

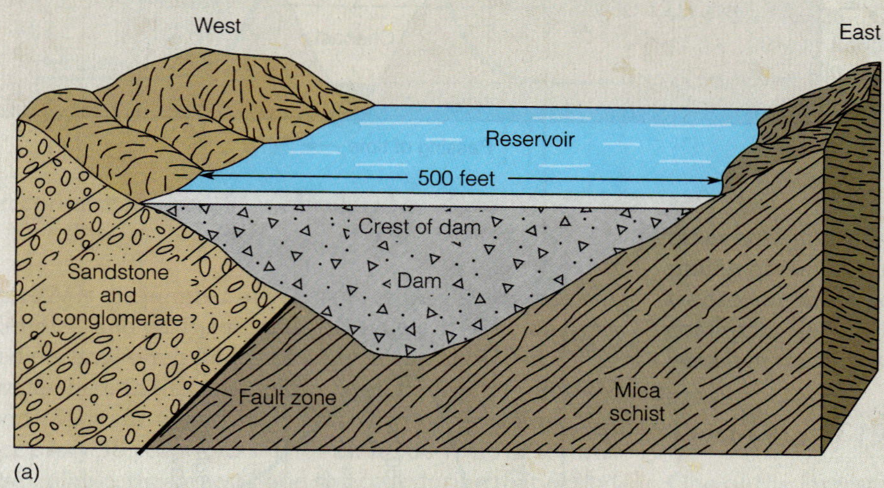

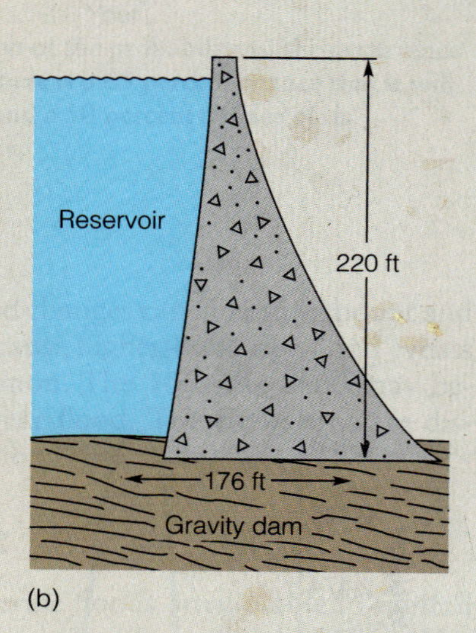

➢ **FIGURE 2** (*a*) **Geologic cross section of the St. Francis Dam through its abutements.** (*b*) **Cross section of a typical gravity dam. The dimensions shown are those of the St. Francis Dam. (Not to scale)**

➢ **FIGURE 3 All that remained of the dam on 13 March 1928.**

➢ **FIGURE 9.29 During the great flood of 1993, workmen sandbagged flood walls and earthen levees along the Mississippi and Missouri Rivers.**

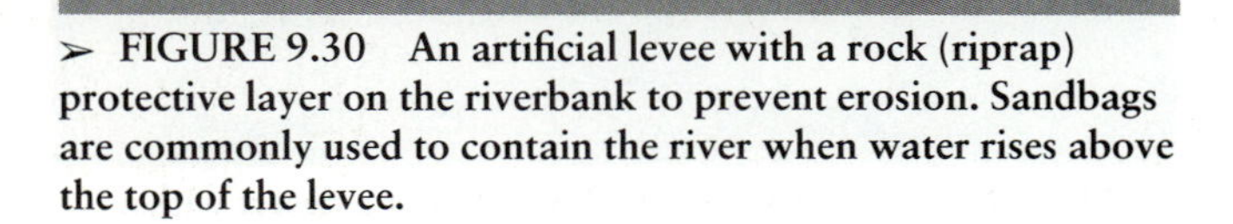

➢ **FIGURE 9.30 An artificial levee with a rock (riprap) protective layer on the riverbank to prevent erosion. Sandbags are commonly used to contain the river when water rises above the top of the levee.**

➢ FIGURE 9.27 Flood-frequency curve for the Red River at Grand Forks, North Dakota. These data are most useful for flood planning and mitigation. Note that the recurrence interval scale

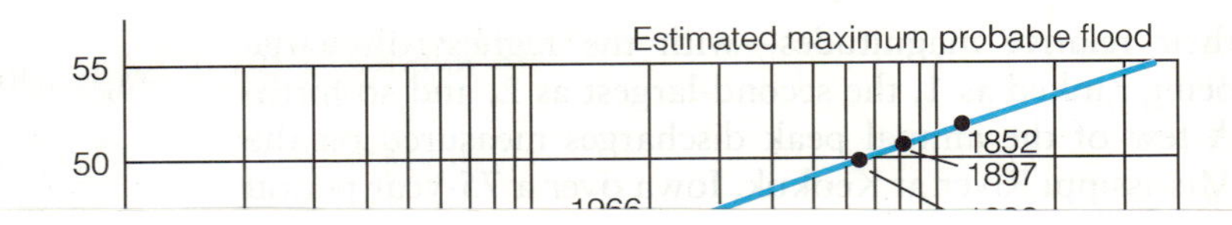

CASE STUDY 9.4

A Leaky River and an Underground Flood

On 13 April 1992 a freight tunnel 15 feet below the Chicago River collapsed, causing flood damage estimated at $1 billion, mostly to underground utility and transportation systems. The flood has come to be called the "Chicago Loop" flood, for the downtown business district in which it occurred. The subsurface facilities of at least a dozen large buildings, including the Chicago Board of Trade, the world's largest commodity market, were flooded.

The cause of the disaster was an error in the installation of a new piling array for the Kinzie Street Bridge, an older structure whose deteriorating foundation was being replaced. The new foundation was to consist of clusters of 37 50-foot-long wooden piles strapped together, each pile driven into the ground individually by a barge-mounted driver. One of these piles cracked the wall of the tunnel beneath the river as it was being driven (see ➢ Figure 1). At first, only a little mud seeped into the tunnel as the riverbed clays and silt plugged the crack. The crack eventually burst open, however, and water from the overlying river gushed into the tunnel. An estimated 250 million gallons of water drained into the tunnel system and basements in the area.

The city of Chicago claims that fault lies with the pile-driving company; the company says it had permission from the city to relocate the pile array three feet south of the planned location, a specification that allowed only one foot of clearance. Both sides agree that permission was given to move the array, but the city maintains the company moved it too far. The city and the construction company are facing legal tests of liability, and years may pass before the cases are settled.

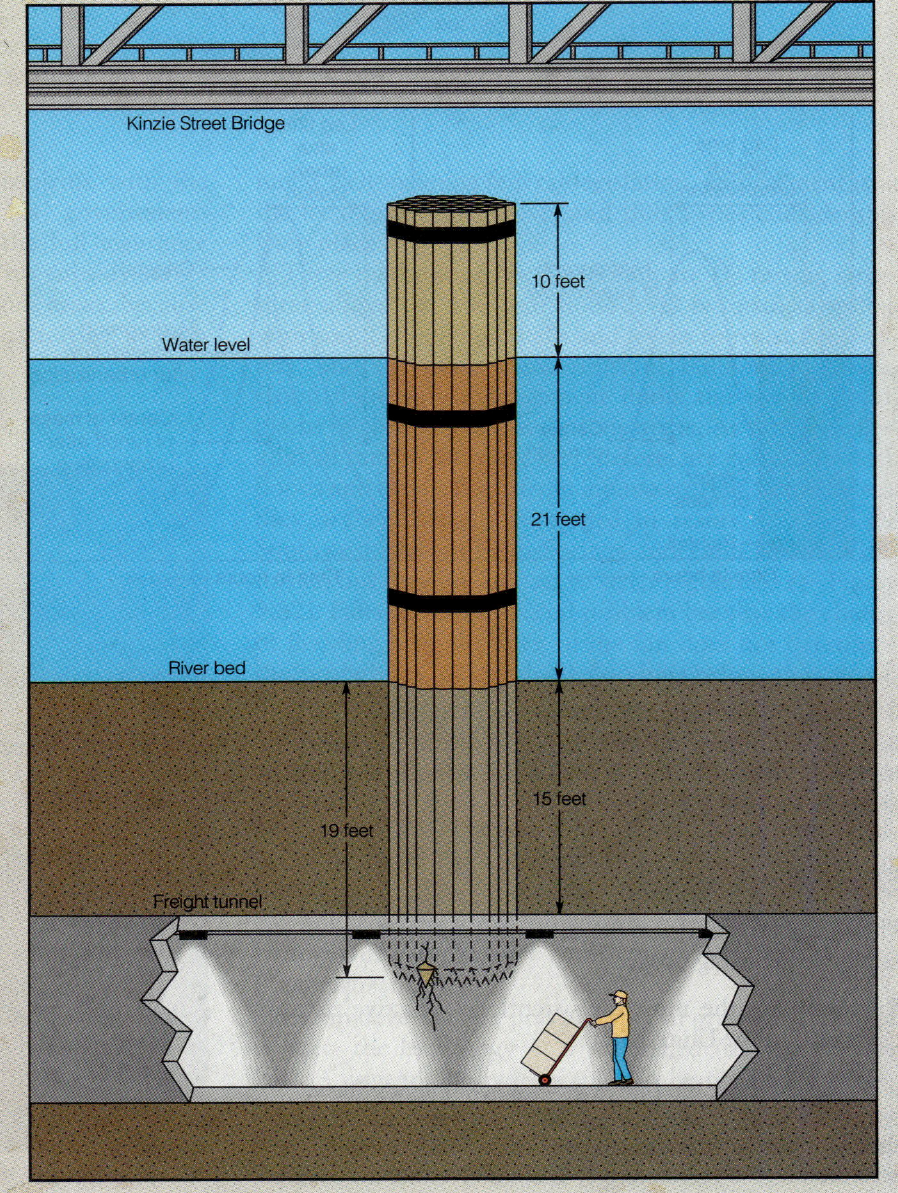

➢ FIGURE 1 The view below Kinzie Street Bridge, Chicago, before the flooding. At least one of the piles on the outside of the cluster penetrated the freight tunnel wall, causing seepage of water and silt into the tunnel. When the wall gave way, a torrent of water from the river above flooded all of the underground structures in the Chicago Loop.

- Flooding cannot be controlled completely, but its damaging effects can be mitigated by restricting floodplain development or by government acquisition of floodplain lands.
- Flood damage costs have been increasing at a rate of 5 percent annually over the past several decades in real dollars.
- Urbanization increases flood peaks on small streams by two to six times.
- Use of the 100-year-flood standard for regulation is effective and should be continued, and the 500-year flood should prevail for such critical facilities as hospitals, water supply, and power plants.

Two towns on the Illinois River, both heavily damaged during the 1993 floods, have totally different flood-planning options. Liverpool, Illinois, a town of 200 residents, has opted to build a levee to protect against future flood events. By 1994 construction will begin on a 1.4-kilometer-long levee at a cost of 2.1 million dollars. Engineers say this should solve Liverpool's flood problem. At Kampsville to the south a levee was not feasible for a variety of reasons and since 1986 the government has been buying homes and businesses on the floodplain. The government spent 1.1 million dollars buying 60 homes, but as many remained and they are now under water. One of the problems, residents say, is that the government doesn't offer enough money to buy land, and government-subsidized flood insurance allows them to rebuild after each flood. Thus, the government pays whether or not they are able to buy floodplain property. However, as noted earlier, if 50 percent of a home is damaged an owner cannot rebuild with flood insurance money. This will probably force the remaining floodplain residents to sell out. From this example one can see the problem with federally subsidized floodplain insurance.

Don Barnett, the mayor of Rapid City, South Dakota, at the time of the disastrous 1972 flood, said "Elected public officials must give the same attention and priority to their drainage problems as they give to their police and fire problems. In the history of Rapid City, perhaps 35 people have died in fires and another 35 have been killed during the commission of crimes, but in just two hours, 238 died in a flood."

GALLERY: RIVER WONDERS

The Grand Canyon and the Falls of the Niagara River are just two of the incredible landscapes fashioned by running water. The Grand Canyon of the Colorado River is truly one of the natural wonders of the world (➢ Figure 9.36). The first sighting of it by Europeans is credited to

➢ FIGURE 9.36 Grand Canyon downstream from Marble Canyon. Flat-lying Paleozoic sedimentary rocks form near-vertical cliffs in this part of the canyon.

the Coronado expedition of 1540, and subsequent Spanish explorers believed the "awesome abyss," as it is sometimes called, was formed by a great earthquake. Established as a national park in 1919, Grand Canyon is visited by more than 2 million people a year. The canyon and adjacent canyonlands are the product of a dry climate, which preserves the beautiful stairstep topography, and a river with a steep gradient and great erosive power (➢ Figure 9.37). The pre-canyon river flowed westerly, and it remained essentially in place, forming the canyon by incessant downward erosion as the plateau was slowly uplifted and tilted southward. This has taken place during the last 5 million years, mostly during late Miocene and Pliocene time, and it explains the west-flowing river on the south-sloping plateau.

The Niagara Falls are of geologic interest because the water falls over a hard rock escarpment, or ledge, that forms a temporary base level for the Niagara River. The river and falls formed when the glaciers covering the area melted about 12,000 years ago, and the escarpment has been retreating ever since. To date, the falls have retreated 12 kilometers (close to 7½ mi). With continued retreat, the river will cut through the hard ledge rock into softer underlying shales. When this occurs the "falls" will fall no more. They will become cascades that will rapidly erode downward to a new base level (➢ Figure 9.38).

➢ FIGURE 9.37 Monument Basin, Canyonlands National Park, Utah. Downcutting by the Colorado River and its tributaries has produced these steep cliffs and isolated "monuments" of sedimentary rock.

➢ FIGURE 9.38 Niagara Falls, on the international boundary between Canada and the United States. Carbon-14 dating indicates the falls came into existence 12,000 years ago and have retreated upstream an average of almost one meter (3.1 ft) per year.

SUMMARY

Water Resources

HYDROLOGIC CYCLE: The continuous flow of water in various states through the environment.

WATER BUDGET: Varies with geologic location but it can be generalized for the 48 contiguous states.

1. *Precipitation* supplies about 75 cm/y
2. *Evapotranspiration* returns 55 cm/y back to the atmosphere by evaporation and transpiration.
3. *Runoff into oceans and infiltration into the ground* account for the remaining 20 cm/y.

USES:

1. Consumptive uses rule out immediate reuse. Agricultural irrigation is consumptive use.
2. Nonconsumptive uses return water to the surface flow via storm drains, sewage treatment plants, and rivers.

SOURCES: Rivers and lakes provide 80% of U.S. off-stream water use. In arid regions where consumption is greater than the local supply, water must be imported or underground water must be "mined" (defined as extraction > recharge).

River Systems

DRAINAGE BASIN: The fundamental geographic unit or tract of land that contributes water to a stream or stream system. Drainage basins are separated by divides.

DISCHARGE: The amount of water flowing in a stream channel depends upon the amount of runoff from the land, which in turn depends upon rainfall or snowmelt, the degree of urbanization, and vegetation.

EROSION: The erosive power of a stream is a function of its velocity; the greater the velocity, the greater the erosion. Stream velocity is determined by discharge, channel shape, and gradient of the stream.

BASE LEVEL: The lowest level to which a stream or stream system can erode. This is usually sea level, but there are also temporary base levels such as lakes, dams, and waterfalls.

GRADED STREAM: A stream that has reached a balance of erosion, transporting capacity, and the amount of material supplied to it. This condition is represented by a concave-upward profile of equilibrium.

Stream Features

ALLUVIUM: Sediment deposited by a stream, either in or outside the channel.

ALLUVIAL FAN: A buildup of alluvial sediment at the foot of a mountain stream in an arid or semiarid region.

DELTA: Forms where a sediment-laden stream flows into standing water. An example is the delta of the Nile River.

FLOODPLAIN: A low area adjacent to a stream that is subject to periodic flooding and sedimentation; the area covered by water during flood stage.

MEANDERS, OXBOW LAKES, AND CUTOFFS: Flowing water will assume a series of S-shaped curves known as meanders. The river may cut through the neck of a tight meander loop and form an oxbow lake.

Flooding and Flood Frequency

FLOODS: Upland floods come on suddenly and move with great energy through narrow valleys. Upland floods are often "flash" floods; the water rises and falls in a matter of a few hours. Riverine floods, on the other hand, inundate broad adjacent floodplains and may take many days, weeks, or even months to complete the flood cycle.

HYDROGRAPH: A graph on which the hydrologist plots measured water level (stage) or discharge over a period of time.

RECURRENCE INTERVAL: The length of time (T) between flood events of a given magnitude. Mathematically, $T = (N+1) \div M$, where N is the number of years of record and M is the rank of the flood magnitude in comparison with other floods in the record. This enables engineers and planners to anticipate how often floods may occur and to what elevation the water may rise.

FLOOD PROBABILITY: The chance that a flood of a particular magnitude will occur within a given year based on historical flood data for the particular location. Using the calculation for T established for the recurrence interval, probability is calculated as $1/T$. Thus the 100-year flood has a 1% chance of occurring in any one year.

Mitigation

STRUCTURES: Dams, retaining basins, artificial levees, and raising structures on artificial fill are common means of flood protection.

FLOOD INSURANCE: The National Flood Insurance Act of 1968 provides insurance in flood-prone areas, provided certain building regulations and restrictions are met.

Urban Development

HYDROLOGY: Urbanization causes floods to peak sooner during a storm, results in greater peak runoff and total runoff, and increases the probability of flooding. Floods, including coastal flooding during hurricanes, are the greatest natural hazard facing humankind.

Key Terms

alluvial fan
alluvium
base level (stream)
channel shape
Continental Divide
cutoff
delta
discharge
drainage basin
drainage divide
evapotranspiration
flood frequency
floodplain
geologic agents
graded stream
gradient
hydrograph
hydrologic cycle
hydrology
meander
natural levee
oxbow lake
point bar
recurrence interval
Regulatory Floodplain
wetted perimeter

Study Questions

1. Distinguish between weathering and erosion.
2. How can we prevent floods or at least minimize the property damage due to flooding?
3. How is the velocity of a stream related to erosive power, discharge, and cross-sectional area?
4. Sketch the hydrologic cycle.
5. What has been the impact of urbanization on flood frequency, peak discharge, and sediment production?
6. What is meant by the "100-year" flood, and how reliable is the statistic?
7. How does the hydrograph record of a flash flood differ from that of a lowland riverine flood? Where do we find flash flooding, and what precautions should be observed in such areas?
8. What is the source of most of the fresh water used in the United States? Explain the difference between consumptive and nonconsumptive use.
9. What has the federal government done to alleviate people's suffering due to flooding? Do you believe this is a good thing, or does it just promote more construction in flood-prone areas? Explain your answer.
10. Artificial levees proved beneficial to some cities and detrimental to others during the great Mississippi River flood of 1993. What are the benefits gained by constructing levees, and how may levees cause damage downstream along a river course?

FURTHER READING

Carrier, Jim. 1991. The Colorado: A river drained dry. *National geographic,* June.

Hays, W. W., ed. 1981. Facing geologic and hydrologic hazards: Earth science considerations. U.S. Geological Survey professional paper 1240-B. Washington, D.C.

Leopold, Luna B. 1968. Hydrology for urban land planning: A guidebook on the hydrologic effects of urban land use. U.S. Geological Survey circular 554.

Leopold, L.; M. G. Wolman; and John P. Miller. 1964. *Fluvial processes in geomorphology.* San Francisco: W. H. Freeman and Co.

McDowell, Jean, and Richard Woodbury. 1991. The Colorado: A fight over liquid gold. *Time,* July 22.

McPhee, John. 1989. Atchafalaya, chapter 1. *The control of nature.* New York: Farrar, Straus & Giroux.

National Geographic. 1993. Water: the power, promise and turmoil of North America's Fresh Water: Special edition, October.

Saarinen, T. F.; V. R. Baker; R. Durrenberger; and T. Maddock, Jr. 1984. *The Tucson, Arizona, flood of October 1983.* Washington, D.C.: National Research Council, National Academy Press.

Waananen, A. O., and others. 1977. Flood-prone areas and land-use planning: Selected examples from the San Francisco Bay region, California. U.S. Geological Survey professional paper 942.

CHAPTER

10

WATER UNDER THE GROUND

Nothing under heaven is softer or more
yielding than water,
but when it attacks things hard and resistant,
there is not one of them can prevail.

THE TAO-TE CHING (800–300 B.C.)

Some of the water in the hydrologic cycle infiltrates underground and becomes **ground water,** one of our most valuable natural resources (see Figure 9.1). In fact, most of the fresh water on earth exists beneath the land's surface (see Table 9.1), and its geologic occurrence is the subject of many misconceptions. For example, it is commonly believed that ground water occurs in large lakes or pools beneath the land. The truth is that almost all ground water occurs in pore spaces and fractures in rocks. Another misconception is that deposits of ground water can be found by people with special skills or powers, people referred to as *water witches* or *dowsers.* They locate underground water using two wires bent into L-shapes, forked branches, or other forked devices called *divining rods.* A dowser locates an underground "tube" or lake of water by walking over the ground, holding the divining rod in a prescribed manner until it jerks downward, pointing to the location of ground water, a mineral deposit, or a lost set of keys. Some dowsers claim they can locate good places to drill water wells from afar, using a map of the land. The divining rod is passed over the map until it jerks downward where presumably a well should be located (X marks the spot). Water dowsers are found

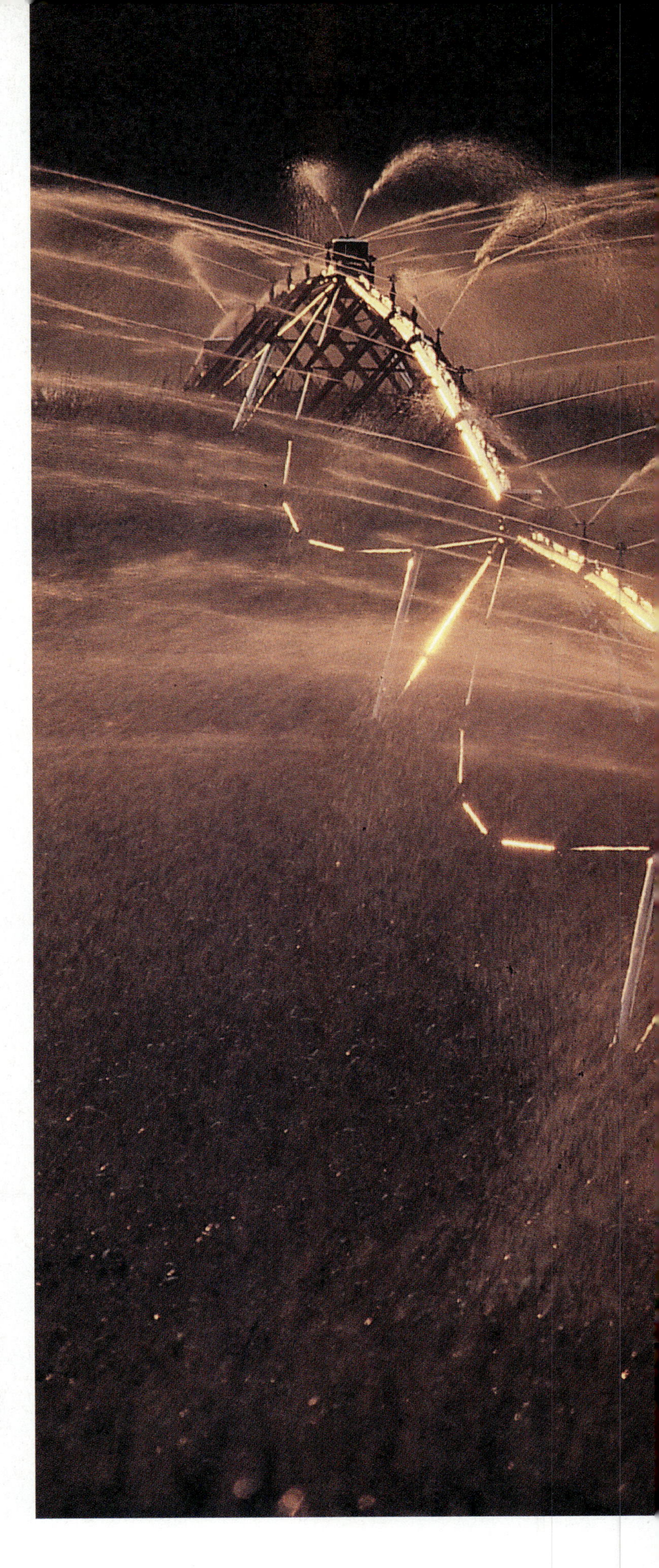

For decades, ground water from the Ogallala Aquifer has irrigated Nebraska cornfields. Seen here is a center-pivot irrigation system that is connected to a well tapping the aquifer that can yield as much as 4,000 liters per minute (about 1,000 gallons per minute), 24 hours a day.

CASE STUDY 10.1

Pores, Flow, and Floating Rocks

Porosity is the ratio of the volume of pore spaces in a rock or other solid to the material's total volume. Expressed as a percentage, it is

$$P = \frac{V_p}{V_t} \times 100$$

where P = porosity
V_p = pore volume
V_t = total volume

If an aquifer has a porosity of 33 percent, a cubic meter of the aquifer is composed of ⅔ cubic meter of solid material and ⅓ cubic meter of pore space, which may or may not be filled with water. Porosities of 20 to 40 percent are not uncommon in clastic sedimentary rocks.

Permeability is a measure of how fast fluids can move through a porous medium and is thus expressed in a unit of velocity, such as meters per day. Knowing the permeability of an aquifer allows us to determine in advance how much water a well will produce and how far apart wells should be spaced. To determine permeability we use **Darcy's Law,** which states that the rate of flow (Q in cubic meters/day) is the product of hydraulic conductivity (K), the **hydraulic gradient,** or the slope of the water table (I in meters per meter), and the cross-sectional area (A in square meters). Thus,

$$\underset{\text{rate of flow}}{Q} = \underset{\text{hydraulic conductivity}}{K} \times \underset{\text{hydraulic gradient}}{I} \times \underset{\text{cross-sectional area}}{A}$$

Hydraulic conductivity, K, is a measure of a rock or sediment's permeability, its water-transmitting characteristics. It is defined as the quantity of water that will flow in a unit of time under a unit hydraulic gradient through a unit of area measured perpendicular to the flow direction. K is expressed in distance of flow with time, usually meters per day.

If we use a unit hydraulic gradient (see ➢ Figure 1), then Darcy's Law reduces to:

$$Q = K \times A$$

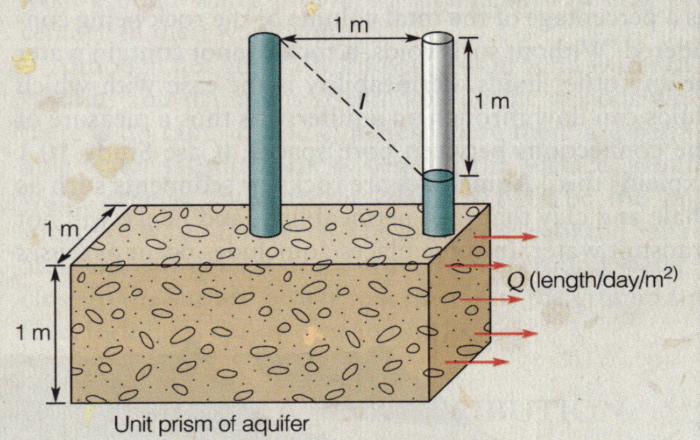

➢ **FIGURE 1 The unit hydraulic gradient (I) and unit prism (1 m^2) of a sandstone aquifer. When the hydraulic conductivity (K) is known, then the discharge (Q) in liters or meters cubed per day per square meter of aquifer can be calculated.**

Unit hydraulic gradient is the slope of 1-unit vertical to 1-unit horizontal (I = 1/1 = 45°), and it allows us to compare the water-transmitting ability of aquifers composed of materials with different permeabilities. Flow velocities may be less than a millimeter per day for clays, 1–100 meters per day for well-sorted sands, and 100–500 meters per day for clean gravels.

Pumice provides an interesting example. It is a porous but almost impermeable rock. It has porosities as high as 90 percent but few connected pore spaces. Pumice will float for a time because it is light-weight and its low permeability prevents immediate saturation. A cubic meter of pumice may weigh as little as 500 pounds, a fifth of the weight of solid rock of the same composition.

Static Water Table

Unconfined aquifers are formations that are exposed to atmospheric pressure changes and that provide water to wells by draining adjacent saturated rock or soil (➢ Figure 10.4). As the water is pumped, a **cone of depression** forms around the well, creating a gradient that causes water to flow toward the well (➢ Figure 10.5). A low-permeability aquifer will produce a steep cone of depression and large lowering of the water table in the well. The opposite is true for an aquifer in highly permeable rock or soil. This lowering, or the difference between the water-table level and the water level in the pumping well, is known as **drawdown**. Stated differently, low permeability produces a large drawdown for a given yield and increases the possibility of a well "running dry." A good domestic (single-family) well is capable of yielding about 11 liters (2.6 gal) per minute, which is equivalent to about 16,000 liters (3,700 gal) per day, although some families use as little as 4 liters per minute (1,350 gal/day). As pointed out, a pumping well in permeable materials creates a small drawdown and a gentle gradient in the cone

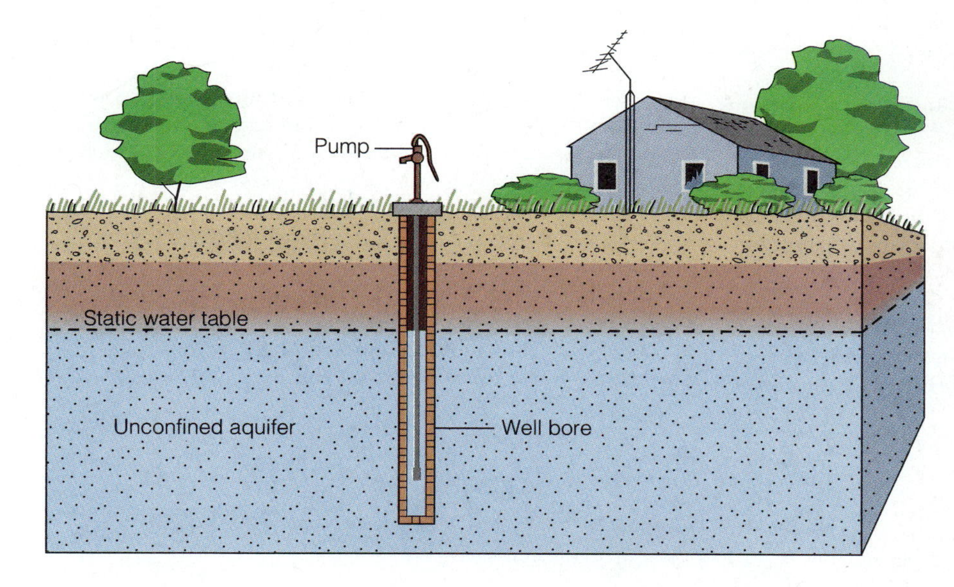

➢ FIGURE 10.4 A horizontal (static) water table penetrated by a well that is not being pumped. The water table remains static until pumping begins.

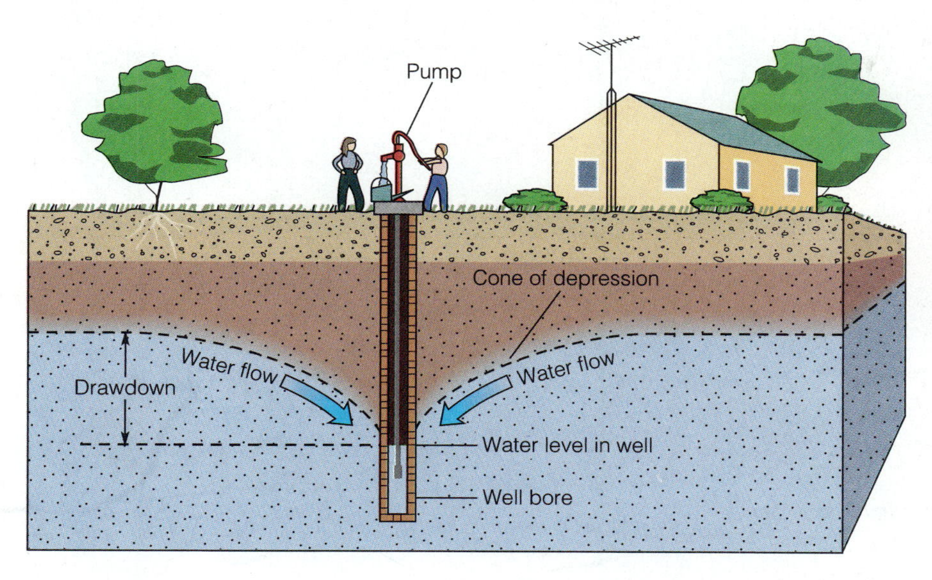

➢ FIGURE 10.5 The water table in a pumping well is drawn down an amount determined by the rate of pumping and the permeability of the aquifer. The development of a cone of depression creates a hydraulic gradient, which causes water to flow toward the well and enables continuous water production. The amount of water the well yields per unit of time pumped depends upon the aquifer's hydrologic properties.

CASE STUDY 10.2

Ground-Water Law

Ownership of ground water is a property right and has traditionally been regulated as such by individual states. Each state applies one or more of four legal doctrines to ground-water rights, and none of the doctrines considers the geological relationship between surface-water infiltration and ground-water flow. The doctrines are

- *Riparian* (Latin *ripa,* river or bank) *doctrine* holds that landowners have absolute rights to the water beneath their properties whenever they choose to exercise this right and with no limitation.
- *Reasonable use doctrine* restricts a landowner's right to ground water to "reasonable use" on the land above that does not deprive neighboring landowners of their rights to "reasonable use." This doctrine results in a more equal distribution of water when there is a shortage.
- *The doctrine of prior appropriation* holds that the earliest water users have the firmest rights; in other words, "first-come, first-served."
- *Correlative rights doctrine* holds that landowners hold shares to the water beneath their collective property that are proportional to their shares of the overlying land.

In the Central Valley of California, for example, surface-water rights are governed by the doctrine of prior appropriation (first-come, first-served), whereas ground-water rights are governed by the doctrine of correlative rights (proportional shares). In the late nineteenth century the Sierra Nevada to the east of the valley and the lower ranges to the west maintained high water tables, which contributed water to the flow of the San Joaquin River (see ➢ Figure 1). By 1966, however, heavy ground-water withdrawals on the west side of the San Joaquin Valley had reversed the regional ground-water flow, so that the river was losing water underground rather than vice versa. The San Joaquin had become an influent stream. Imagine a court-room scenario in which hundreds of people with first-come, first-served rights to *surface* water sue thousands of people holding proportional shares rights to *underground* water for causing their surface water to go underground.

As explained throughout this chapter, it is impossible to treat ground water and surface water as separate entities. Surface water contributes to underground water supplies, and ground water contributes to streamflow and springs.

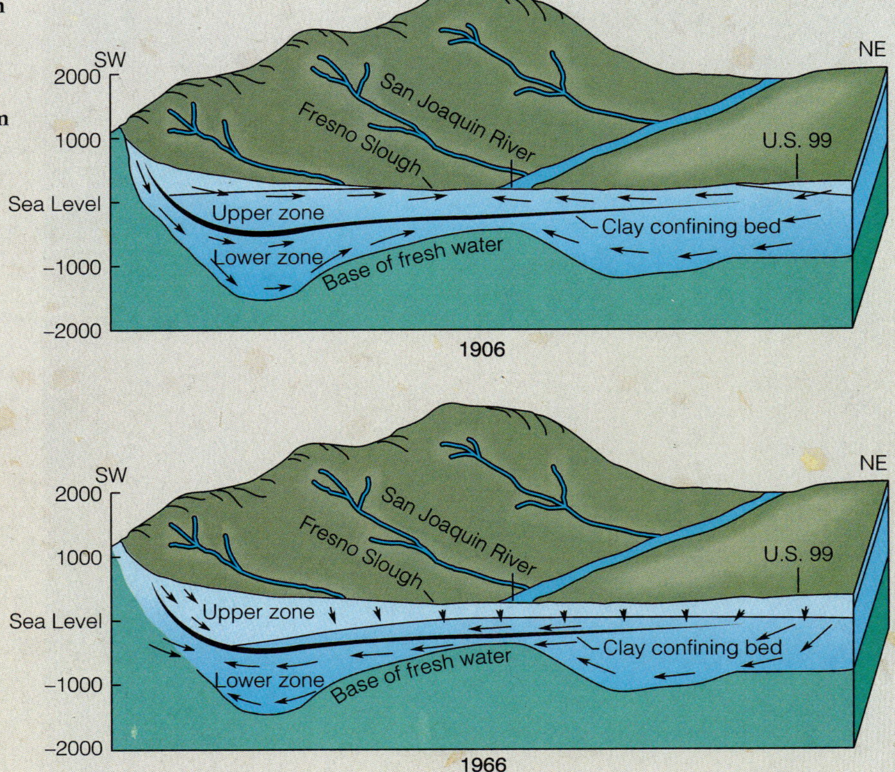

➢ FIGURE 1 Ground-water flow in the central San Joaquin Valley, 1906 and 1966. Note that the San Joaquin River was an effluent (gaining) stream in 1906 and an influent (losing) stream in 1966. The change reflects intensive development and extraction of underground water.

of depression. Some closely spaced wells, however, have overlapping "cones," and excessive pumping in one well lowers the water level in an adjacent one (➢ Figure 10.6). Many lawsuits have arisen for just this reason. Water law is briefly discussed in Case Study 10.2.

The water table rises and falls with consumption and the season of the year. Generally, the water table is lower in late summer and higher during the winter or the wet season. Thus water wells must be drilled deep enough to accommodate seasonal and longer term, climatic fluctuations of the water table (➢ Figure 10.7). If a well "goes dry" during the annual dry season, it definitely needs to be deepened.

Streams that are located above the local water table are called **influent** or "losing" streams, as they contribute to the underground water supply (➢ Figure 10.8a). Beneath influent streams there may be a mound of water above the local ground-water table known as a *recharge mound.* A stream that intersects the water table and which is fed by both surface water and ground water is known as an **effluent** or "gaining" stream (➢ Figure 10.8b). A stream may be both a gaining stream and a losing stream, depending upon the time of year and variations in the elevation of the water table.

Where water tables are high, they may intersect the ground surface and produce a spring. Springs also are subject to water-table fluctuations. They may "turn on and off" depending upon rainfall and the season of the year. ➢ Figure 10.9 illustrates the hydrology of three kinds of springs.

Pressurized Underground Water

Pressurized ground-water systems cause water to rise above aquifer levels and sometimes even to flow to the ground surface. Pressurized systems occur where water-

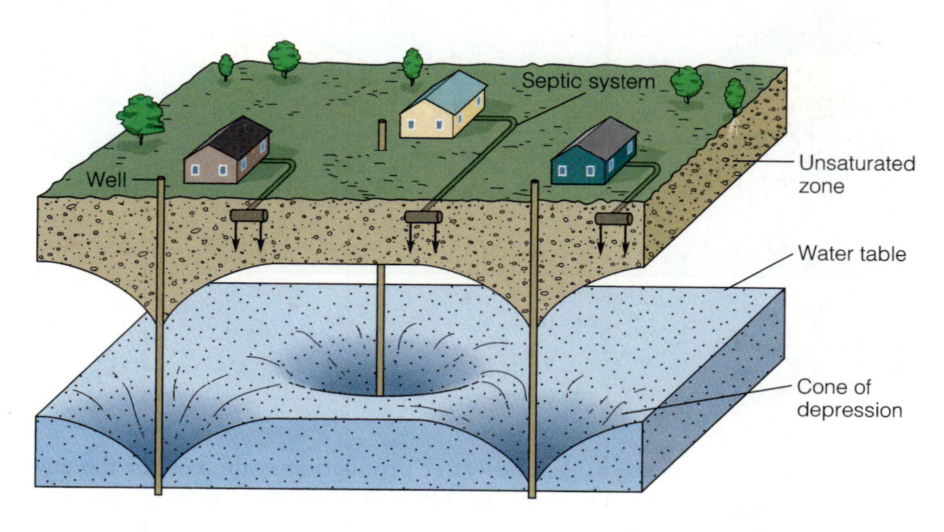

➢ **FIGURE 10.6 Overlapping cones of depression in an area of closely spaced wells. Note the contributions to ground water from septic tanks.**

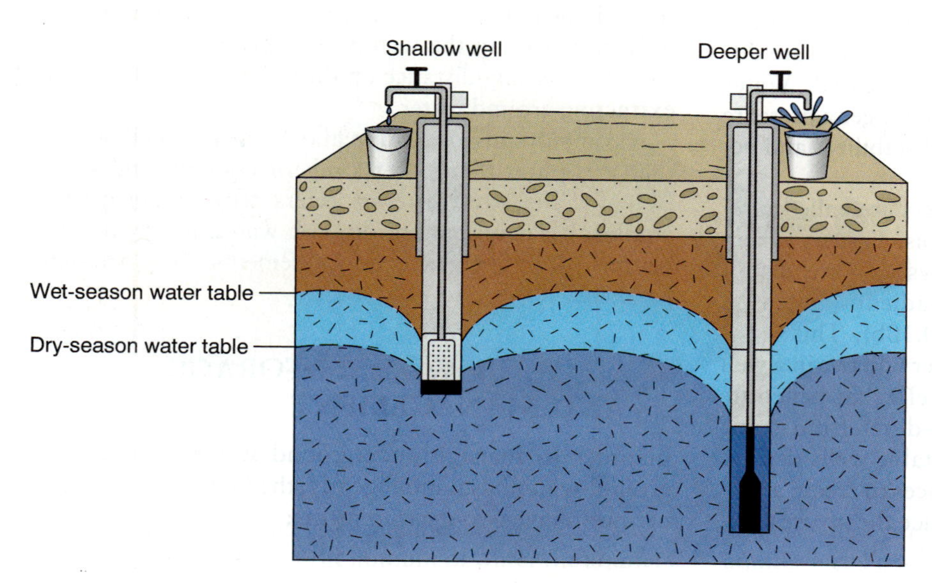

➢ **FIGURE 10.7 The effect of seasonal water-table fluctuations on producing wells. The shallow well runs dry in the summertime. Either it must be deepened, or a deeper well must be drilled.**

A ground-water basin consists of an aquifer or a number of aquifers that have well-defined geological boundaries

gravity flow to the total volume of material, expressed as a percentage. Thus a saturated clay with a porosity of 25

FIGURE 10.8 (a) An influent

CASE STUDY 10.3

Long Island, New York—Saltwater or Fresh Water for the Future?

Long Island is an informative case study of the cause and remedies of saltwater encroachment. By far, the most important source of fresh water on Long Island is its aquifers, which are estimated to hold 10–20 trillion gallons of recoverable water (➢ Figure 1). This ground water infiltrated underground over centuries before development, and the excess was naturally discharged to the sea. Ground water was exploited as the island was urbanized, and the water table in Kings County, the county closest to the mainland, dropped below sea level. By 1936 saltwater had invaded the freshwater aquifer (➢ Figure 2a). Pumping wells were then abandoned or were converted to recharge wells using imported water, and by 1965 the water tables had recovered to acceptable elevations above sea level (➢ Figure 2b).

Meanwhile, in adjoining Queens County, pumping had been increasing and recharge was diminishing due to the construction of sewers and paving of streets. Treated wastewater was thus being carried to the sea, rather than infiltrating underground as it would with cesspools and septic tanks. With lowered water tables, saltwater invaded the aquifers below Queens County (Figure 2), and the State of New York in 1970 asked the U.S. Geological Survey to conduct ground-water studies.

In order to preserve the fresh water below Long Island, it was necessary to determine the *safe yield,* the amount of water that could be withdrawn from Long Island aquifers without causing undesirable results. It was also necessary to achieve a balance between total freshwater outflow and recharge. Several methods of managing the water resource and preventing saltwater encroachment were developed, and they are offered here as examples.

The basic challenge was to offset ground-water withdrawals of 1.7 million cubic meters per day with equal amounts of surface infiltration or injection. A line of injection wells was proposed for replacing the decreased natural recharge. These wells would inject treated sewage effluent or imported water and build a freshwater barrier against saltwater intrusion (➢ Figure 3a). Spreading highly treated sewage effluent and natural runoff into recharge basins could be carried out along with injection into the deeper aquifer. In both the injection-well and the spreading-basin alternatives, the treated sewage is purified to drinking-water standards so that it does not affect local shallow wells. Recharge, like injection, builds a freshwater ridge that keeps saltwater at bay. An unusual but viable method is to allow controlled saltwater intrusion so that a true Ghyben–Herzberg lens of fresh

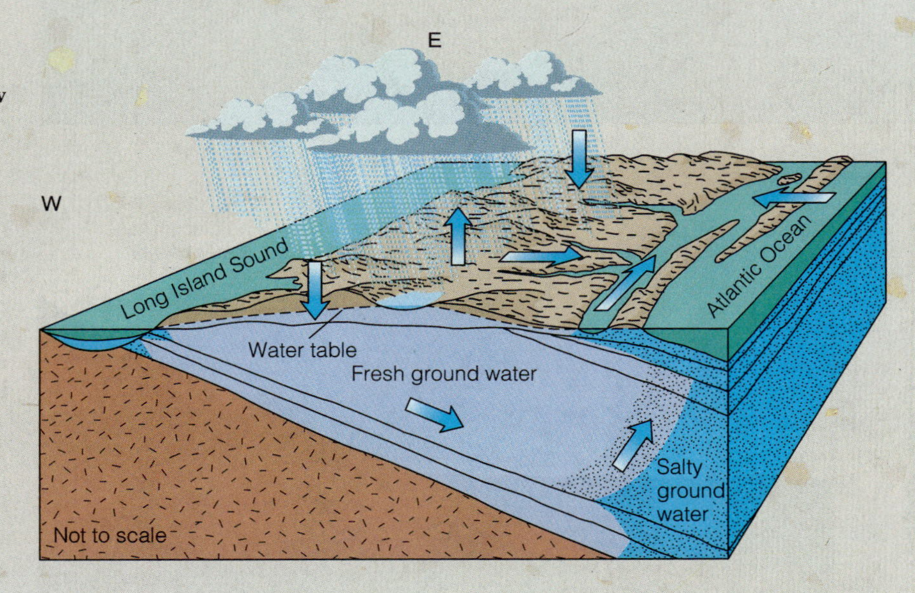

➢ **FIGURE 1 Hydrologic relationship between fresh and salty underground water; Long Island, New York. Arrows indicate direction of water movement in the hydrologic cycle.**

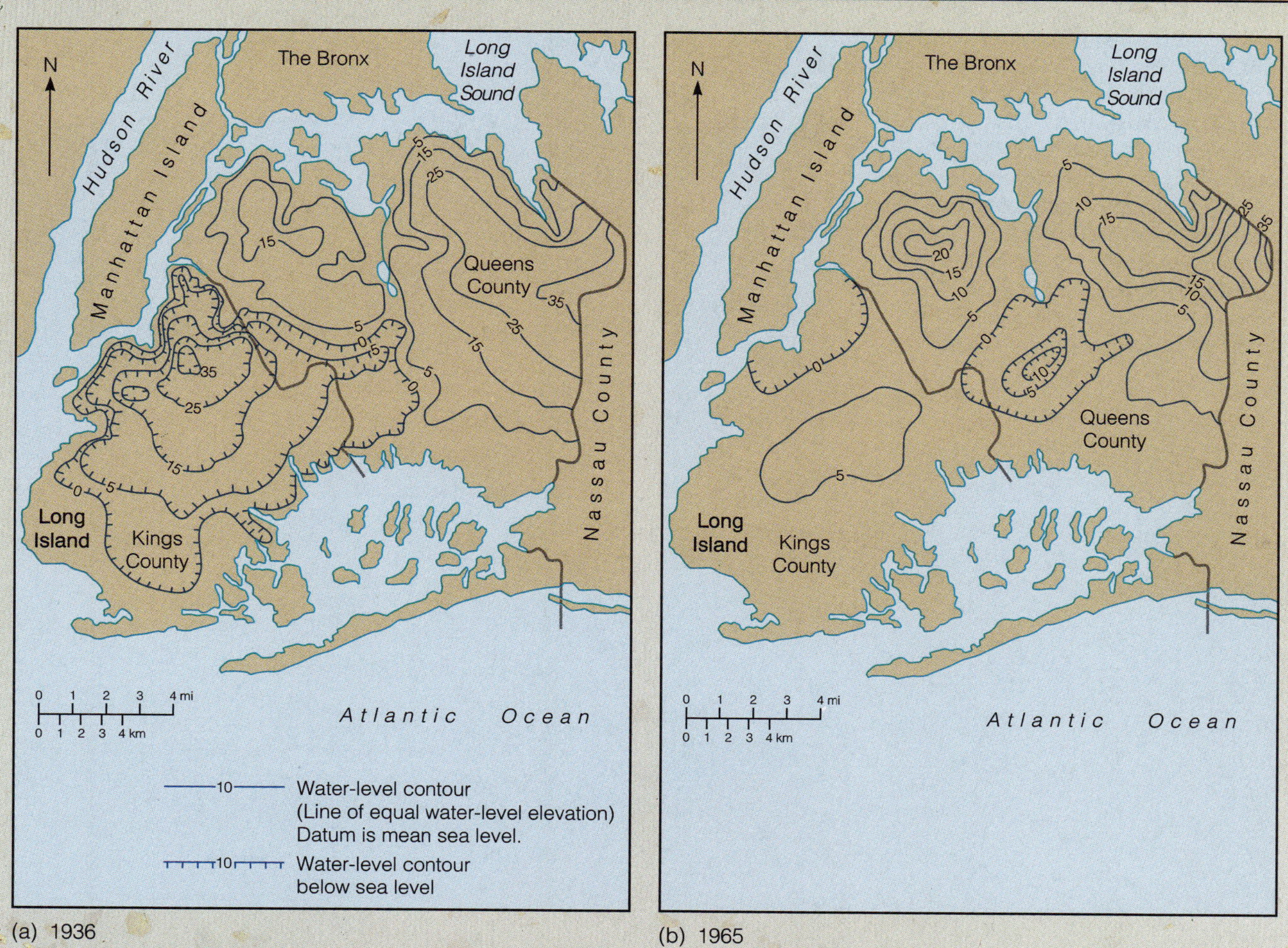

➤ **FIGURE 2 Water-table levels in (*a*) 1936 and (*b*) 1965 for Kings and Queens Counties, Long Island, New York.**

water floating on saltwater develops (➤ Figure 3b). Treated sewage effluent would be discharged into the ocean or injected underground as the need arose. This method has the distinct advantage of salvaging much of the fresh water that otherwise would flow through aquifers into the ocean. In addition, this alternative would increase the yield of the aquifer by several hundred million cubic meters per day, although it would decrease the total volume of fresh water in the reservoir.

Each method of balancing outflow and inflow—injection wells, spreading basins, and controlled saltwater intrusion—has advantages and disadvantages. As is the case with many geological problems, applying a combination of methods yields the most fruitful and cost-effective solution. King and Queens Counties now rely entirely upon importation for their freshwater needs. Public water in Nassau and Suffolk Counties is entirely underground water and only a very small amount of seawater intrusion occurs in the southwest corner

continued

CASE STUDY 10.3 *Continued*

➢ **FIGURE 3 Methods of balancing freshwater outflow and seawater inflow; Long Island, New York. (*a*) The aquifer is recharged with highly treated wastewater (*T*) using injection wells. This method reverses saltwater intrusion and also improves the outlook for long-term water yield. (*b*) Saltwater is allowed to move inland and a lens of fresh water forms, floating on the saltwater. This serves to establish a new equilibrium between fresh water and seawater.**

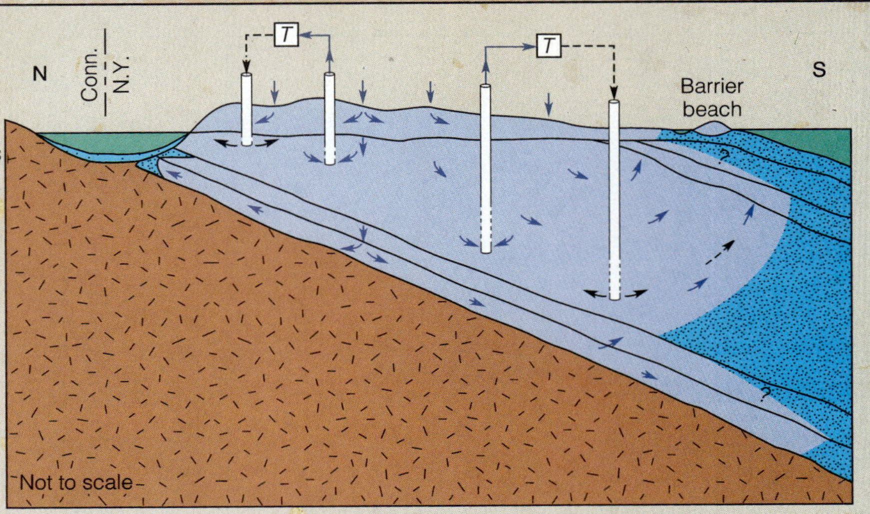

(a)

Treated waste water discharged to the sea
N
Conn.
N.Y.
Barrier beach
S
Not to scale
Permit saltwater intrusion

(b)

of Nassau County. Injection of tertiary treated waste water (see Chapter 15) was tested in the 1980s; however, chemical differences between the injected water, natural fresh water, and saltwater caused the injection wells to clog and require excessive maintenance. Waste-water injection is feasible but at present is too expensive. Currently, there is a network of recharge basins that collect surface water runoff and allow it to percolate underground into the shallow aquifer. The counties require developers to dedicate land for recharge basins within their housing or commercial projects. This method of recharge has proven effective in maintaining the balance between seawater and fresh water.

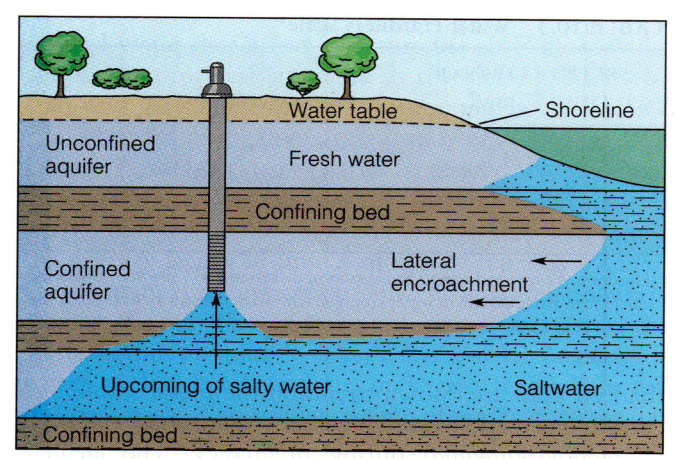

➢ **FIGURE 10.16 Two geologic examples of saltwater encroachment: An unconfined aquifer above and a confined bed at depth. Note the upconing of saltwater from the deepest aquifer, where pressure has been depleted because of the pumping well.**

by good management along with artificial recharge using local or imported water (water spreading), or with injection of imported water into the aquifer through existing wells.

■ WATER QUALITY

Dissolved Substances

Dissolved in natural waters are varying concentrations of all the stable elements known on earth. Whereas some of these elements and compounds, such as arsenic, are poisonous, others are required for sustaining life and health; common table salt is one and it occurs in almost all natural waters. Years ago the U.S. Public Health Service established **potable** (drinkable) **water** standards for public water supplies based on purity, which is defined as allowable concentrations of particular dissolved substances in natural waters. These concentrations are expressed in weight per volume, which is milligrams per liter (mg/L) in the metric system. The standards can also be expressed in a weight-per-weight system. For example, in the English system, one pound of dissolved solid in 999,999 pounds of water would be one part per million (ppm). Clean water standards allow only minuscule amounts of pollutants. For example, 1 ppm or 1 mg/L is equal to about 1 ounce of a dissolved solid in 7,500 gallons of water, or 0.0001 percent. The Public Health Service standard for "sweet" water is less than 500 ppm of *total dissolved solids* (tds), and the standard for "fresh" water is less than 1,000 ppm of total dissolved solids. Table 10.1 lists the Public Health Service classification of water based upon total dissolved solids. Secondary standards, that is, substances that are in low concentrations but affect water taste, odor, and clarity, are also presented.

The quality of ground water can be impaired either by a high total amount of various dissolved salts or by a small amount of a specific toxic element. High salt content may be due to solution of minerals in local geological formations, seawater intrusion into coastal aquifers, or the introduction of industrial wastes into ground water.

TABLE 10.1 Secondary* Drinking Water Standards

WATER CLASSIFICATION		TOTAL DISSOLVED SOLIDS
Sweet		<500 mg/L (ppm); preferable for domestic use
Fresh		<1,000 mg/L (ppm)
Slightly saline		1,000–3,000 mg/L (ppm)
Moderately saline		3,000–10,000 mg/L } unfit for humans and animals
Very saline		10,000–35,000 mg/L } unfit for humans and animals
Brine		>35,000 mg/L } unfit for humans and animals
DISSOLVED SUBSTANCE	**LEVEL (mg/L)**	**EFFECT**
Calcium and magnesium	—	principal cause of water hardness; boiler scale (Ca and Mg salts)
Chloride	250	unpleasant taste; corrosive to pipes
Copper	1.0	metallic taste; staining clothes and containers
Fluoride	2.0	may stain tooth enamel
Sulfate	250	bitter taste; laxative effect
Iron	0.3	unpleasant taste
Zinc	5.0	unpleasant taste
Manganese	0.05	unpleasant taste
Total dissolved solids (tds)	500	possible relationship to cardiovascular disease; corrosive to pipes

* Secondary standards are nonenforceable and apply to substances that occur in concentrations below U.S. Public Health Service standards and affect such water qualities as taste and odor. For example, although it is desirable that drinking water contain no iron, sulfate, or zinc, some is permissible.

CHAPTER

11

OCEANS AND COASTS

Man marks the earth with ruin;
his control stops with the shore.

LORD BYRON, *CHILDE HAROLD'S PILGRIMAGE* (1812)

Although Lord Byron's insight into human degradation of the planet was prophetic, environmental pollution no longer stops at the shoreline, and no relationship is more important to humans than that between the two most common substances on earth, water and air. The exchange of heat from the ocean to the atmosphere feeds energy-hungry hurricanes. The reverse exchange causes evaporation, which puts moisture into the atmosphere and begins the hydrologic cycle. Heat exchanged in either direction between the ocean and the atmosphere creates maritime climates and modifies day-to-day weather along coasts and adjacent inland areas. Much of the intense weather along the U.S. Gulf Coast is the result of moisture-laden air flowing northward from the Gulf of Mexico mixing with cold air flowing southward from the northern states and Canada.

Coastal oceans are important because they are the location of estuaries and beaches. **Estuaries** are semi-enclosed bodies of water where fresh water and saltwater mix, forming brackish waters that are attractive to a host of marine and terrestrial life. A significant estuary is Chesapeake Bay, the site of many important seaports and centers of commercial fishing activities. Beaches are geologically important because they are the last rampart protecting the land from the sea. In addition, they are unexcelled as areas of recreation and relaxation.

"Watch that first step!" might well be said to one exiting these stilt houses that once rested on the beach. Strong waves reduced the beach level almost four meters during a recent storm at Westhampton, New York.

CHAPTER

12

ARID REGIONS, DESERTIFICATION, AND GLACIATION

I would hasten my escape
from the windy storm and tempest.

PSALMS 55:8

In this chapter we examine features and processes that result from the interaction of the solid earth and the atmosphere in two areas of climatic extreme: extreme cold and extreme dryness. The natural processes acting in these environmentally sensitive areas are responsible for a variety of desert, coastal, alpine, polar, and subpolar landforms, together constituting about 40 percent of the earth's land area. As the global population grows, humans are forced to occupy less-desirable lands, some of which are in desert regions and others in subarctic and possibly arctic climates. We will first examine how arid regions and wind can impact people who occupy these regions. Then we will go on to examine glaciers, their landforms, their effects on humankind, and the clues they offer about climate changes in the future.

WINDS AND THE ORIGIN OF DESERTS

Wind is movement of air from a region of high pressure to a region of low pressure due to unequal heating of the earth's surface. Because the sun's rays strike the earth more directly at the equator than at other places on the earth, more radiation (heat) per unit area is received there

McBride Glacier, Glacier National Park, Alaska. The glacier is retreating and a lake has formed between the ice front and the sediment deposited by the glacier at the mouth of the valley. The prominent dark band of sediment along the center of the glacier is known as a medial moraine.

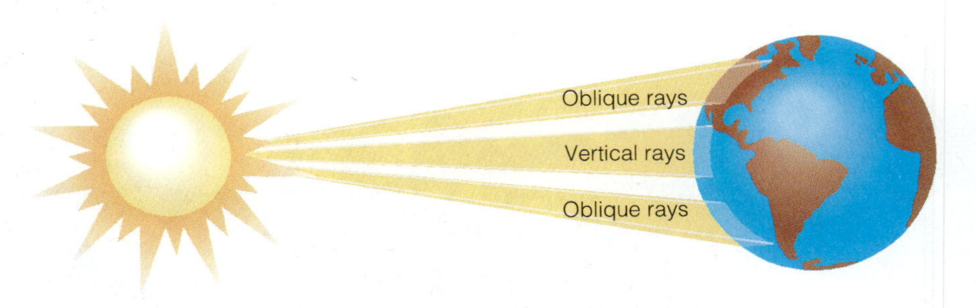

➢ **FIGURE 12.1** The intensity of the sun's heat depends upon the angle at which the sun's rays hit the earth. The intensity and heating is greatest where the rays hit vertically.

than elsewhere (➢ Figure 12.1). Solar heating at the equator warms the air, and it rises, creating a zone of low atmospheric pressure known as the **equatorial low.** In polar regions, the sun rays strike the earth at a low angle, resulting in less-intense heating per unit area. There the colder, drier, denser air creates a zone of high pressure known as the **polar high** (➢ Figure 12.2a). The cold, dry, polar air sinks and moves laterally into areas of lower pressure at lower latitudes. Upper-level air flow from the equatorial regions moves toward the poles. When it reaches approximately 30 degrees north and south of the equator, it sinks back to the earth's surface as dense, dry air to form high-pressure belts known as the **subtropical highs,** or the **horse latitudes.** Winds, then, are driven by differences in air pressure that are caused by unequal heating, but as they move they do not follow straight paths. Instead, they are deflected by the Coriolis effect.

The **Coriolis effect** is caused by the rotation of the earth. A simple analogy of the Coriolis effect is to consider two people playing catch on a merry-go-round (➢ Figure 12.3). If the merry-go-round is stationary, the ball will follow a straight path. An apparent paradox occurs when the merry-go-round is turning, however. As seen by stationary observers not on the merry-go-round, the thrown ball still travels in a straight path, but to the people on the turning merry-go-round, the ball appears to follow a curved path. The reason for this paradox is that the merry-go-round and its passengers are turning below the ball as it travels from one person to the other. To the people on the merry-go-round, the path of the ball appears to be deflected as though some force is acting on it. The curved path will be to the right if the merry-go-round is turning counterclockwise, the same direction the earth rotates if observed from above the North Pole. The deflection of freely moving objects also occurs on the earth. Rifle bullets, ballistic missiles, ocean currents, and

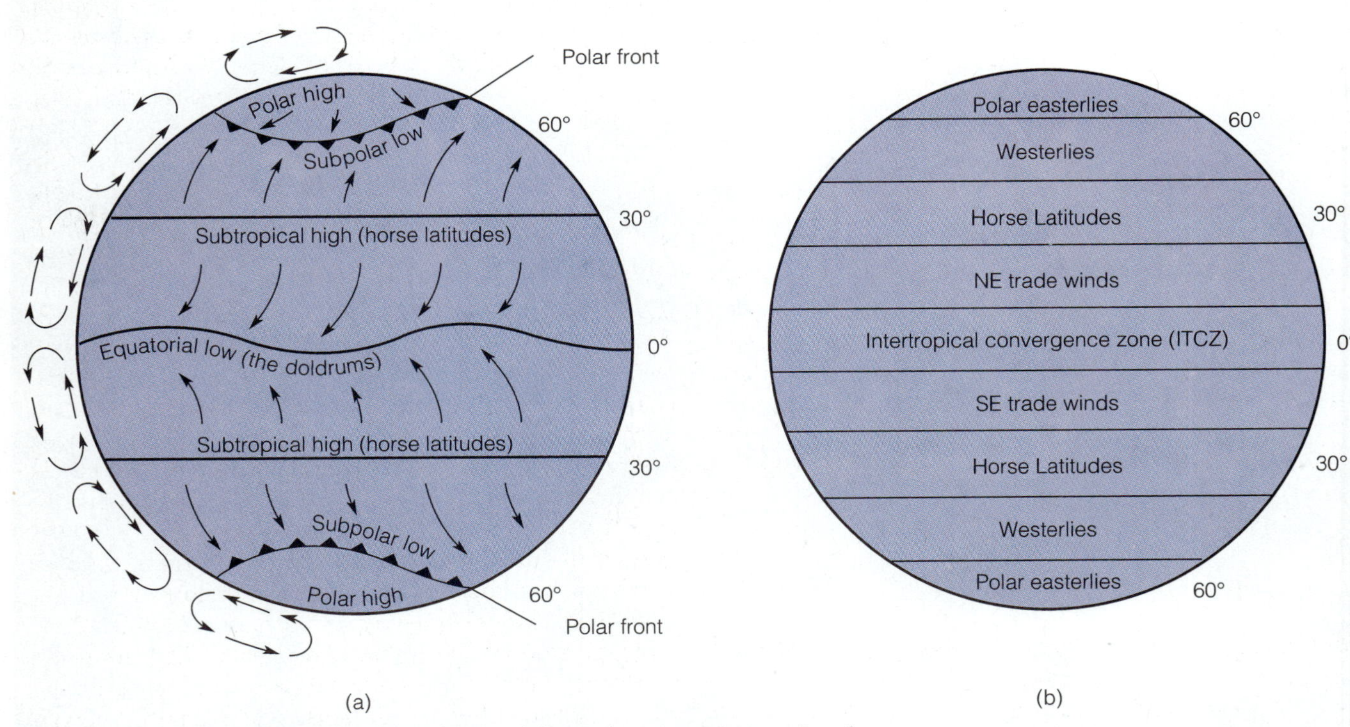

➢ **FIGURE 12.2** (*a*) Generalized circulation of the earth's atmosphere and surface air-pressure distribution. (*b*) Atmospheric pressure zones and surface wind belts.

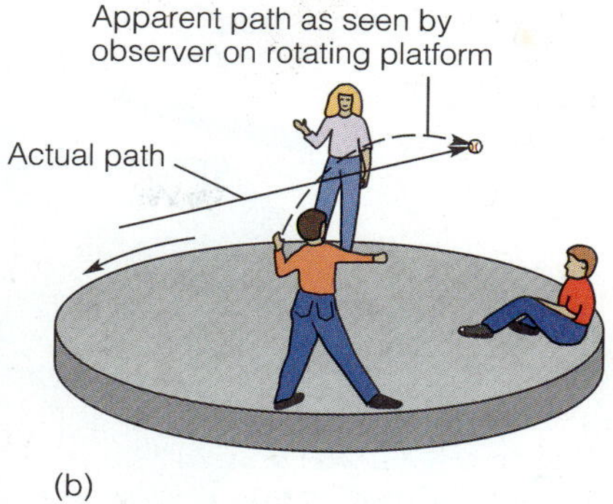

➢ **FIGURE 12.3 The Coriolis effect in a game of catch. (*a*) The path of the ball is straight when the game is played on a stationary platform. (*b*) On a rotating platform the ball still follows a straight path, but to observers on the platform, the path appears to be deflected to the right. This is the Coriolis effect.**

winds that travel long distances would all follow apparent curved paths as seen from a great distance by observers on the earth because of the earth's rotation. The path is deflected to the right in the Northern Hemisphere and to the left in the Southern Hemisphere.

The rising warm air in the equatorial belt expands and cools as it rises. Cooling reduces the air's ability to hold moisture, resulting in the formation of clouds and heavy precipitation. For this reason, the equatorial low-pressure region—called the *intertropical convergence zone (ITCZ)*, and sometimes *the doldrums*—is one of the rainiest parts of the world. We generally do not find deserts there. Air also rises at latitudes of approximately 60° north and south, where cold polar surface air and mild-temperature surface air converge but tend not to mix. At this boundary, called the **polar front,** surface air rises, forming a zone of low pressure called the **subpolar low.** It is along the polar front that rising air causes storms to develop. In the subtropics, in contrast, cool, dry air sinks, forming high-pressure regions. As the air sinks, it is compressed, which causes its temperature to increase. This warming enables the air to absorb water and precludes the formation of clouds. This is the reason that the regions of subtropical highs, from about 30° to 35° north and south of the equator, the so-called *horse latitudes,* experience warm, dry winds with generally clear skies and sunny days. It is here that we find most of the world's hot deserts.

Kinds of Deserts

A **desert** (Latin *desertus,* "deserted, barren") is a region with mean annual precipitation of less than 25 centimeters (10 in), with a potential to evaporate more water than falls as precipitation, and that is so lacking in vegetation as to be incapable of supporting abundant life. Notice that high temperatures are not necessary for a region to be called a desert; only low precipitation, a dry climate, and limited biological productivity. On this basis we can identify four kinds of desert regions: polar, subtropical, mid-latitude, and coastal.

Polar deserts are marked by perpetual snow cover, low precipitation, and intense cold. The most desertlike polar areas are the ice-free dry valleys of Antarctica. The Antarctic continent and the interior of Greenland are true deserts. They are very dry, even though most of the fresh water on earth is right there in the form of ice.

Tropical deserts are the earth's largest dry expanses. They lie equatorward of, the subtropical zones of subsiding high-pressure air masses in the western and central portions of continents. The Saharan, Arabian, Kalahari, and Australian deserts are of this type.

Mid-latitude deserts, at higher latitudes than the subtropical deserts, exist primarily because they are positioned deep within the interior of continents, remote from the tempering influence of oceans due to distance or to a topographic barrier. The Gobi Desert of China, characterized by low rainfall and high temperatures, is an example. Marginal to many middle-latitude deserts are *semiarid grasslands,* extensive treeless grassland regions. Drier than prairies, they are especially sensitive to human intervention. In North America and Asia the mid-latitude and subtropical arid regions merge to form nearly unbroken regions of moisture deficiency. Also in the mid-latitudes are mountain ranges that act as barriers to the passage of moisture-laden winds from the ocean, a situation leading to the formation of **rain-shadow deserts.** As water-laden maritime air is drawn across a continent, it is forced to rise over a mountain barrier, which causes the air to cool and to lose its moisture as rain or snow on the windward side of the mountains. By the time the air descends on the mountains' leeward side, it has lost most of its moisture, and there we find desert conditions

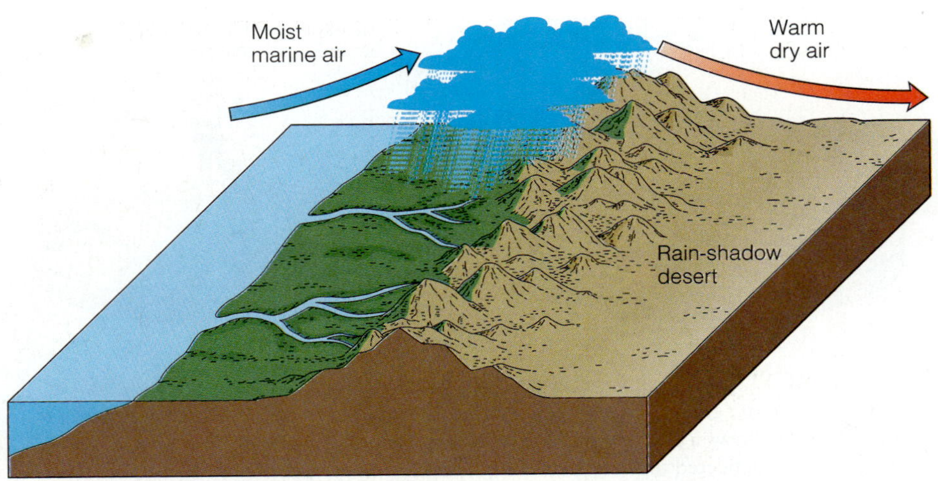

➢ **FIGURE 12.4 The origin of rain-shadow deserts in middle- and high-latitude regions. Moist marine air moving landward is forced upward on the windward side of mountains, where cooling causes clouds to form and precipitation to fall. The drier air descending on the leeward side warms, producing cloudless skies and a rain-shadow desert.**

(➢ Figure 12.4). The Mojave Desert and the Great Basin, leeward of the high Sierra Nevada in California, are examples of rain-shadow deserts. Death Valley, the lowest and driest place in North America, is in the rain shadow of the Sierra Nevada. The Sierra Nevada receives heavy precipitation, and Mount Whitney, the highest point in the range and the highest point in the adjacent 48 states, is a scant 100 kilometers (60 mi) from arid Death Valley.

Another type of subtropical deserts are the **coastal deserts.** These deserts lie on the coastal sides of land or mountain barriers in the horse latitudes and are tempered by cold, upwelling oceanic currents (see Chapter 11, Case Study 11.1). Their seaward margins are perpetually enshrouded in gray coastal fog, but inland, where the air temperatures are higher, the fog evaporates in the warmer dry air. The Atacama Desert in Chile, a coastal desert that may be the driest area on earth, lies in a double rain shadow. The Andes on the east block the flow of easterly winds, and air from the Pacific Ocean on the west is chilled and stabilized by the Humboldt Current, the ocean's most prominent cold-water current.

Collectively, mid-latitude deserts and semiarid grasslands cover about 30 percent of the earth's land area and constitute the most widespread of the earth's geographic realms. It is in these dry regions where the effects of winds are the most prevalent (➢ Figure 12.5).

Deserts and Desertification

Deserts resulting from human intervention may be called *deserts of infertility*. Such deserts are generally variants of middle-latitude deserts, but they are not controlled by climate as much as by human misuse. An understanding of the causes of this class of desert is important, because much of humankind's future in many parts of the world is dependent on the proper and careful use of arid and semiarid lands.

Human activities, grazing of domestic animals, cutting trees and brush for fuel, certain farming practices, and other uses beyond the carrying capacity of the land can so deplete the natural vegetation in the semiarid grasslands bordering arid lands that true deserts can easily expand into these regions (see Case Study 12.1). The expansion of deserts into formerly productive lands because of excessive human use is called **desertification,** and it is occurring at alarming rates in countries that can least afford to lose any productive land. Globally, deserts are estimated to be advancing by about 70,000 square kilometers (27,300 mi^2) per year, an area a bit larger than that of the state of West Virginia, and they are expanding over grazing lands, villages, and agriculturally productive lands. In the past few decades millions of people have died of starvation or have been forced to migrate because of drought in these misused lands where malnutrition is a way of life. All desert regions of the world are expanding (➢ Figure 12.6), but beginning in the 1970s the expansion of the Sahel region of Africa became especially severe.

Natural long-term climate variations cause gradual expansion and contraction of deserts, but recent human overuse of semiarid lands has greatly accelerated the rate of desertification. The clearing of natural vegetation from semiarid lands for cultivating crops and grazing domestic

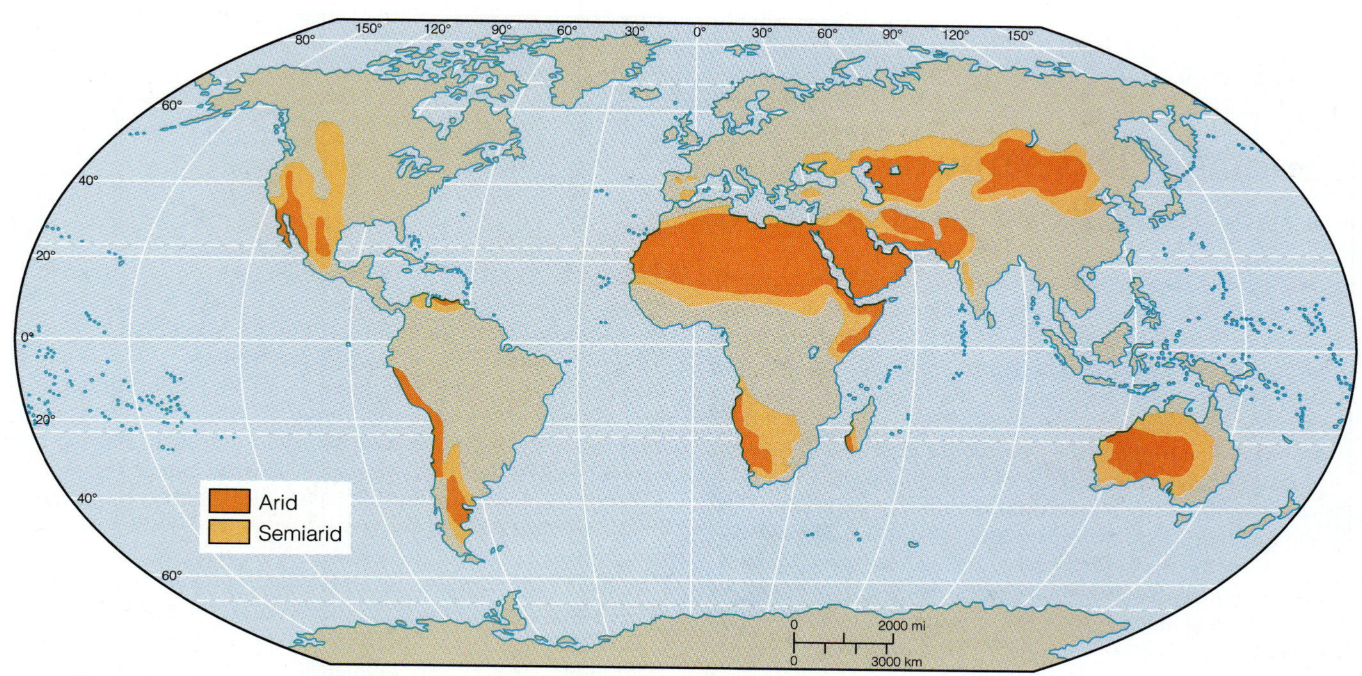

➢ **FIGURE 12.5 Distribution of the earth's arid and semiarid regions.**

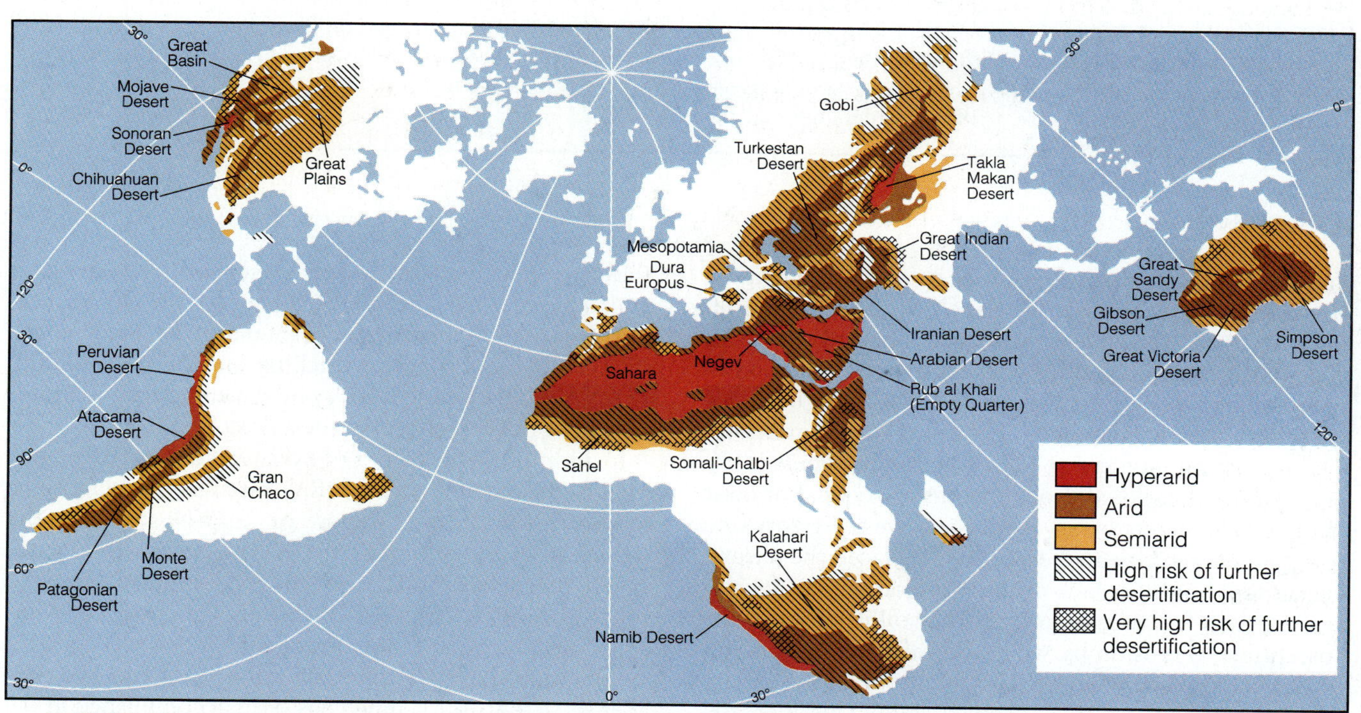

➢ **FIGURE 12.6 Arid and semiarid regions of the world currently experiencing desertification.**

CASE STUDY 12.1

Somalia, Desertification, and Operation Restore Hope

The famine and political problems in Somalia that necessitated the United Nations-sponsored Operation Restore Hope are indeed rooted in desertification due to the interplay of factors explained in this chapter. From a long-term perspective, however, the Somalia crisis is not the result of drought or politics; it is the result of nonsustainable "development" that began in the nineteenth century when Italy and Great Britain imposed colonial economic programs on Somalia's traditional nomadic pastoral (herding) society. These nations redesigned the subsistence culture and developed an economy for the Somalis based on the export of livestock. These changes were expanded after 1960, when a united Somalia gained independence from the Italians and the British. Since then, exports of cattle, goats, and sheep have increased tenfold and exports of camels twentyfold. The increase in livestock production has caused overgrazing, excessive soil losses from erosion, and a general degradation of the Somalian rangelands.

Ethiopian expansionism at the turn of the century also changed the nomadic life of the Somalis. Ethiopia, Italy, and Great Britain established new national borders at that time that split a region of seasonally rich grazing land between Ethiopia and Somalia. These artificial borders ignored the cultural and ecological aspects of the region and restricted the Somalis' traditional use of lush grasslands.

Additional water wells financed by the World Bank were drilled in Somalia after it became independent. This encouraged nomads to settle near the wells, which resulted in overgrazing and increased desertification in these areas. The settled nomads further stressed the environment by excessive wood cutting for fuel. This depleted the already sparse woodlands of Somalia, which in turn depleted soil moisture and increased soil erosion. Furthermore, the settling of the nomads led to population growth, such that the country's population growth rate had increased to about 3 percent a year by the early 1990s. At this rate, the population will double by about the year 2020, well in excess of the country's sustaining capacity (see Chapter 1).

The UN troops in Somalia may well restore order and peace among the clans, but addressing the root causes of the problems—population growth, loss of a grazing economy, and misuse of the land—is more important. Furthermore, the focus of international aid on the cities and agriculture is probably misplaced. The focus should be on the rural sector and the traditional nomadic culture.

animals seemed necessary in order to feed an increasing population in places like the Sahel, the 300- to 1,500-kilometer (180- to 940-mi)-wide belt along the southern edge of the Sahara Desert. But these semiarid grasslands, marginal to true deserts, are part of a delicately balanced ecosystem that is adapted to natural climate variations. The vegetation indigenous to these marginal lands is tolerant of the occasional droughts that are typical of these climates, the plants being able to survive the rigors of a few dry years. Cultivated crops that replaced the natural vegetation, however, lack drought tolerance, and for this reason, crop failure is common. The soil is left bare and susceptible to erosion by wind and rain. Water used for irrigating crops may worsen the situation, because evaporation concentrates the dissolved salts. This increases the salinity of the soil and decreases fertility (see Chapter 6). Salinization of soil is contributing to desertification in the Southwestern United States (especially in the Imperial, Coachella, and southern San Joaquin Valleys of California), North Africa, parts of Asia, and the Middle East. Once salinized, it is difficult to correct the condition, although some success has been achieved in the Imperial and Coachella Valleys. Extensive subsurface systems of drainage pipes have been placed beneath the agricultural lands there. Excessive water is intentionally used in irrigating for the purpose of leaching out the salt buildup. The water is then carried away by the underground drainage network for disposal in the Salton Sea.

In addition to the causes of overuse it is clear that lands used for off-road vehicles (ORVs) are also showing signs of desertification. The report of a 1974 study by the American Association for the Advancement of Science stated that recreational pressure on semiarid lands of southern California by ORVs is "almost completely uncontrolled," and that this area had experienced greater degradation than any other arid region. In the two decades since that study, much has been accomplished in the regulation and control of ORVs, but they still exert severe impacts on areas established, or sacrificed, for their use. A committee of scientists convened by the Geological Society of America in 1977 concluded that "many delicate interdependencies between organisms and their habitats, having been obliterated by ORVs, can never be restored" (➢ Figure 12.7).

➢ FIGURE 12.7 Motorcycle damage to the desert. This view of Jawbone Canyon, California, is a subtle but important example of how human activities contribute to desertification. Because the vehicles compact and alter the surface, they cause reductions in vegetation and permeability that reduce soil moisture content and thus affect local biological systems. Desert rains have eroded the tracks down the slope, transforming them into gullies, which increases runoff and sediment yield.

Reversing desertification certainly is a desirable goal. The simplest program for reversing the trend is to halt exploitation and to adopt measures that will help re-establish natural ecosystems. It might take many centuries for misused semiarid regions to recover fully, but some natural recovery has already occurred in Kuwait as an unexpected consequence of the 1991 Gulf War. Thousands of unexploded land mines and bombs are scattered across the desert in western Kuwait. These have discouraged off-road driving by hunters who formerly entertained themselves by shooting the region's desert animals for recreation. Recent studies show that the native vegetation is becoming re-established, the desert is now resembling a U.S. prairie, and the number of birds has increased immensely since before the war.

Wind as a Geologic Agent

Erosion

Wind erodes by deflation and by abrasion. **Deflation** occurs when things are blown away. This produces deflation basins where loose, fine-grained materials are removed, sometimes down to the water table (➢ Figure 12.8 and Case Study 12.2). *Desert pavement,* an interlocking cobble "mosaic," results where winds have removed fine-grained sediments from the surface (➢ Figure 12.9). **Abrasion** occurs when mineral grains are blown against each other and into other objects. Among the wind-abraded geological features are fluted and grooved bedrock outcrops (➢ Figure 12.10) and **ventifacts,** cobbles or boulders bordering a dune field that have been sandblasted and polished by wind action.

➢ FIGURE 12.8 The Devil's Cornfield, a deflation basin that has been lowered to the capillary fringe at the top of the water table; Death Valley National Monument, California. The rugged plants growing here—arrow weed, which resemble shocks of corn—are able to tolerate heat and saline soil, and their roots can withstand the effects of exposure to episodic sandblasting.

CASE STUDY 12.2

Deflation Basins, Winds, and the North Africa Campaign of World War II

The Qattara Depression in northwestern Egypt is an immense, hot wasteland of salt marshes and drifting sand that covers an area of about 18,000 square kilometers (10,800 mi^2) at an elevation nearly 135 meters (443 ft) below sea level. Its origin is partly due to deflation by wind. Wind scour has lowered the land surface to the capillary fringe at the top of the water table. This has led to evaporation of the saline water and crystallization of the salt as a sparkling white salt crust on the depression floor.

This impressive and forbidding quagmire played an important role during World War II in a decisive Allied victory in 1942. The impassable trough served to block German Field Marshal Erwin Rommel's Afrika Korps' attempt to turn the flank of the British Eighth Army led by General Bernard Montgomery. The geography forced Rommel to attack the Eighth Army's heavily fortified positions on El Alamein, a 64-kilometer (40-mi)-wide strip of passable land between the Mediterranean on the north and the hills to the south that mark the edge of the impassable Qattara (➢ Figure 1). Rommel outran his lines of supply, sustained heavy losses, and was forced into retreat. This pushed him out of Egypt, and eventually westward across Libya.

Throughout the two-year North Africa campaign, the Allied and Axis forces had a common enemy: the *khamsin*, the searing hot winds that blow out of the Sahara. These winds raised temperatures as much as 19°C (35°F) in a few hours and sent sand grains swirling across hundreds of square kilometers. The onslaught of a *khamsin* dictated the actions of men and machines, preempting military tactics and strategies. Airplanes had to be grounded and battlefield action stopped for days as the sun was blotted out and visibility dropped to zero. Soldiers wore tight-fitting goggles to protect their eyes from the clouds of stinging sand and gasped for air through gas masks or makeshift cloth masks. Winds of hurricane strength that tore up camps and turned over vehicles were common. Swirling clouds of sand generated static electricity that rendered the magnetic compasses needed for navigating across the desert useless. Storm-generated static electricity during one *khamsin* detonated and destroyed an ammunition dump.

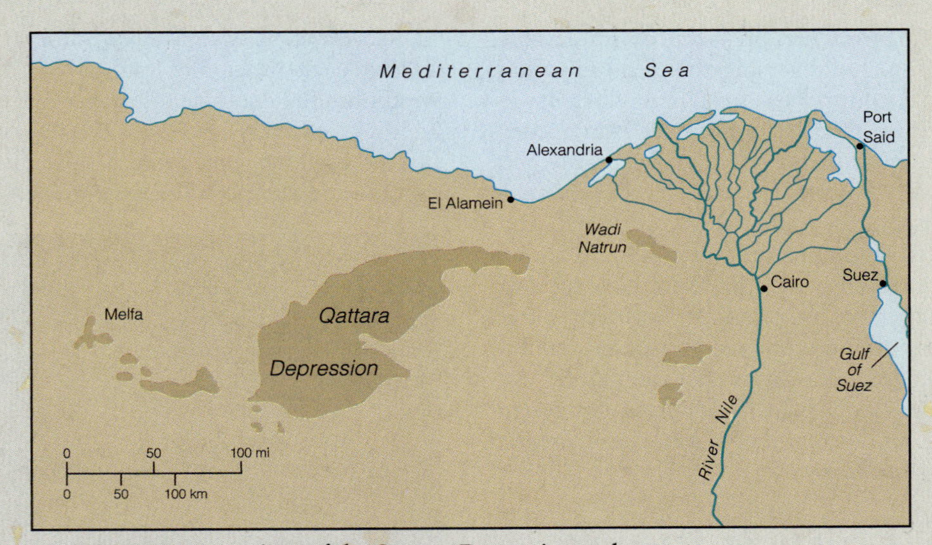

➢ **FIGURE 1 Locations of the Qattara Depression and El Alamein. The decisive battle between the forces of Montgomery and Rommel was fought north of the Qattara Depression.**

➢ FIGURE 12.9 Desert pavement, an interlocking cobble mosaic that remains after winds have removed the surficial fine-grained sediments. The desert floor is left armored with a tightly interlocked mosaic of gravel. An unsorted mixture of coarse- and fine-grained sediments underlies the armor.

➢ FIGURE 12.10 Granitic rocks and a juniper tree near Palm Springs, California, that have been sculptured by intense sand abrasion. The strong prevailing wind moves from left to right relative to this photo. (Width of view in foreground is about 8 meters, or 26 ft.)

Wind transports erosion products by saltation, rolling, and suspension. **Saltation** is the movement of sand grains by impact-caused jumps (➢ Figure 12.11). That is, one grain hits another and causes it to bounce; when that grain hits another grain, it also moves, and so on. The saltation layer is only about 30 centimeters (12 in) high, but bouncing sand grains have great erosive power and can be responsible for considerable destruction within those 30 centimeters. Wind carries fine-grained material, mostly silt- and clay-sized particles, into the atmosphere by **suspension,** creating clouds of dust and even dust storms. Once suspended in the atmosphere, the fine particles may be carried thousands of kilometers from their source (see Chapter 6).

Deposition

LOESS. Rarely is wind able to carry coarse-grained sediment very far. Fine-grained silt and dust, however, may be suspended in the atmosphere for long periods and be carried great distances. Buff-colored, wind-blown silt deposits composed of angular grains of feldspar, quartz, calcite, and mica are called **loess** (see Chapter 6). Loess (pronounced as "loss," or "luss") is relatively unconsolidated and easily eroded, but it is commonly stabilized by vegetation and moisture. Though unconsolidated and relatively uncompacted, the interlocking, angular grains give loess cohesive strength. This cohesion accounts for some unusual properties. Loess is easily excavated, but because of its cohesion, it can stand in remarkably steep cliffs without falling (➢ Figure 12.12). These properties served the Union forces well in Mississippi during the Civil War. Pennsylvania coal miners were utilized to excavate trenches, revetments, and other earthworks prior to the important Battle of Vicksburg. In Shansi Province, northern China, people reside in rather elaborate cave houses dug in loess, which unfortunately are subject to collapse during earthquakes. Loess is also subject to collapse and subsidence due to hydrocompaction (see Chapter 6).

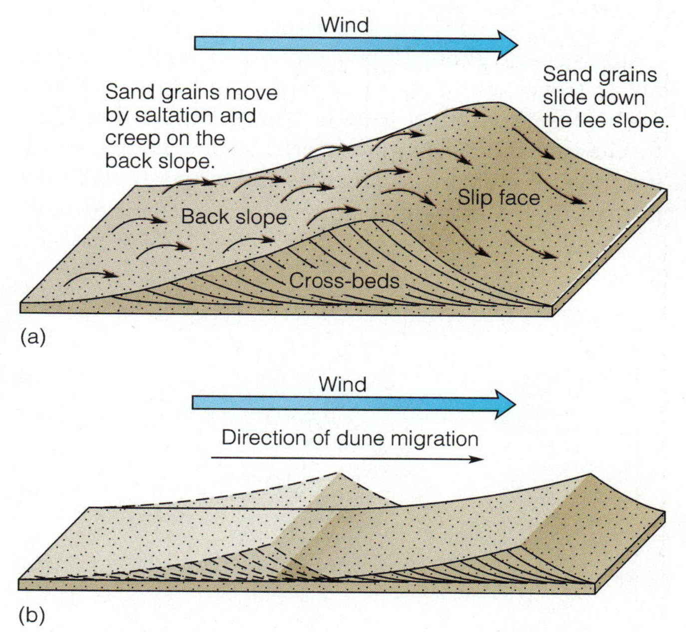

➢ FIGURE 12.11 (*a*) Profile of a sand dune. (*b*) Dune migration occurs by sand ripples migrating up the windward side and grains toppling down the leeward side. The migration produces a series of inclined layers called cross-beds.

➢ **FIGURE 12.12 Remarkably steep roadcut excavated in loess; People's Republic of China.**

Loess covers about 20 percent of the United States and about 10 percent of the earth's total land area (➢ Figure 12.13). The sources of loess in North America are Pleistocene glacial deposits and deserts. Strong Ice Age winds swept loose silt from arid regions and outwash plains deposited by the rivers that drained melting ice and redeposited it as blanket deposits, filling valleys and covering ridge tops. Loess in the United States has produced especially rich agricultural lands in eastern Oregon and Washington and the central Great Plains. Important agricultural soils in the Ukraine are developed on loess. So good are these Ukranian soils that Adolf Hitler had quantities

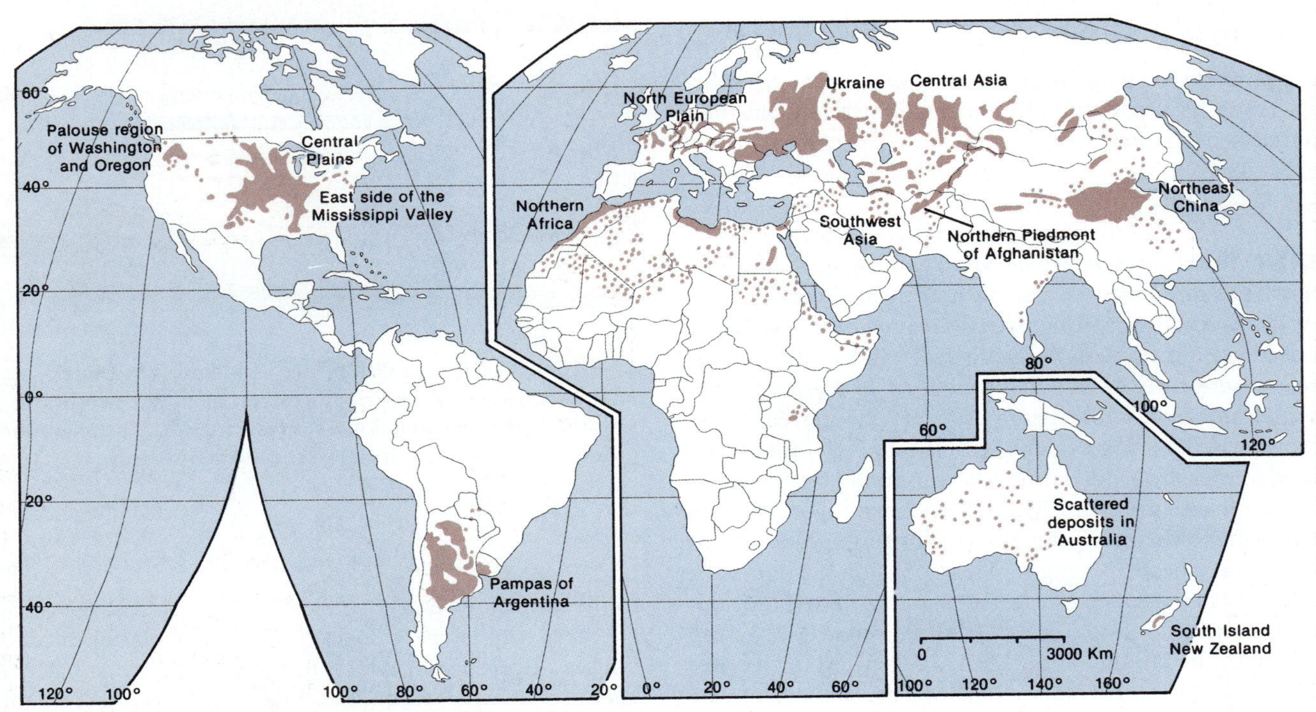

➢ **FIGURE 12.13 Loess-covered regions of the world.**

of it moved to Germany during World War II. Loess in eastern Sudan was carried there by winds from the Sahara Desert.

The most impressive loess deposits are in northeastern China, where they cover an area of 800,000 square kilometers (312,500 mi^2) and in places attain thicknesses of hundreds of meters. The loess of northern China was deposited there by winds from the Gobi Desert. Even today, airborne sediment from the Gobi is noticeable sometimes on windy days in Beijing, hundreds of miles away. Loess is easily eroded, and loess sediment carried by the Huang Ho River in China colors its waters, giving it its English name, the *Yellow River*. Furthermore, the body of water into which the Huang Ho flows is known as the *Yellow Sea*.

SAND DUNES. Hills of wind-blown sand, sand **dunes,** are the most familiar wind deposits. Winds blow sand grains as series of migrating *sand ripples* that move upward along the dune's gentle **back slope** to its crest, where the grains tumble down the steeper **slip face** (Figure 12.11). Successive layers of sand sliding down the slip face create a slanting depositional pattern called *cross-bedding* (see Chapter 2). Sand accumulates on the slip face at the *angle of repose,* which is 34° for dry sand. At any steeper angle, the slope will fail, and a sand slide will cascade down the slip face. Cross-bedding may be preserved in sandstone layers, which allows geologists to interpret ancient wind directions.

Environmental Effects and Mitigation of Blowing Sand

Migrating dunes and drifting sand may adversely affect desert highways and communities all over the world, but the effects of migration are not limited to arid regions. Strong sea breezes in humid coastal regions also cause dune migration (see Chapter 11 and ➢ Figure 12.14). The problems of migrating sand have been mitigated by building fences, by planting windbreaks of wind- and drought-tolerant trees such as tamarisk (➢ Figure 12.15), and by oiling or paving the areas of migrating sand. Dune stabilization in humid coastal areas of Europe, the eastern United States, and the California coast has been accomplished by planting such wind-tolerant plants as beach grass (*Elymus*) and ice plant "Hottentot's fig" (*Carpobrotus*).

Oddities of Wind Erosion and Deposition

Arid regions of the world are a storehouse of intriguing and bizarre features, some examples of which are shown in ➢ Figure 12.16. Among the curiosities produced by wind are teetering rocks that have been nearly severed at the base by wind abrasion and playa rock scrapers, rocks that leave complicated tracks on mud-surfaced playas as they slide in high winds.

➢ FIGURE 12.14 Coastal dune field near Pismo Beach, California. In the 1960s these dunes began migrating leeward onto agricultural lands due to the destruction of stabilizing plant cover by uncontrolled off-road vehicle use in the dune field.

➢ FIGURE 12.15 Windbreaks of salt-tolerant *tamarisk* (salt cedar) keep drifting sand from burying railroad tracks; Palm Springs area, California.

GLACIERS AND THE ENVIRONMENT

The Importance of Glaciers

About 97 percent of the earth's water is in the oceans, and three fourths of the remainder is in **glaciers.** A tenth of the earth's land area is covered by glaciers at present, the same amount that is cultivated globally. During the Pleistocene Epoch, which radioactive age-dating places as beginning 1.6 million years ago and lasting until about 10,000 years ago, as much as thirty percent of the earth was covered by glaciers (➢ Figure 12.17). Today's glaciers store enormous amounts of fresh water, more than

(b)

(a)

(c)

(d)

➢ FIGURE 12.16 (*a*) Mushroom rock; Death Valley National Monument, California. Rigorous wind erosion in the abrasion zone just above the ground surface may produce unusual landforms such as this. (*b*) Remnants of a 1920s plank road, an early attempt to maintain a road in a region of drifting sand; Algodones dune field between El Centro, California, and Yuma, Arizona. (*c*) Playa lake scraper; Racetrack Playa, Death Valley National Monument, California. The boulder's tracks are preserved on the mud-cracked surface. The boulder moved in the winter due to just the right combination of rain-soaked mud and gusty high winds. (*d*) Devil's Golf Course, Death Valley, California, the lowest elevation in the Western Hemisphere. The salt-encrusted surface forms at the capillary fringe on the top of the underlying water table.

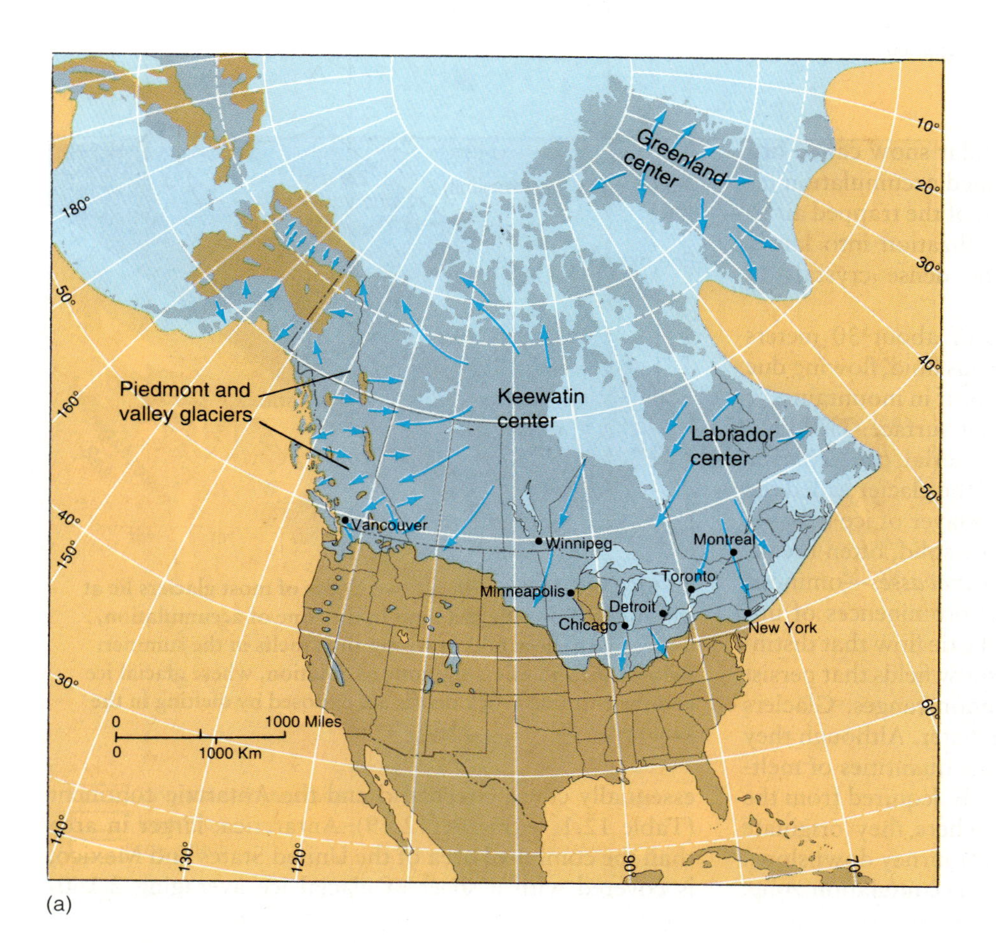

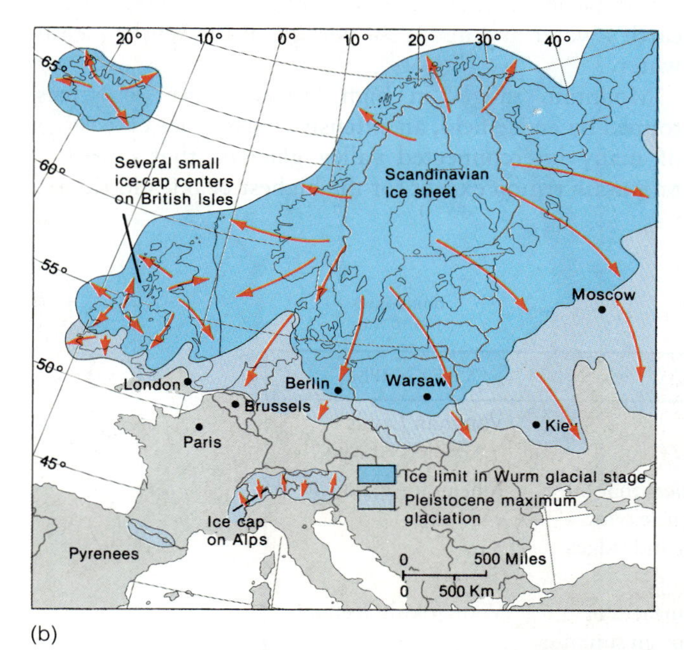

➢ **FIGURE 12.17 Major centers of ice accumulation and the maximum extent of Pleistocene glaciation in (*a*) North America and (*b*) Europe.**

exists in all of the lakes, ponds, reservoirs, rivers, and streams of North America. Farmers in the Midwest grow corn and soybeans in soils of glacial origin, Bunker Hill in Boston is a glacial feature, and the numerous smooth, polished, and grooved bedrock outcroppings in New York City's Central Park were eroded by a glacier that flowed out of Canada (Case Study 12.3 in the next section has more information on this). Glaciers in Norway, Switzerland, Alaska, Washington state, Alberta, and British Columbia exert a vital effect on regional water supplies during dry seasons. They do this by supplying a natural base flow of meltwater to rivers that helps to balance the annual and seasonal variations in precipitation. Meltwater from glaciers serves as the principal source of domestic water in Switzerland. The Arapaho Glacier is an important water source for Boulder, Colorado, so much so that Boulder residents take pride in telling visitors that their water comes from a melting glacier. (The glacier, however, is only one part of a drainage basin of several tributaries that together supply the city's water.)

Origins and Distribution of Glaciers

Glaciers form on land where for a number of years more snow falls in the winter than melts in the summer. Such climatic conditions exist at high latitudes and at high altitudes. The increasing pressure as the thickness of the snow pack increases causes the familiar six-sided snow-

The Juneau Ice Field in Alaska is an example (➢ Figure 12.20). An **ice cap** is a dome-shaped or platelike cover of perennial ice and snow that enshrouds the summit area of a mountain mass so that no peaks emerge through it, spreading under its own weight radially outward in all directions. Continental and ice-cap glaciers may nurture **valley glaciers** (also called *alpine glaciers)*, which are confined on both sides by steep valley walls (➢ Figure 12.21). A valley glacier also may originate in a *cirque,* an armchair-shaped basin at the head of a steep-walled glacial valley. Where a valley glacier emerges from the mouth of a mountain valley onto a plain or a gentle piedmont slope, it may spread laterally to form a **piedmont glacier** (➢ Figure 12.22). When the climate warms and a valley glacier recedes (melts back), a straight, steep-walled U-shaped canyon remains (➢ Figure 12.23).

➢ FIGURE 12.20 Juneau Ice Field, Alaska. Ice fields are extensive mountain glaciers. Only the highest peaks and ridges remain above the ice surface.

➢ FIGURE 12.21 A valley glacier with lateral and medial moraines; Vaughan Lewis Glacier, Alaska.

➢ FIGURE 12.22 The Malaspina, a 2,400 square-kilometer (1,500 square-mile) piedmont glacier formed on the coastal plain of Alaska by the discharge of about two dozen valley glaciers. The glacier is the largest in North America; the contorted pattern of medial moraines is the result of differential flowage from episodic surging of valley glaciers that nurture the Malaspina.

➢ **FIGURE 12.23 A U-shaped glacial valley; Kern Canyon in Sequoia–Kings Canyon National Park, California.**

Should the seaward end of a coastal U-shaped valley be drowned by an arm of the sea, it is called a **fjord** (➢ Figure 12.24).

Glacier Budget

When the rate of glacial advance equals the rate of wasting, the front (terminus) of the glacier remains stationary, and the glacier is said to be in equilibrium. Glaciers **ablate,** or waste away, by melting or, if they terminate in the ocean, by calving. **Calving** is the breaking off of a block of ice from the front of a glacier that produces an iceberg. If the rate of glacier advance exceeds the rate of melting, the terminus advances. Conversely, if the rate of advance is less than the rate of wasting, the terminus retreats. It is important to realize that glacier ice always flows toward its terminal or lateral margins, irrespective of whether the glacier is advancing, is in equilibrium, or is retreating. Glaciers move at varying speeds. Where slopes are gentle, they may creep along at rates of a few centimeters (inches) to a few meters (feet) per day; where slopes are steep, they may move eight to ten meters (24 to 30 ft) a day.

Probably the best-documented and most dramatic retreat of a glacier is found at Glacier Bay National Park, Alaska. When George Vancouver arrived there in 1794, he found Icy Bay, at its entrance, choked with ice and Glacier Bay only a slight dent in the ice-cliffed shoreline. By 1879, when John Muir visited the area, the glacier, by then named the *Grand Pacific,* had retreated nearly 80 kilometers (50 mi) up the bay. The Grand Pacific Glacier had retreated another 24 kilometers (15 mi) by 1916, and

➢ **FIGURE 12.24 A fjord; Gastineau Channel near Juneau, Alaska.**

CASE STUDY 12.3

Continental Glaciers and the New York Transit Authority

A superb example of a major continental end moraine, the Harbor Hill moraine, extends along Long Island's north shore from Orient Point southwestward through Queens and into Brooklyn, where the rolling ground surface in Prospect Park betrays the glacial origin of the park grounds. The last pulse of the advancing Canadian ice sheet reached this point about 20,000 years ago, leaving the moraine as a kind of glacial footprint. South of the Harbor Hill moraine is the older and less obvious Ronkonkoma moraine. Much of the Harbor Hill moraine on Long Island is inaccessible because it is on private property, but from Port Jefferson to Orient Point, the moraine is easily visible, because its edge is hugged by state highways 25, 25A, and 27 (➢ Figure 1). Sagamore Hill National Historic Site, at Oyster Bay, Long Island, the family home of Theodore Roosevelt, twenty-sixth President of the United States, sits on a recessional moraine that is younger than the Harbor Hill. The cliffs along the bay at Garvies Point, next to the Nassau County Museum of Natural History at Glen Cove, expose a one to two m (3 to 7 ft) cover of glacial till overlying Cretaceous sands and clays. The most striking evidence of Pleistocene glaciation are the erratic boulders, large rock fragments on the beach and near the museum, carried here by the glacial ice and deposited with the till (➢ Figure 2). These erratics consist of granites and gneisses, unlike any bedrock that crops out nearby, which illustrates the variety of rocks eroded and carried here from distant areas far north of Long Island. At Prospect Park in Brooklyn, which is easily reached by subway, you may stroll along the undulating ground surface of the moraine across terrain described as **knob-and-kettle topography,** the *kettles* being depressions that mark former locations of chunks of glacial ice that were buried in the sand and gravel of the moraine and left behind as the ice retreated. Eventually the trapped ice blocks melted, and the ground surface collapsed into the rounded depressions. The moraine continues southwest from Prospect Park to Staten Island, where the moraine gravels form Todt Hill, the highest point in the region at 125 meters (410 ft), and then on into New Jersey and Pennsylvania.

When the Harbor Hill moraine extended across the area that is now the entrance to New York Harbor, it dammed the meltwaters coming down the valley of the modern Hudson River, forming a lake. Eventually the lake waters broke through the moraine dam and eroded the channel to the sea that is now called the Verrazano Narrows. Passing near the Verrazano Narrows Bridge on the New York Transit Authority's Staten Island Ferry or driving across the bridge, one might ponder whether the breakthrough of the moraine was a slow process or a catastrophic one some 20,000 years ago.

In addition to Pleistocene glaciation being responsible for the development of the Long Island landscape, it also set the stage for the region around Hempstead to become the early cradle of American aviation. Before urbanization, the moraines and till of Long Island, like much of New York State, were heavily forested. However, the stratified drift of the glacial outwash plain south of the moraines supported a mat of

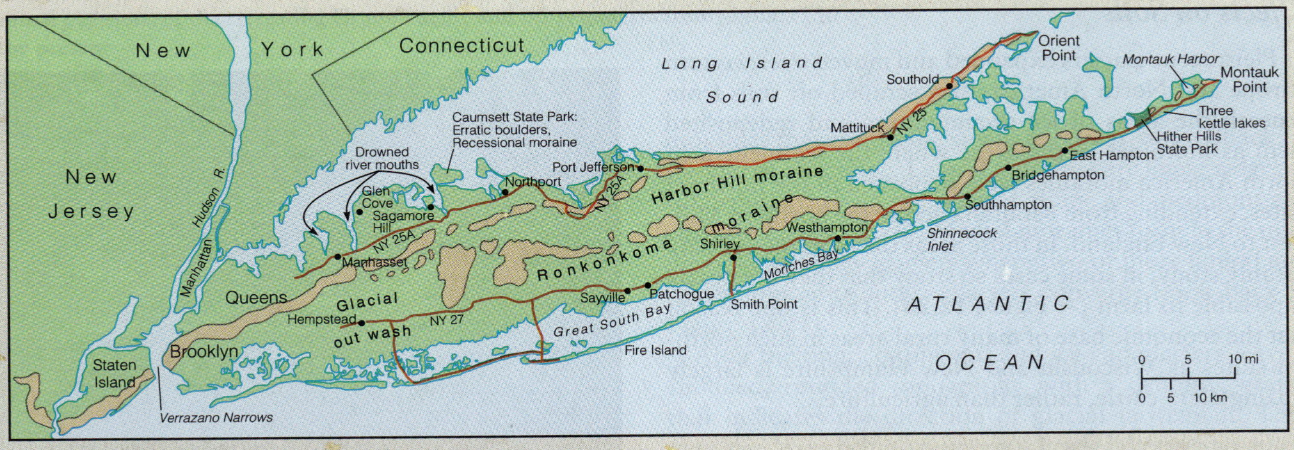

➢ **FIGURE 1 Locations of the Harbor Hill and Ronkonkoma moraines and other features mentioned in the case study.**

grasses so dense that it excluded trees. It was this treeless plain around Hempstead that Glenn Curtiss and other aviation pioneers recognized as the most suitable site for a landing field for their early flying exploits. Thus, the fledgling American aircraft industry got its start here about 1910, and Mitchell and Roosevelt airfields were built soon afterward; it was from Roosevelt Field that Charles Lindbergh began his epic transatlantic flight in 1927.

Central Park in Manhattan, accessible by bus or subway, reveals its share of glacial footprints. There are numerous **roche moutonnée** (French, "rock sheep"), rounded knobs of bedrock so sculptured by a large glacier that their long axes are oriented in the direction of ice movement. Some of these show *glacial grooves* that were rasped out by boulders. An especially good set of grooves and glacial polish is found on a large roche moutonnée in the southwestern part of the park immediately south of Heckscher baseball fields numbers 3 and 4 (➢ Figure 3). The orientation of the grooves shows that the glacier scraped across Manhattan Island from the northwest to the southeast.

➢ **FIGURE 2 Glacial erratic boulders on the beach at Garvies Point, Glen Cove, Long Island, New York.**

(a)

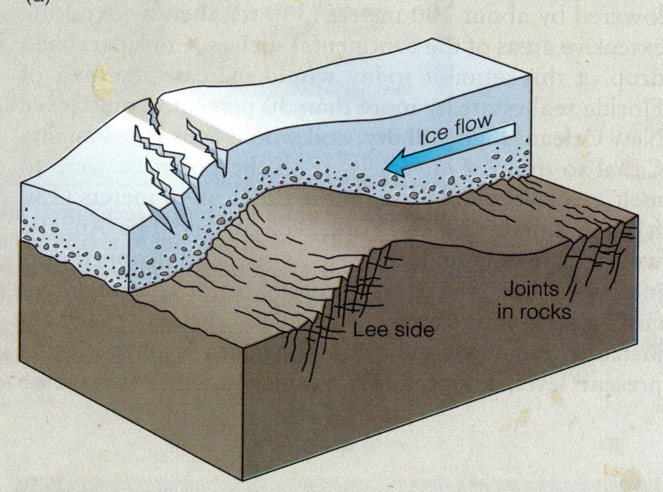

(b)

➢ **FIGURE 3 (*a*) Smoothed roche moutonnée (rock knob) in Central Park, New York City. (*b*) The manner of ice flow and the influence of jointing on the development of a roche moutonnée. The glacier flows upward and over the unjointed part of the rock, smoothing and polishing it to form the gentle side, and then plucks out and removes blocks from the jointed part of the rock to form the steep side.**

CASE STUDY 12.4

Glaciers Have a Mind of Their Own

Most of the world's glaciers advance and retreat synchronously as a result of global climatic changes. Glaciers began advancing worldwide in the thirteenth century during a generally cooler period known as the "Little Ice Age." World temperatures warmed in the late nineteenth century, and glaciers began a general retreat and thinning. These regular, synchronous variations in glacier behavior are virtually restricted to glaciers lying totally on land. **Tidewater glaciers,** those that terminate in an ocean, on the other hand, may exhibit episodes of advance or unusually rapid retreat that are unrelated to climate change. In the last century, for example, several Alaskan tidewater glaciers have retreated at extremely rapid rates. Furthermore, tidewater glaciers located in the same region and experiencing the same climatic conditions, may independently exhibit advance and retreat.

When a tidewater glacier advances, it may do so for a great distance due to the forward movement of a protective submarine end moraine it is bulldozing along at its terminus (➢ Figure 1). Most of the base of the glacier lies in deep water, but it will remain stable (that is, will not calve) as long as the front is grounded on the end moraine. However, if a small area of the terminus retreats off its moraine shoal, ocean water will occupy the space formerly occupied by ice, and calving will begin. Because glaciers thicken in the up-glacier direction, the ocean water at the calving site becomes progressively deeper as calving continues. Once started, the active calving face retreats into increasingly deeper water, which in turn promotes additional calving. A positive-feedback mechanism is thus established that causes irreversible and rapid retreat until the glacier terminus reaches shallow

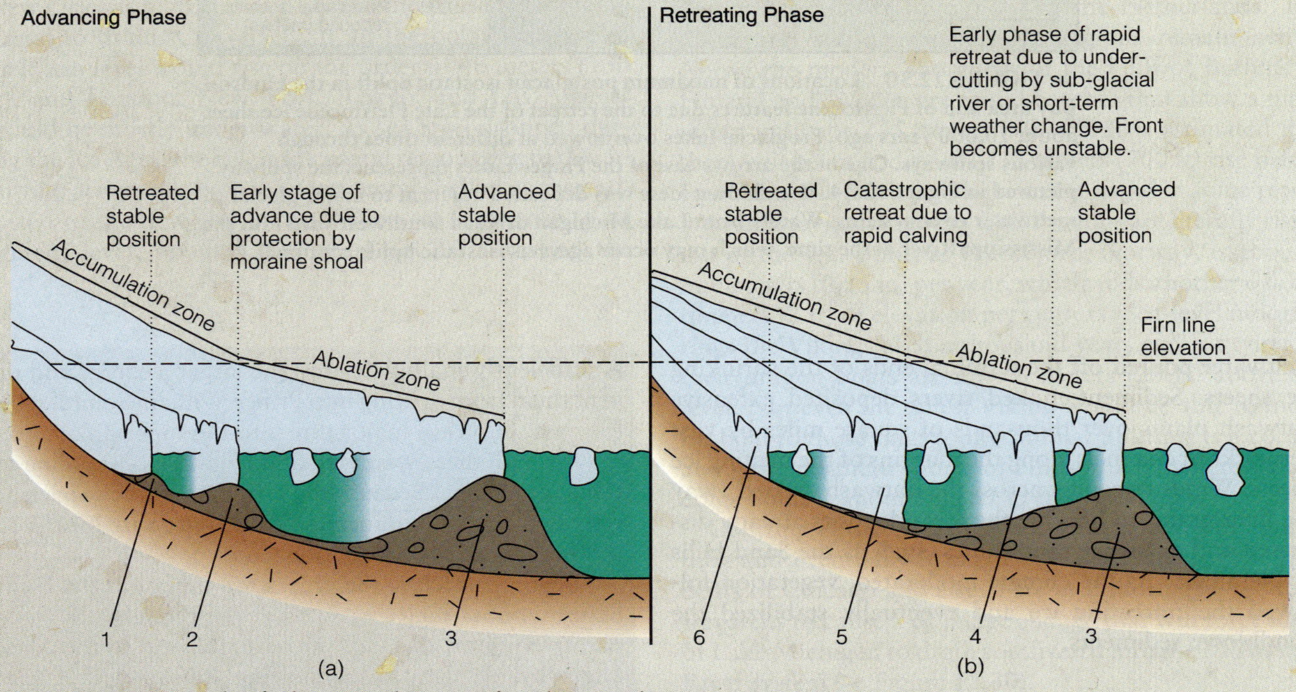

➢ FIGURE 1 Cycle of advance and retreat of a tidewater glacier. (*a*) Points 1, 2, and 3 mark changing positions of the protective moraine shoal as it is shoved seaward at the front of an advancing tidewater glacier. (*b*) Points 4, 5, and 6 mark changing positions during retreat. Such changes may occur without climatic change, with the average elevation of the firn line remaining unchanged throughout the cycle.

water near the head of what has by then become a fjord. Such asynchronous behavior is well-illustrated by the Columbia Glacier near Valdez, Alaska, which is now in rapid retreat, while the Grand Pacific and other nearby glaciers are advancing.

A peculiar episodic or periodic rapid advance of a glacier is observed in a few of the world's glaciers. These glaciers normally flow for decades at steady rates of a few meters per day until suddenly they exhibit flow rates that may exceed 30 meters (100 ft) per day. This is called **surging.** During surging, a large volume of ice from the accumulation area is displaced rapidly down-glacier into the ablation area, where the surface bulges and becomes chaotically crevassed. Surging usually lasts a few months to a year, following which the glacier slows and resumes its normal behavior. Surging appears to be caused by the slow accumulation of a large amount of water beneath the glacier, which exerts a buoyant lifting force on the overlying ice. When lifted enough, the friction between the glacier and the rock is reduced sufficiently that the glacier surges until the film of water is depleted.

About 20 glaciers in Alaska exhibit surging, two of them—the Variegated and Hubbard Glaciers—near Yakutat, Alaska surged in the 1980s. The surge of the Hubbard attracted national attention when its advanced snout blocked a fjord, trapping a number of marine mammals that would have been doomed if the condition had continued for very long. Studies revealed that the Hubbard Glacier began its surge when about 95 percent of its area lay in the accumulation zone. Only about 60 to 70 percent of the total area of a normal, fully advanced glacier flowing in equilibrium is within the accumulation zone.

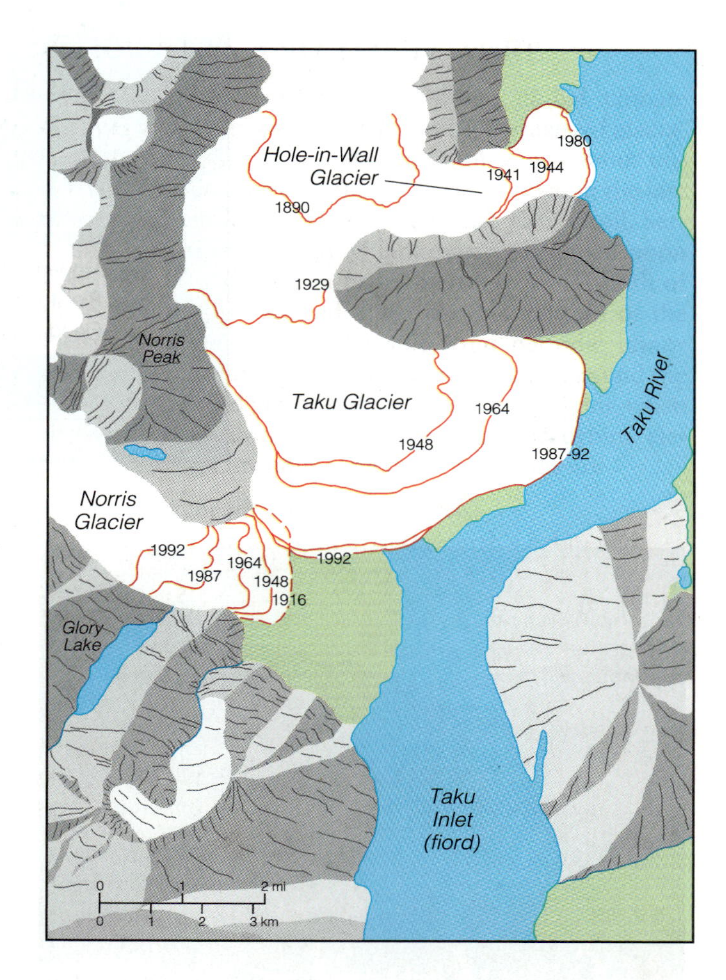

➢ FIGURE 12.31 Map recording the behavior of the retreating Norris Glacier and the advancing Taku Glacier, Alaska, from 1890 to 1990. The reason for the different behavior of the neighboring glaciers is that the Norris Glacier has a much smaller and lower-elevation accumulation zone than the Taku. Only two percent of the Norris' accumulation zone is above 1,370 meters (4,500 ft), while about 40 percent of the Taku's is above that elevation.

flowed in ice tunnels in or beneath a retreating glacier. Hence, especially in Maine, the eskers were called *horsebacks* (Figure 12.27).

Pleistocene Lakes

The cooler climates of the Pleistocene brought increased precipitation, slower evaporation, and the runoff of glacial meltwater. Lakes dotted the basins of North America (➢ Figure 12.32). Accumulations of water, mostly in valleys between the fault-block mountain ranges in the Great Basin, formed hundreds of lakes far from continental glaciers. Today many of these valleys contain aquifers that were probably charged by waters of the Ice Age lakes. California had dozens of lakes, now mostly dried up; among them were Owens Lake, Lake Russell (the ancestor of Mono Lake, ➢ Figure 12.33), and China Lake. Lake Manly covered the floor of what is now Death Valley National Monument, being named after the Manly party that suffered great hardships there in 1849. The Great Salt Lake in Utah is all that remains of the oncevast Pleistocene Lake Bonneville. The Bonneville Salt

possible weather variations that would lead to climate change is so complex that is impossible at this time to develop clear-cut models of the effects of global warming. Regardless of whether the greenhouse effect causes global warming, cooling, or both, the changes could last many thousands of years. Eventually, however, even in a best-case scenario, the variations in the earth's orbital geometry will change, and the earth may be thrust into yet another ice age.

GALLERY: WONDERS OF GLACIATION

Some of the world's most-visited scenic areas are popular because of the spectacular landscapes that remain as superlative legacies of the work of Pleistocene glaciers. ➢ Figure 12.39 is a collage of photographs illustrating features of glaciers and glaciation that may enrich your appreciation of glaciers and glacial landscapes.

(a)

(b)

(c)

(d)

➢ FIGURE 12.39 (*a*) **Glacial polish and grooves on rock outcropping near the Provincial Parliament Building, Victoria, British Columbia, Canada.** (*b*) **Tuckerman's Ravine, a cirque that is well-known as a challenge for expert skiers, on the flanks of Mount Washington, New Hampshire.** (*c*) **A glacial erratic transported by the Pleistocene valley glacier that occupied Lee Vining Canyon, California. This location is more than 25 kilometers (15 mi) from the nearest bedrock outcropping of the particular rock type that could have yielded this erratic boulder.** (*d*) **Dark layers of volcanic ash and dust on a tidewater glacier on Livingston Island off the Antarctic Peninsula preserve a record of volcanic eruptions.**

Summary

Wind

DESCRIPTION: Movement of air due to unequal heating of the earth's surface.

CAUSES:

1. Variations in the heating of the earth at different latitudes because they receive varying amounts of solar heat energy due to the differing angles of the sun's rays.
2. Due to the earth's rotation, the Coriolis effect deflects winds to the right in the Northern Hemisphere and to the left in the Southern Hemisphere.

Air Pressure Differences and Climate Belts

POLAR HIGHS: Polar regions of high pressure; subsiding air produces variable winds and calms.

WESTERLIES: Zones of generally consistent winds lying between 35° and 60° north and south latitudes.

SUBTROPICAL HIGHS: Zones of high pressure and subsiding air masses lying between 30° and 35° north and south latitudes; have variable winds and calms and, commonly, clear and sunny skies. These zones, often called the *horse latitudes,* are the regions in which most of the earth's deserts are located.

TRADE WINDS: Belts of generally consistent easterly winds lying between 50° and 30° north and south latitudes.

EQUATORIAL LOW: Zone of variable winds and calms lying between latitudes 5° north and 5° south.

Deserts

DESCRIPTION: A region where annual precipitation averages less than 25 centimeters (10 in) and that is so lacking in vegetation as to be incapable of supporting abundant life.

CAUSES:

1. High-pressure belts characterized by subsiding and warming air that absorbs water and precludes cloud formation.
2. Isolation from moist maritime air masses by position in the deep continental interior.
3. Windward mountain barrier that blocks the passage of maritime air.

CLASSIFICATION: Deserts may be classed into four categories determined by climate belts and a fifth category due to human misuse.

TYPES:

1. *Polar deserts* are regions of perpetual cold and low precipitation.
2. *Mid-latitude deserts* are found at higher latitudes than the subtropical deserts, primarily because they are within the interior of continents, remote in distance from the influence of oceans.
3. *Subtropical deserts,* the earth's largest realm of arid regions lying in, and on the equatorial side of, the subtropical zones of subsiding high-pressure air masses in the western and central portions of continents.
4. *Coastal deserts* occur on the coastal sides of land or mountain barriers in the subtropical latitudes. Because they are bordered by the ocean, they are cool, humid, and often foggy.
5. *Deserts of infertility* are generally variants of mid-latitude deserts, but they are not controlled by climate as much as from human misuse.

Wind as a Geologic Agent

EROSION AND TRANSPORTATION:

1. *Deflation,* lifting up and blowing things away.
2. *Abrasion,* blowing mineral grains against each other and into other objects.
3. *Suspension,* carrying fine-grained material, mostly silt- and clay-sized particles, into the atmosphere. Once suspended in the atmosphere, the particles may be carried thousands of kilometers from their source.
4. *Saltation,* the movement of sand grains by impact-caused jumps; that is, one grain hits another, causing it to bounce, and when that grain hits another grain, it also moves, and so on. Bouncing sand grains have great erosive power and can cause considerable destruction within 30 centimeters (12 in) of the earth's surface.

DEPOSIT TYPES:

1. *Loess:* Relatively unconsolidated, buff-colored, wind-blown silt deposits composed of angular grains of feldspar, quartz, calcite, and mica.
2. *Sand dunes:* Hills of wind-blown sand that migrate as a series of *sand ripples* moving upward on the dune's gentle back slope to the crest, from where the grains tumble or slide down the steeper slip face.

CONTROL OF MIGRATING SAND DUNES: Accomplished with sand fences, paving, windbreaks, and stabilizing plants.

Glaciers

DESCRIPTION: Glaciers are large masses of ice that form on land where, for a number of years, more snow falls in winter than melts in summer, and which deform and flow due to their own weight because of the force of gravity.

LOCATION: Glaciers form at high latitudes and at high altitudes.

CLASSIFICATION: The major types are continental, ice field, ice cap, valley or alpine, and piedmont.

Important Glacial Features

MORAINES: Landforms composed of till and named for their site of deposition. The principal types are terminal, recessional, lateral, and medial.

CIRQUE: Semiround, amphitheater-shaped depression at the head of a glaciated canyon.

U-SHAPED CANYONS remain after the melting of a valley glacier.
STRATIFIED DRIFT: Sediment deposited by a meltwater river originating at the melting edge of a glacier.

Effects of Pleistocene Glaciation on Modern Society

Ice Age glaciers are directly or indirectly responsible for:

1. Some soil conditions.
2. Local ground water conditions.
3. Shore-line configuration due to sea-level changes and isostatic rebound.
4. Some areas of dunes.
5. Loess deposits resulting from transportation by Ice Age winds.

Ice Age Climate and Causes of Ice Ages

DESCRIPTION: Temperatures 4° to 10°C cooler than at present.
CAUSES: A number of factors that may contribute to changes of climate interrelate with each other in a variety of ways that are not clearly understood. These factors include:

1. Obliquity of the earth's axis.
2. Eccentricity of the earth's orbit.
3. Precession of the earth's axis.
4. Variations in content of atmospheric gases.
5. The "greenhouse effect."
6. Changes in landmass positions due to plate tectonics.
7. Tectonic changes in the elevation of continents.
8. Volume and temperature of the oceans.
9. Changes in ocean currents.

Climate Change and the Future

MOST PROBABLE WARMING SCENARIO:

1. Increase in volume of ocean water due to expansion from warming and melting of polar ice.
2. Worldwide drowning of low-level coastal plains.
3. Changes in global climate and weather patterns.
4. Decreasing soil moisture in the Northern Hemisphere, making farming nearly impossible in much of North America and Europe.
5. Declining summer runoff of glacial meltwater, which would reduce base flow volumes to rivers.
6. Thawing of much permafrost with consequent release of methane gas presently stored in subarctic ice, which would increase greenhouse effect.

ALTERNATE WARMING SCENARIO:

1. Increasing snowfall in higher northern latitudes because of increased evaporation from a warmer ocean.
2. Expansion of glaciers much as occurred 120,000 years ago at the close of the last interglacial epoch.

KEY TERMS

ablate
ablation zone
abrasion
accumulation zone
back slope
calving
chlorofluorocarbons (CFCs)
coastal desert
continental glacier
Coriolis effect
deflation
desert
desertification
drift, stratified
dune
eccentricity
end moraine
equatorial low
erratic
esker
firn
firn limit
fjord
glacial outwash
glacier
global warming
greenhouse effect
horse latitudes
ice cap
ice field
isostatic rebound
kettle lake
knob-and-kettle topography
lateral moraine
loess
medial moraine
mid-latitude desert
moraine
obliquity
piedmont glacier
polar desert
polar front
polar high
precession
proglacial lake
rain-shadow desert
recessional moraine
roche moutonnée
saltation
slip face
subpolar low
subtropical desert
subtropical high
surging, glacial
suspension
tidewater glacier
till, glacial
valley glacier
ventifact
wind
zone of flowage

STUDY QUESTIONS

1. What makes the wind blow? Why do winds in the Northern Hemisphere turn to the right, and winds in the Southern Hemisphere turn to the left?
2. Why are most of the world's deserts in the so-called desert belts located at or near the horse latitudes?
3. Contrast the relative importance of desert winds as agents of (*a*) erosion, (*b*) transportation, and (*c*) deposition. In other words, which process is the most significant?
4. Is it simply the recent drought that is responsible for the desertification and starvation so prevalent in Africa and elsewhere? (*Hint:* Consider carrying capacity, as discussed in Chapter 1.)
5. Why is it that in some of the driest places in the world—for example, the Devil's Golf Course, at about 75 meters (250 ft) below sea level in Death

Valley, and the Qattara Depression, at about 134 meters (475 ft) below sea level in the Sahara Desert—ground water is found only a few centimeters to a meter (3 ft) or so below the salt-encrusted surface?

6. What measures may be taken to halt drifting sand dunes?
7. Most coastal dunes, even in humid climates such as those of Long Island and the Northwestern United States, are located at the head of a large embayment. What is the origin of the sand that nurtures these dune fields? (*Hint:* Refer to Chapter 11.)
8. Explain the generalization "Glaciers are found at high latitudes and at high altitudes."
9. Glacial ice begins to flow and deform plastically when it reaches a thickness of about 30 meters (100 ft). What is the driving force that causes the ice to flow?
10. Describe the behavior of a glacier in terms of its *budget;* that is, the relationship between accumulation and ablation. How does a glacier behave when it has a balanced budget, a negative budget, and a positive budget? What kind of a budget does a surging glacier have?
11. How do stratified drift and unstratified drift differ? Which would be a more likely source of ground water? Why?
12. Areas far beyond the limits of the ice sheets were affected by Pleistocene continental glaciation. What are some of the effects? (*Hint:* Consider coastlines, today's arid regions, and modern agricultural regions.)
13. What is isostatic rebound? What effect has the melting of the great ice sheets had on the crustal regions that were covered by thick masses of ice until about 11,000 to 8,000 years ago? What are the effects of this rebound on society?
14. What is the greenhouse effect? What causes greenhouse warming? What can be done to modify or even control greenhouse warming? What are the potential effects of greenhouse warming to society?
15. What causes ice ages? Will there be more ice ages? Explain.

FURTHER READING

Burger, Jack. 1992. New York: Take a geological field trip. *Earth* 1, no. 2, pp. 60–67.

Byers, Bruce. 1992. The roots of Somalia's crisis. *The Christian Science Monitor,* 24 December, p. 18.

Collier, Richard. 1977. *The war in the desert.* Chicago: Time-Life Books.

Fairbridge, Rhodes. 1960. The changing level of the sea. *Scientific American* 202, no. 5, pp. 70–79.

Geological Society of America. 1977. *Impacts and management of off-road vehicles.* Report of the Committee on Environment and Public Policy. Boulder, Colo.: Geological Society of America.

Holstrom, David. 1992. Cooling that won't heat up the globe. *The Christian Science Monitor,* 9 December, p. 12.

Huber, N. King. 1987. The geologic story of Yosemite valley. *U.S. Geological Survey Bulletin* 1595. Washington: U.S. Government Printing Office.

Kirk, Ruth. 1983. Of time and ice. *Glacier Bay: Official National Park handbook.* Washington: Division of Publications, National Park Service, U.S. Department of the Interior, pp. 22–103.

Mayo, L. 1988. Advance of Hubbard Glacier and closure of Russel Fjord, Alaska: Environmental effects and hazards of the Yakutat area. *Geologic studies in Alaska by the United States Geological Survey during 1987,* U.S. Geological Survey Circular 1016, pp. 4–16.

Monroe, James S., and Reed Wicander. 1991. *Physical geology: Exploring the earth.* St. Paul: West Publishing Co.

National Research Council. 1989. *Ozone depletion, greenhouse gases, and climate change: Proceedings of a joint symposium by the Board of Atmospheric Sciences and Climate and the Committee on Global Change, National Research Council.* Washington: National Academy Press.

Trent, D. D. 1983. California's Ice Age lost: The Palisade Glacier: *California Geology* 36, no. 12, pp. 264–267.

Van Diver, Bradford B. 1992. *Roadside geology of New York,* 5th ed. Missoula, Mont.: Mountain Press.

Webb, Robert H. and Howard G. Wilshire, eds. 1983. *Environmental effects of off-road vehicles: Impacts and management in arid regions.* New York: Springer-Verlag.

Williams, Richard S., Jr. 1983. *Glaciers: Clues to future climate?* Washington: U.S. Geological Survey.

CHAPTER

13

ENERGY

I suspect the energy crisis is over
until we have our next energy crisis.

JAMES SCHLESINGER, 1982.
FORMER SECRETARY, DEPARTMENT OF ENERGY

From the simplest pond scum to the most complex ecosystem, energy is essential to all life. Derived from the Greek word *energia* meaning "in work," *energy* is defined as the capacity to do work. The units of energy are the same as those for work, and the energy of a system is diminished only by the amount of work that is done.

Prosperity and quality of life in an industrialized society such as ours depend in large part on the society's energy resources and its ability to use them productively. We may illustrate this in a semi-quantitative fashion with the equation:

$$L = \frac{R + E + I}{\text{population}}$$

where L represents quality of life, or "standard of living," R represents the raw materials that are consumed, E represents the energy that is consumed, and I represents an intangible we shall call *ingenuity*. As the equation expresses, when high levels of raw materials, energy, and ingenuity are shared by a small population, a high material quality of life results. If, on the other hand, a large population must share low levels of resources and energy, a low standard of living would be expected. Some highly ingenious societies with few natural resources and little energy can and do enjoy a high quality of life. Japan is a prime example. Some other countries that are self-sufficient in resources and energy—such as Argentina—are having difficult times. Thus we see that ingenuity,

Direct utilization of solar energy by reflecting sunlight onto an energy receiver, Mojave Desert near Barstow, California.

CHAPTER

14

MINERAL RESOURCES AND SOCIETY

Our entire society rests upon—and is dependent upon—our water, our land, our forests, and our minerals. How we use these resources influences our health, security, economy, and well-being.

JOHN F. KENNEDY, 23 FEBRUARY 1961

Every material used in modern industrial society is derived from the earth's natural mineral resources. These mineral resources are usually classified as either metallic or nonmetallic, as shown in Table 14.1. Production and distribution of all metals, the thousands of manufactured products, and the food we eat is dependent upon the utilization of metallic and nonmetallic mineral resources. The per capita U.S. consumption of mineral resources used directly or indirectly in providing shelter, transportation, energy, and in the manufacture of clothing is enormous (➢ Figure 14.1). The availability and cost of mineral and rock products influences our nation's standard of living, our Gross National Product, and our position in the world.

With the value of nonfuel minerals produced in the United States running 30–35 billion dollars per year, it is amazing that the general public has relatively little knowledge of where these minerals naturally occur, the methods by which they are mined and processed, and the extent to which we are dependent on them. It is important that the public recognize that exploitable mineral resources occur only in particular places, having formed

Argyle Diamond Mine, 120 kilometers southwest of Kununurra, Western Australia. This open-pit mine is the world's largest producing diamond mine.

➢ **FIGURE 14.6 A disseminated porphyry copper deposit.**

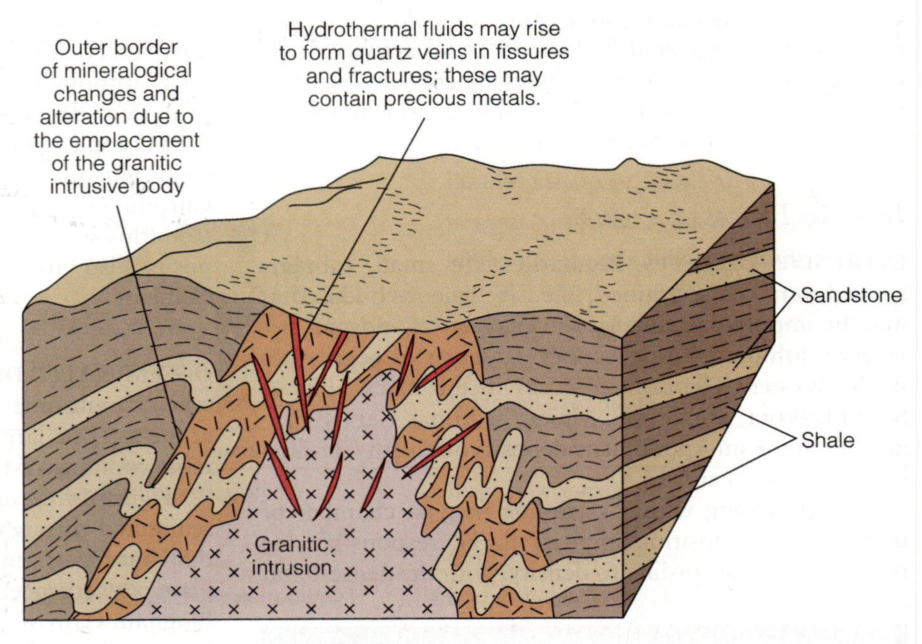

➢ **FIGURE 14.7 Kennecott Corporation's Bingham Canyon mine south of Salt Lake City, Utah, the world's largest open-pit mine and largest human-made excavation. The pit is now more than a half-mile deep and nearly 2½ miles wide.**

will vary with the composition of the particular hydrothermal fluids, but they commonly include such suites (associations) of elements as gold–quartz and lead–zinc–silver (➢ Figure 14.8).

VOLCANOGENIC DEPOSITS. When volcanic activity vents fluids to the surface, sometimes associated with ocean-floor hot-spring **black smoker** activity, **volcanogenic deposits** are formed (➢ Figure 14.9). These deposits are so-named because they occur in marine sedimentary rocks that are associated with basalt flows or other volcanic rocks, and the ore bodies they produce are called **massive sulfides.** The rich copper deposits of the island of Cyprus,* which have supplied all or part of the world's needs of copper for more than 3,000 years, are of this type. Cyprus's copper sulfide ore formed millions of years ago adjacent to sea-floor hydrothermal vents that were near a sea floor spreading center. Warping of the copper-rich sea floor due to convergence of the European and African plates brought the deposits to the surface when the island formed.

➢ **FIGURE 14.8 Detail of a mineralized quartz vein; Colosseum Mine, San Bernardino County, California.**

Sedimentary Processes

SURFICIAL PRECIPITATION. Very large and rich mineral deposits may result from evaporation and direct **precipitation** of salts in ocean water, usually in shallow marine basins. Minerals formed this way are called **evaporites.** Evaporites may be grouped into two types: marine evaporites, which are primarily salts of sodium and potassium, gypsum, anhydrite, and bedded phosphates; and nonmarine evaporites, which are mainly calcium and sodium carbonate, nitrate, sulfate, and borate compounds. Other important sedimentary mineral deposits are the **banded-iron ores** (➢ Figure 14.10). Banded-iron ore deposits formed in Precambrian time, more than a billion years ago, when the earth's atmosphere lacked free oxygen. Without free oxygen, the iron that dissolved in surface water could be carried in solution by rivers from the continents to the oceans, where it precipitated with silica to form immense deposits of red chert and iron ore. Banded-iron deposits are found in the Great Lakes region, northwestern Australia, Brazil, and elsewhere. These deposits are enormous; they provide two hundred years or more of reserves even without substantial conservation measures.

* Copper, from Latin *Cyprium,* "Cyprian metal."

➢ **FIGURE 14.9 A "black smoker" hydrothermal vent on the Endeavor segment of the Juan de Fuca mid-oceanic ridge photographed during a dive of the submersible *Alvin*. The "smoke" is hot water saturated with dissolved metallic sulfide minerals, which precipitate into black, particulate material upon contact with the cold seawater.**

➢ **FIGURE 14.10 Banded-iron ore specimen. The red bands are chert, and the gray bands are iron oxide minerals that were deposited as sedimentary layers in a shallow marine basin bordering a deeply weathered landmass.**

DEEP-OCEAN PRECIPITATION. **Manganese nodules** (see ➢ Figure 14.11) are formed by precipitation on the deep ocean floor. The nodules are mixtures of manganese and iron oxides and hydroxides, with small amounts of cobalt, copper, nickel, and zinc, that grow in onionlike concentric layers by direct precipitation from ocean waters. Any commercial recovery of this resource appears to be many decades away because of technological, economic, international political, and environmental limitations, but eventually it may be necessary to exploit the deposits.

PLACER DEPOSITS. Mineral deposits concentrated by moving water are called **placer deposits** (Spanish *placer,* "reef"; pronounced "plass-er"). Dense, erosion-resistant minerals such as gold, platinum, diamonds, and tin are readily concentrated in placers by the washing action of moving water. The less-dense grains of sand and clay are carried away, leaving gold or other heavy minerals concentrated at the bottom of the stream channel. Such deposits formed by the action of rivers are **alluvial placers;** and ancient river deposits that are now elevated as stream terraces above the modern channels are called **bench placers** (➢ Figure 14.12 and Case Study 14.1). Some placer deposits occur on beaches, the classic example being the gold beach placers at Nome, Alaska.

Sand and gravel concentrated by rivers into alluvial deposits are important sources of aggregate for concrete. River deposits formed by **glacial outwash**—that is, resulting from the runoff of glacial meltwater—are another form of placer deposit.

Weathering Processes

The deep chemical weathering of rock in hot, humid, tropical climates promotes mineral enrichment by the solution and removal of the more soluble materials, which leaves a residual soil of less-soluble minerals. Because iron and aluminum are relatively insoluble under these conditions, they tend to remain behind in laterite, a highly weathered, red subsoil or material that is rich in oxides of iron and aluminum and lacking in silicates (see Chapter 6, Figure 6.13). However, when the iron content of the parent rock is low or absent, lateritic weathering produces rich deposits of **bauxite,** the principal ore mineral of aluminum.

Ground water moving downward through a disseminated sulfide deposit may dissolve the dispersed metals from above the water table to produce an enriched deposit below the water table by **secondary enrichment** (➢ Figure 14.13). At Miami, Arizona, for example, the primary disseminated copper ore body is of marginal to submarginal grade, containing less than one percent copper, but secondary enrichment has improved the grade to more than three percent.

➢ **FIGURE 14.11 Manganese nodules, an example of deep-ocean precipitation. The nodules consist mainly of manganese, iron, and lesser amounts of nickel and copper.**

Metamorphic Processes

The high temperatures, high pressures, and ion-rich fluids that accompany the emplacement of intrusive igneous rocks produce a distinct metamorphic halo around the intrusive body. The result, in concert with the accompanying hydrothermal mineralization, is known as a **contact-metamorphic deposit.** If, for example, granite intrudes limestone, a diverse and colorful group of contact-metamorphic minerals may be produced, such as a tungsten–molybdenum deposit (➢ Figure 14.14). Asbestos and talc originate by **regional metamorphism,** metamorphism that affects an entire region.

MINERAL RESERVES

Metallic Mineral Reserves

The metallic minerals are often grouped on the basis of their relative crustal abundance.

THE ABUNDANT METALS. The geochemically abundant metals—aluminum, iron, magnesium, manganese, and titanium—have abundances in excess of 0.1 percent by weight of the average continental crust. Economically valuable ore bodies of the abundant metals—such as iron and aluminum, for example—need only comparatively small concentration factors for profitable mining (Table 14.2) and are recovered mainly from ore minerals that are oxides and hydroxides. Even though they are abundant, these metals require large amounts of energy for production. Because of this, it is not surprising that the world's industrialized nations are the greatest consumers of these metals.

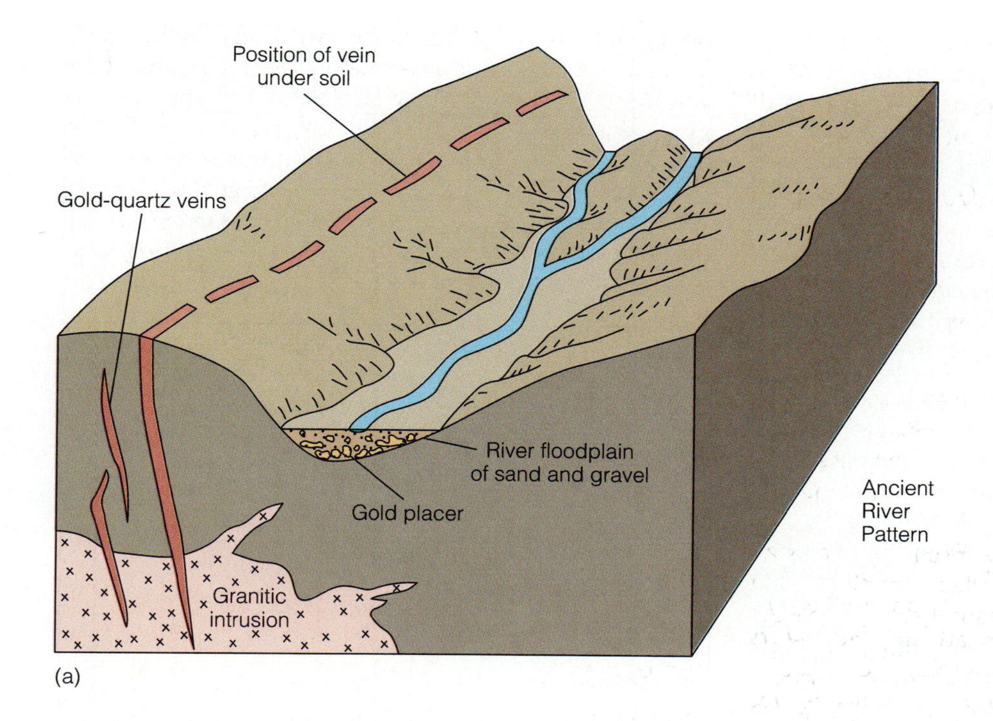

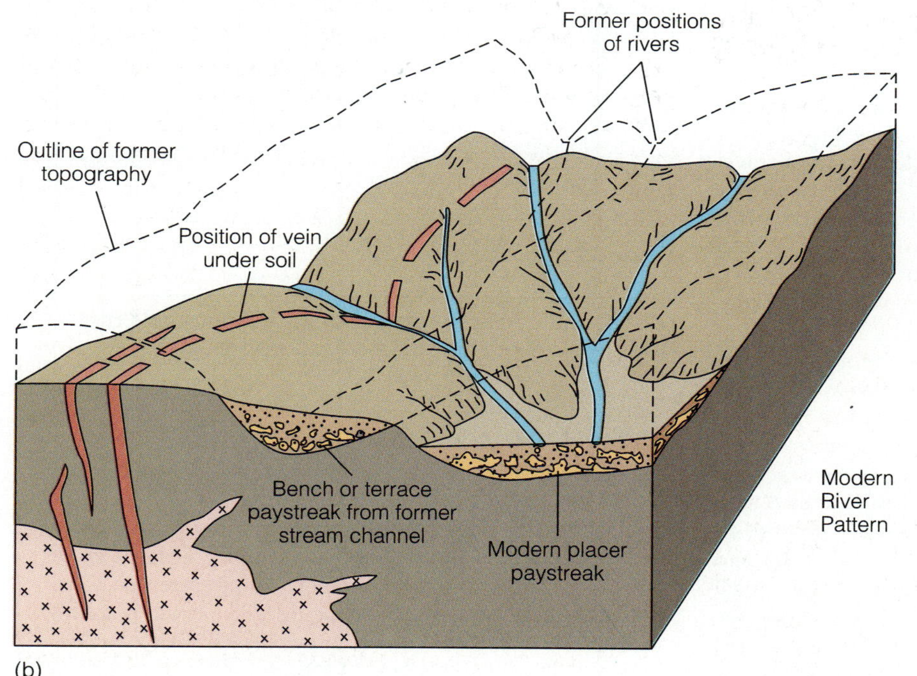

➢ **FIGURE 14.12 The origin of gold placer deposits. (*a*) An ancient landscape with an eroding gold–quartz vein shedding small amounts of gold and other mineral grains that eventually become stream sediments. The heavier gold particles settle to the bottom of the sediments in the channel. (*b*) The same region in modern time. Streams have eroded and changed the landscape and now follow new courses. The original placer deposits of the ancient stream channels are elevated as bench placers above the modern river valley.**

CASE STUDY 14.1

The California Gold Rush and the Mother Lode

Although rumors of gold in Spanish-held California were circulating as early as 1816, the first rush for California's gold did not begin until 1842 following the discovery of gold by ranchero Francisco Lopez in what is now Placerita Canyon on Rancho San Fernando near Pueblo de Los Angeles. The lack of water at the site resulted in mining by only the crudest of methods until the arrival of experienced miners from Sonora, Mexico, who introduced the method of dry washing. The spurt of interest quickly waned, however, because the small amount of gold available was soon extracted, and the lack of water for washing the gravels was discouraging to all but the most experienced miners. In 1848, however, James Marshall's recognition of gold at Sutter's Mill on the American River electrified the world and triggered the greatest gold rush in history, ultimately altering the development of Western North America and perhaps of the United States (➢ Figure 1).

The crude early mining methods by which knives and spoons were used to extract gold from placer deposits soon gave way to the gold pan and "rocker," an open-topped wooden box mounted on curved boards. Using a rocker was a two-person operation: one miner shoveled sand and gravel and poured water into the box, while the second rocked the box back and forth. The dirt would wash through the box and the gold would be caught behind "riffles," slats placed across the bottom of the box perpendicular to the flow of water. Rockers in turn led to the development of the elongated rocker, the sluice, and the "Long Tom," a large version of the sluice with similar transverse riffles (➢ Figure 2). The Long Tom was positioned in a stream so that water would flow down the length of the box. Miners shoveled in gold-bearing sand and gravel, and the gold was caught by the riffles. Cocoa matting was sometimes used to catch very small particles of gold, and oftentimes a copper plate coated with mercury was used to recover the very fine particles more efficiently by forming an *amalgam,* an alloy of mercury and gold—or of silver, hence the reason that mercury is called *quicksilver*. When "cleaning up," the miners would heat the amalgam in a vessel to separate the dissolved gold from the amalgam by vaporizing and condensing the mercury, which could then be reused. The miners were careful to avoid breathing the fumes in recovering the vaporized mercury, because they knew of its toxicity.

The easily obtainable gold was soon exhausted, and by 1853 the method of hydraulic mining—extracting gold by blasting a jet of water through a nozzle, called a *monitor,* against hillsides of ancient alluvial deposits—was developed (➢ Figure 3). Once the necessary ditches, reservoirs, penstocks (vertical pipes), and pipelines had been installed, it was possible to wash thousands of cubic yards of gold-bearing gravels without hand labor. An attempt that same year to mine alluvial gold by dredging failed; it was not until 1898 that the first real success was achieved with a bucket-conveyor dredge (text Figure 14.19).

Experienced miners immediately began searching for the *mother lode*—the bedrock source veins from which the placer deposits originated. The first gold–quartz lode vein was discovered in 1849 on Colonel John C. Fremont's grant near Mariposa. Discoveries of more gold–quartz veins of the lode system soon followed. In contrast to the simplicity of mining

➢ **FIGURE 1 California's nineteenth-century gold rush belt, also known as the Mother Lode district.**

➢ **FIGURE 2 American and Chinese gold miners working a sluice box at Auburn, California, in 1852.**

➢ **FIGURE 3 Hydraulicking a bench placer deposit.**

placer deposits, lode mining requires extensive underground workings (which require drilling, blasting, and tunneling) and a complex surface plant for hoisting men and ore from underground, for pumping water out of the mine, and for milling the raw ore. A few banks of old stamp mills (➢ Figure 4), batteries of heavy weights that were alternately lifted and dropped to pulverize the ore, remain today in the gold country. The gold was recovered by gravity separation from a slurry of powdered rock and water as it flowed over riffles and by amalgamation with quicksilver. In the 1890s, cyanide extraction was introduced into the milling process to make recovery even more efficient. It had been learned that gold dissolves readily in a dilute solution of sodium or potassium cyanide. Hence the new method mixed crushed ore with cyanide and removed the gold later by chemical methods.

The influx of thousands of Forty-Niners, as the fortune-seekers were known, to the Mother Lode district led to widespread prospecting and to the rapid and complete exploration of the previously little-known, sparsely populated region. Within a year of James Marshall's gold discovery at Sutter's Mill, California had attained statehood and become, almost overnight, the most important Western state. California's gold rush also served as the impetus for widespread prospecting rushes in the 1850s that led to important gold strikes in Nevada, Colorado, Montana, Australia, and Canada. A conservative estimate places California's annual gold production at $750,000,000 by 1865. The major world impacts of the California gold rush were the widespread redistribution of population, with its resulting problems and challenges, and the general increase of money in circulation.

➢ **FIGURE 4 Interior of a large stamp mill, like those used in the Mother Lode gold belt in the nineteenth century. Mills like this ran 24 hours a day six days a week with a noise that must have been deafening. It is said that one could travel the length of the Mother Lode in the 1880s and never be out of earshot of the pounding stamp mills as they crushed the ore.**

CASE STUDY 14.2

Mining in the Mother Lode in the 1990s

Gold mining in the Mother Lode of California continued into the twentieth century, with activity fluctuating as inflation, innovations in recovery technology, increases in gold values, and the interruptions of two world wars occurred. The U.S. Treasury Department's suspension of gold purchases in 1968 left mining companies free to sell their gold on the open market, and President Nixon's 1974 order ended the ban on private ownership of gold and terminated the convertibility of U.S. dollars into gold. Because of these factors, plus the initiation of gold trading on U.S. commodities markets in 1975, the price of gold had risen dramatically by 1980. This renewed interest in development, coupled with improvements in the technology of gold recovery, increased gold production greatly in California—to the extent that by 1991 the state output was more than 300 times that of 1980. In spite of the improved technology and gold's higher value, however, new environmental constraints and mining-permit application requirements dampened the enthusiasm of the mining industry. Many mining companies were reluctant to develop and explore California properties because of their perception that permit approvals were difficult or impossible to obtain. Nevertheless, by 1990, sixteen major gold mines in California had been successfully permitted and were in production. Gold production in the state rose from 102 kilograms (3,195 troy oz) in 1979 to 23,449 kilograms (729,272 troy oz) in 1988, by which time California had become the second-largest producer in the United States. The state's gold production in 1991 was 33,362 kilograms (1,072,733 troy oz) with an estimated value of $396,866,000.

The Yuba Gold Dredge at Marysville, California, was retrofitted in the 1970s to enable it to dig 40 feet deeper and thus rework the old dredged waste rock that it began working in the 1880s. Still operating in the 1990s, it is the only active gold dredge in California, yielding about 30,000 ounces of gold a year valued at about $10,500,000. The present operation not only recovers the deeper placer gold missed when first dredged, but it also recovers mercury that escaped from the early operations, thus removing its potential for environmental pollution (see text Figure 14.19).

Typical of several "new " Mother Lode gold mines of the 1990s is the Jamestown Mine near Sonora, in Tuolumne County (➢ Figure 1). By 1990 this mine had become the third-largest producer in California, yielding approximately 110,000 ounces per year with a value of about $37,000,000. Its reserves were estimated at about 19 million tons of ore averaging 0.063 ounce of gold per ton. Mining is by open pit and metallurgical recovery by standard crushing and milling procedures at the mine plant. The final gold separation is carried out near Yerington, Nevada, where a weak cyanide solution dissolves the gold within a closed system of tanks. The gold is removed chemically from the cyanide, and the cyanide is recycled (➢ Figure 2). Each day about 6,300 tons of ore are crushed and fed into the mill circuit for processing, with about three and a half tons of barren rock being moved for each ton of ore.

Environmental safeguards are a primary concern at the Jamestown mine. Its environmental program monitors water and air quality, noise, and seismic activity. Measures have been taken on the mine property to enhance the wildlife hab-

AGRICULTURAL MINERALS. With the world's human population doubling every 40 years and the global population expected to reach 7 billion by the end of this century (see Chapter 1), it seems obvious that agricultural food production will continue to expand and that the corresponding need for fertilizer and agricultural chemicals also will increase. Therefore, nitrate, potassium, and phosphate compounds will continue to be in great demand. Nearly all agricultural nitrate is derived from the atmosphere, but phosphate and potassium come only from the crust of the earth. Phosphate reserves occur in many parts of the world, but the major ones are in the United States, Morocco, Turkmenistan, and Buryat (the latter two are republics of the former Soviet Union). The primary sources of phosphate in the United States are marine sedimentary rocks in North Carolina and Florida (see Figure 14.22 in the following section), and there are other valuable deposits in Idaho, Montana, Wyoming, and Utah. The main U.S. supply of potassium comes from widespread nonmarine evaporite beds beneath New Mexico, Oklahoma, Kansas, and Texas, with the richest beds being in New Mexico. Large potential reserves of potassium salts are also found in Canada.

CONSTRUCTION MATERIALS. Nonmetallic minerals used as construction materials include aggregate, the sand, gravel, and crushed stone used in concrete or for making roadbeds and asphalt road surfaces; limestone and shale, which are used in making cement; clay, used for tile and bricks; and gypsum, the primary component

itat and to replace bird-nesting sites lost through mining activities. A 46-acre wildlife-management area with a three-acre pond has been established, and more than 100 bluebird nesting boxes have been placed on the property. A study shows that the bluebird population has increased in the area since the mining operations began. Waste-rock piles are contoured, planted, and seeded with native vegetation.

➢ **FIGURE 1 The Jamestown Mine, Jamestown, California. The tailings pond is at the left, and the waste rock is piled behind it. The structures are the mill complex. Settling ponds below the mill trap suspended sediment in order that the water can be reused in the milling process. Reclamation of the site has already begun, with grasses and shrubs planted on the tailings dam and on the face of the waste-rock dump.**

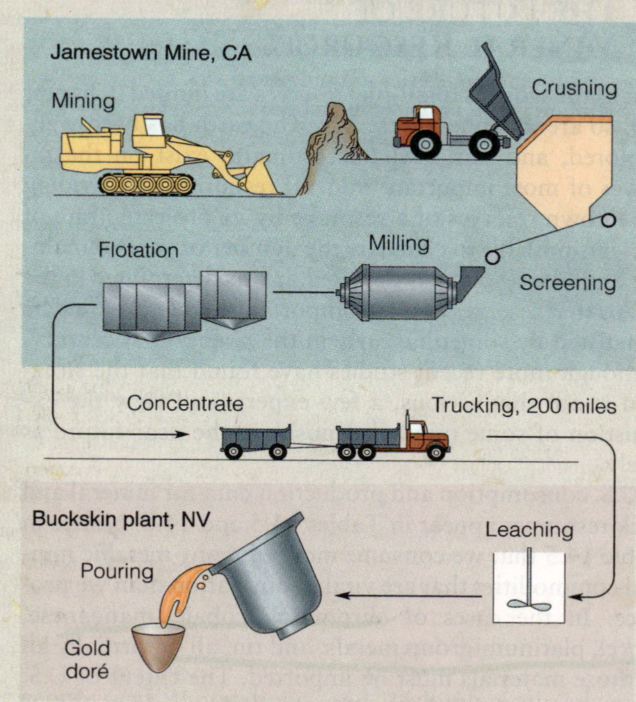

➢ **FIGURE 2 Flowchart of the Jamestown Mine's operation. The final leaching and gold pour is at the company's Buckskin plant near Yerington, Nevada. Only 0.06 ounce of gold is produced from each ton of ore mined.**

of plaster and wallboard. It is the mining of aggregate that is probably the most familiar to urban dwellers, because quarries customarily are sited near cities, the major market for the product, in order to minimize transportation costs. The quantity of aggregate needed to build a 1,500-square-foot house is impressive: 67 cubic yards are required, each cubic yard consisting of one ton of rock and gravel and 0.7 ton of sand. This amounts to 114 tons of rock, sand, and gravel per dwelling, including the garage, sidewalks, curbings, and gutters. Thus, it is no wonder that aggregate has the greatest commercial value of all mineral products mined in most states, ranking second only in those states that produce natural gas and petroleum. The annual per capita production of sand, gravel, and crushed stone in the United States amounts to about 6 tons, the total value of which exceeded $15 billion in 1991. The principal sources of aggregate are open-pit quarries in modern and ancient flood plains, river channels, and alluvial fans. In areas of former glaciation, aggregate is mined from glacial outwash and other deposits of sand and gravel that remained after the retreat of the great Pleistocene ice sheets in the northern United States, Canada, and Northern Europe (see Chapter 12).

Concrete, in addition to aggregate, contains cement. The major raw materials for manufacturing cement are large amounts of limestone or marble, from which is produced lime (CaO), and lesser amounts of gypsum or anhydrite, shale, clay or sand, and iron-bearing minerals such as hematite (Fe_2O_3) or magnetite ($FeO{\cdot}Fe_2O_3$). The principal raw materials are obtained by open-pit mining,

➢ **FIGURE 14.19 Placer mining with a bucket-conveyer dredge near Platinum, Alaska, 1958. The dredge operated from the early 1930s until the late 1970s, during which time it was the major producer of platinum-group metals in North America.**

➢ **FIGURE 14.20 Placer gold mining with a bucket-conveyor dredge on the Yuba River, California, in 1993. Surrounding the dredge operation is the scarified landscape of waste rock piles remaining after mining.**

against hillsides of ancient alluvial deposits (Case Study 14.1). Hydraulicking requires the construction of ditches, reservoirs, a penstock (vertical pipe), and pipelines. Once constructed, it is possible to wash thousands of cubic yards per day of gold-bearing gravel efficiently. The gold is recovered in sluice boxes, where mercury (the liquid metal that dissolves gold and silver to form an amalgam; Case Study 14.1) usually is added to aid recovery at a low cost. Hydraulicking is efficient but highly destructive to the land (➢ Figure 14.21). Because nineteenth-century hydraulicking was found to create river sediment that increased downstream flooding, clogged irrigation systems, and ruined farmlands, court injunctions stopped most hydraulic mining in the United States before 1900 (although it still continues in Alaska on a limited scale). Hydraulicking is still being used in the Baltic region of Russia for mining amber, and in 1979 thousands of gold seekers began invading remote areas of Brazil's Amazon basin and using it for extracting placer gold. The danger of using mercury is that when it escapes from the mining activity, it can accumulate in the food chain; consumption of mercury-tainted food causes birth defects and neurological problems in humans and animals. Its current use in gold recovery in Brazil as well as in Indonesia and the Philippines is causing extensive contamination. It is estimated that 100 tons of mercury annually works its way into the ecosystem of the Amazon basin.

Strip-mining is used mainly for coal, most commonly when the resource lies parallel and close to the surface. The phosphate deposits of North Carolina and Florida are shallow, horizontal beds that are strip-mined by excavating to a depth of about 8 meters (26 ft) (➢ Figure 14.22). After these phosphate beds are removed, the excavated area is back-filled to return the surface to its original form.

Open-pit mining is the only practical way to extract many minerals when a very large low-grade ore body is located near the surface. The process requires processing enormous amounts of material, and it is devastating to the landscape. The epitome of open-pit mining is the Bingham Canyon copper mine in Utah, where about 3.3 billion tons of material—about 7 times the volume moved in constructing the Panama Canal—have been removed since 1906 (Figure 14.7). Now a half-mile deep, the pit is the largest human excavation in the world.

The environmental consequences of open-pit mining are several. The mine itself disrupts the landscape, and the increased surface area of the broken and crushed rocks from mining and milling sets the stage for erosion and the leaching of toxic metals to the environment. This is especially true of sulfide ore bodies that produce acid mine waters, because the waste rocks and tailings are highly susceptible to chemical weathering (see Case Study 14.4).

Impacts of Mineral Processing

Except for some industrial minerals, excavating and removing raw ore are only the first steps in producing a marketable product. Once metallic ores are removed from the ground, they are processed at a mill to produce an enriched ore, referred to as a *concentrate*. The concentrate is then sent to a smelter for refining into a valuable commodity.

Concentration and smelting are complex processes, and a thorough discussion of them is well beyond the

➢ FIGURE 14.21 A monument to the destructive force of hydraulic mining of bench placer deposits; Malakoff Diggins State Park, California. When the hydraulic mining operations here ended in 1884, Malakoff Diggins was the largest and richest hydraulic gold mine in the world with a total output of about $3.5 million. More than 41 million cubic yards of earth had been excavated to obtain gold. The site is now marked by colorful, eroded cliffs along the sides of an open pit that is some 7,000 feet long, 300 feet wide, and as much as 600 feet deep.

➢ FIGURE 14.22 Rich phosphate beds are strip-mined in Florida by a mammoth drag-line dredge. The bucket, swinging from a 100-meter boom (that's longer than a football field), can scoop up nearly 100 tons of phosphate in a single bite. The phosphate is then piped as a slurry to a plant for processing. Fine-grained gypsum, a by-product of processing, is pumped to a settling pond (*upper left*), where it becomes concentrated by evaporation.

CASE STUDY 14.5

The Mother Lode's Toxic Legacy

The Penn Mine began operation in 1861, two years after copper was discovered in the volcanic rocks beneath gold-bearing gravels about 40 miles southeast of Sacramento. The deposit is classed as a volcanogenic copper–zinc sulfide ore body enriched by hydrothermal activity, and the Penn Mine was one of the largest in the "Foothill" copper–zinc belt. Mine operation was sporadic from 1861 to 1958, when it was finally abandoned (➢ Figure 1). Initial development at the Penn Mine was by underground methods, but in later years it became an open-pit mine. Total production included 80 million pounds of copper, 12 million pounds of zinc, 1,900,000 ounces of silver, and 62,000 ounces of gold. Perhaps tens of miles of abandoned tunnels lace the ground beneath the Penn Mine area, some of the shafts reaching depths of 3,000 feet. Many of the shafts were flooded during periods of inactivity and were later dewatered when mining resumed. All of the workings are now flooded, and most of their entrances are closed.

This once-productive mine left two legacies: the open pit and contaminated water. The latter appears to be the more serious threat to the environment. Water from the Penn Mine is highly acidic with a pH of 2.2 and sufficient concentrations of copper and zinc to kill fish and other aquatic organisms in the surface waters of the area. Drainage water from the mine eventually reaches the Mokelumne River, where the fish population has varied over time in response to operations at the mine. Dewatering of the mine has always produced fish kills.

➢ FIGURE 1 The Penn Mine, one of the great producers of copper and zinc in the Foothill Belt, near Campo Seco, California, in the 1940s.

The mine is within the source area of domestic water for the populous East Bay area. The East Bay Municipal Utility District (EBMUD) initially thought that abatement procedures to control the acid mine drainage were unnecessary. It was reasoned that constructing a dam to create Camanche Reservoir downstream from the mine would dilute the mine waters sufficiently to protect the lower Mokelumne River from toxic contamination. Nevertheless, a fish hatchery located immediately below the dam experienced repeated fish kills between the dam's completion in 1964 and 1978, when 170,000 king salmon fingerlings were killed by toxic water at the hatchery. The poisoned water was traced back to the Penn Mine. It was discovered that the denser, cooler, sediment-laden river water carrying the toxic mine drainage enters the reservoir and continues to follow the old river channel beneath the reservoir water. Little mixing of water occurs, and concentrations of copper and zinc in the stream channel remain high, even at the dam. These toxic waters entered the hatchery through a siphon emplaced near the bottom of the dam.

The problem was temporarily resolved in 1978, when the EBMUD, assisted by funds from the Central Valley Regional Water Quality Board (CVRWQB) and the California Department of Fish and Game, constructed a number of reclamation ponds for trapping the contaminated surface runoff flowing from the tailings piles and a ditch for diverting the natural creek that drained the mine area (➢ Figure 2). A circulating pump was installed to increase evaporation of the ponds. Unfortunately, the construction efforts were poorly conceived: the ponds were unlined. Thus, the ponded waters now percolate downward and promote the solution of sulfide minerals, *increasing* the acidity and heavy-metal contamination of the ground water beneath the old mine site. Furthermore, although the ponds were supposedly designed to prevent overspills during the normal winter rainy season, they have overflowed every winter since their construction, except during the drought years of the late 1980s and early 1990s. In an average year, 19 million gallons of toxic mine waters overflow into the Camanche Reservoir, about twice the volume of the 1989 oil spill of the *Exxon Valdez* in Alaska (see Chapter 13).

The utility district and the CVRWQB claim that their abatement procedures have mitigated the situation. Others believe they have made it worse, however, and have filed a lawsuit charging the utility district and the CVRWQB with violating the Clean Water Act.

SOURCE: Gregory R. Wheeler, Department of Geology, California State University–Sacramento.

➢ FIGURE 2 The site of the Penn Mine today. Reclamation ponds are supposedly trapping the acidic mine waters contaminated with heavy metals. Actually, the unsealed ponds are a continuous supply of acidic water that leaches through the mine wastes beneath them, contributing sulfuric acid and heavy metals to the ground water.

begin without prior approval of an adequately funded reclamation plan specifying that the waste and tailings piles will be restored to resemble the surrounding topography to the extent that it is possible, that the disturbed ground and the tailings will be replanted with native species, that all buildings and equipment will be removed, and that the open pits will be fenced and posted with warning signs.

It is claimed by some that enforcement of federal regulations pertaining to mining has been weak. In fact, the EPA has done little to regulate mining wastes, even though mining wastes are the single largest category of industrial waste. Further, Congress specifically exempted hardrock mining wastes and tailings from regulation as hazardous wastes in the Resource Conservation and Recovery Act (1976). Complicating the issues is the fact that instead of the federal government, it is the individual states that play the role of regulator, and the level of enforcement varies greatly from state to state.

Further reform of the 1872 Mining Law is expected. Beginning in the 1980s public concern grew over the problems of protecting nonmineral values on public lands, the lack of meaningful federal reclamation standards, limited environmental protection, and the lack of royalty collection for exploitation of public land—issues of public-resource management that are not addressed in the current mining law. These issues were being addressed in two bills before Congress (December 1993). A point of departure between the two bills is the imposition of a royalty on the mine income (8 percent in the House bill, 2 percent in the Senate bill). The royalty income would finance reclamation of tens of thousands of abandoned mines in the western states.

The future may also bring some changes in the basic attitudes and assumptions that underlie our capitalistic society. The economic assumption that prosperity is synonymous with mineral production is now being questioned. Environmental deterioration from today's unprecedented rate of mineral production will, if continued, eventually overwhelm the benefits gained from increased mineral supplies. As the world's developing nations strive to achieve the economies and life-styles currently enjoyed by the developed countries, there will be severe demands on, and competition for, the world's remaining mineral and fuel resources. By the middle of the next century when world population has reached 10–15 billion (see Chapter 1), new technologies and economic strategies must be in place. The goals of protecting and managing the environment while at the same time exploiting and expanding the mineral resource base in an environmentally responsible manner are not necessarily mutually exclusive. But how can both goals be achieved? Society will need to replace the current materials-intensive, high-volume, planned-obsolescent manufacturing processes—and consumption patterns—with those that use raw materials and fuels more efficiently, that generate little or no waste, and that recycle most of the waste that is generated. Whether technology can modernize fast enough to conserve natural resources at the same time that the world's wealth is increasing is the big question. This is the formidable challenge that must be faced jointly by industry, governments, and society.

GALLERY: THE LEGACY OF MINES AND MINING

➢ Figure 14.27 is a collage of pictures illustrating various aspects of past and present mining, primarily in the Western United States.

➢ FIGURE 14.27 (*Opposite page*)(*a*) Toxic legacies left for perpetuity by turn-of-the-century Colorado silver mining; waste-rock piles from abandoned mines above Silver Plume, Colorado. Although the mines have been shut down for decades, the barren waste-rock piles still lack vegetation because of the accumulation of acid salts (indicated by the white coloration) due to evaporation of acidic seepage water that has percolated through the material. There are thousands of such abandoned mines in Colorado and perhaps hundreds of thousands in the Western states. (*b*) A modern rotary-ball mill; Colosseum Mine, San Bernardino County, California. Ball mills have replaced the antiquated stamp mills. Two ball mills are usually required for grinding ore to a consistency finer than talcum powder. The powdered ore is mixed with water to form a slurry, which is then processed chemically and mechanically to remove the metal. The residual slurry becomes tailings, which are disposed of at the tailings pond. (*c*) The legendary "Mother Lode" gold–quartz vein is exposed in the wall of an open-pit mine; Jamestown Mine, Jamestown, California. (*d*) Creek near the open-pit copper mine at Morenci, Arizona. The water's distinct blue color reveals a high concentration of dissolved copper leached from mine wastes.

(a)

(b)

(c)

(d)

large-scale recycling project, say at the local level? How would you educate the general public to the urgency of recycling? What effects would recycling have on the economy? Would it decrease or increase employment?

Further Reading

Allgood, Genne M. 1991. The Jamestown mine: its history, geology, and operations. In Barabas, Arthur H., ed. *Geology, gold deposits, and mining history of the southern Mother Lode.* National Association of Geology Teachers field conference guidebook, Far Western Section, Fall Meeting, October 11–13, Fresno, California, pp. 38–46. (Originally published Littleton, Colo.: Society for Mining, Metallurgy, and Exploration, Inc., preprint number 90-400, pp. 1–9.)

Averill, Charles V. 1948. History of placer mining for gold in California. In Jenkins, O. P., ed. Geologic guidebook along Highway 49: Sierran gold belt, the Mother Lode country, centennial ed. *California Division of Mines bulletin* 141, p. 19.

Baum, Dan, and Kay Chernush. 1992. In Butte, Montana, A is for arsenic, Z is for zinc. *Smithsonian* 23, no. 8, pp. 46–56.

Cole, Kenneth A., and Ann Kirkpatrick. 1983. Cyanide heap leaching in California: Methods used plus health and safety aspects. *California geology* 36, no. 9, pp. 187–194.

Craig, James R.; David J. Vaughan; and Brian J. Skinner. 1988. *Resources of the earth.* Englewood Cliffs, N.J.: Prentice-Hall.

Cutter, D. C. 1948. The discovery of gold in California. In Jenkins, O. P., ed. Geologic guidebook along Highway 49: Sierran gold belt, the Mother Lode country, centennial ed. *California Division of Mines bulletin* 141, pp. 13–17.

Henning, Robert A., ed. 1982. Alaska's oil/gas and minerals industry. *Alaska geographic* 9, no. 4, p. 216.

Logan, C. A. 1948. History of mining and milling methods in California. In Jenkins, O. P., ed. Geologic guidebook along Highway 49: Sierran gold belt, the Mother Lode country, centennial ed. *California Division of Mines bulletin* 141, pp. 31–34.

McPhee, John. 1992. Annals of the former world: Assembling California, part 1. *New Yorker,* 7 September, pp. 36–68.

Silva, Michael A. 1988. Cyanide heap leaching in California. *California geology* 41, no. 7, pp. 147–156.

Wheeler, G. R. 1984. Environmental geology and mining in the Mother Lode southeast of Sacramento. In Wheeler, G. R., ed. *Field trips guidebook.* National Association of Geology Teachers, Far Western Section, fall meeting, Sacramento, California, pp. 39–67.

Young, John E. 1992. Mining the earth. In Brown, Lester R., et al. *State of the world, 1992: A Worldwatch Institute report.* New York: W. W. Norton & Co., pp. 100–118.

Video

Poison in the Rockies (video). 1992. NOVA, PBS.

Gifts from the earth (video). 1986. The Planet Earth Series, Annenberg/CPB Project. (Available from Films Incorporated, 5547 N. Ravenswood, Chicago, IL 60640-1199; (312) 878-2600, ext. 43.)

CHAPTER

15

WASTE MANAGEMENT GEOLOGY

Garbage isn't generic gunk;
it's specific elements of our behavior
all thrown together.

WILLIAM RATHJE, PROFESSOR OF ANTHROPOLOGY,
UNIVERSITY OF ARIZONA

Getting rid of trash is a major environmental problem in industrialized nations. Growing populations of consumers have created an explosion of solid, liquid, and hazardous wastes, much of which requires special handling. At risk due to careless waste disposal are ground- and surface-water purity, air quality, public health, and less threatening but still important scenic and land-surface degradation. Municipal solid waste generated in the United States in 1985 amounted to 2.1 kilograms (4.6 pounds) per person per day. This includes garbage, yard cuttings, construction materials, and the like (➢ Figure 15.1), and does not include agricultural or industrial waste products. Annually, the United States disposes of an estimated 1.3 billion metric tons of wastes that fall broadly into four categories (Table 15.1). Just one day's total U.S. waste would cover 15 square kilometers (almost 6 mi^2) to a depth of 3 meters (10 ft). If it were loaded into 10-ton trucks lined up bumper-to-bumper, the trucks would stretch around the world 20 times. Those trucks containing only the day's *municipal solid waste* would circle the earth almost 3 times. The spectrum of solid-waste disposal problems ranges from

Greenpeace garbage barge looking for a home in the harbor of New York City. Disposal of solid waste is a world-wide problem for large urban centers in both developed and developing nations.

NEXT TIME... TRY RECYCLING
GREENPEACE

CASE STUDY 15.2

The Highest Point between Maine and Florida

Fresh Kills landfill on Staten Island, New York, is the world's largest landfill. It contains 25 times the volume of one of the Seven Wonders of the World, the Great Pyramid of Giza, Egypt (➢ Figure 1). Fresh Kills (Dutch, *kil,* "stream") serves New York City and its suburbs and parts of New Jersey. It accepts about 13,000 tons of trash per day, and by 1998 its surface will be 165 meters (500 ft) above sea level—the highest elevation on the eastern shore of the United States. The fill was established in 1948 on a salt marsh with no provisions for constraining leachate. For years, more than 4.2 million liters (a million gallons) of leachate leaked into the marsh and its nearby waters each day. Recent cleanup has rectified many of the pollution problems, but Fresh Kills remains a great mountain of an eyesore to Staten Islanders.

What makes Fresh Kills unique, aside from being built upon marshland, is that trash is tranported to the fill by city-

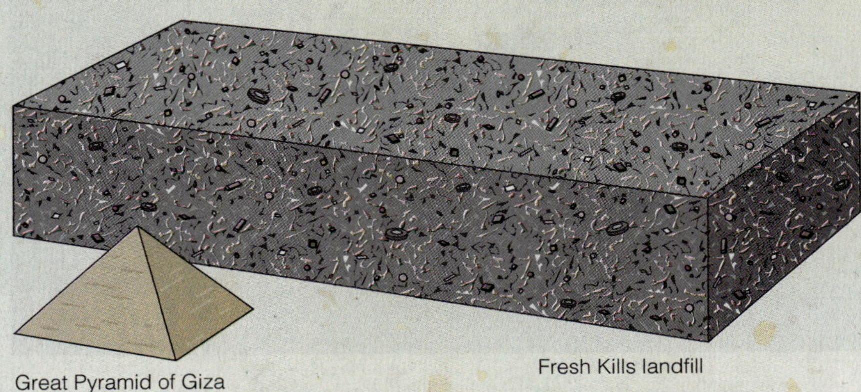

(a)

(b)

➢ FIGURE 1 (*a*) Comparison of the volumes of the Great Pyramid of Giza and the Fresh Kills landfill. (*b*) The Great Pyramid (*right*) built by King Cheops at Giza on the west bank of the Nile River in the third century B.C.. The length of each side at the base averages 756 feet, and its present height is 451 feet (original height, 481 feet). Automobile gives an idea of the immensity of these structures.

owned barges, each with a capacity of 600–700 tons. The barges are loaded from garbage trucks at eight stations located throughout the city and towed to Staten Island by tugboats. The tugboats are followed by specially built "skimmer" boats that pick up trash that falls from the barges or drops into the water at the unloading facilities. At the docks the refuse is unloaded by huge cranes and placed into large, custom-made vehicles that carry the trash to a fill site. After dumping, the refuse is sprayed with a deodorant, compacted by bulldozers, covered by clean earth, and eventually landscaped to "reclaim" the land.

New York City also has three active municipal trash incinerators. In the past the ash residue from these facilities was transported offshore and dumped into the ocean. Now the ash is barged to Fresh Kills and deposited along with raw garbage. Because incinerator ash is so loaded with toxic metals, construction of specially designed ash-only fill areas is planned. The ash-only fill will have a double liner and a double leachate-collection system to keep toxic substances within the fill. In addition, a clay slurry "wall" will be excavated around the fill to further confine leachate should there be leaks. Also planned is a waste-to-energy incinerator, and ash from this incinerator also will be dumped in Fresh Kills (➢ Figure 2).

Major efforts are underway to extend the life of the landfill by recovering resources contained in the trash. Methane is extracted and used for heating and cooking in nearby homes, construction debris is recycled, and yard waste is composted. Grass and leaves from Staten Island and the Borough of Queens are composted and mixed with soil to promote healthy plant growth on the finished fill. Two more recycling facilities with debris-crushing and screening plants are planned, as is an intermediate processing center that will sort and process recyclable glass, cans, paper, and plastics. At the present rate of delivery, Fresh Kills landfill will be filled to capacity by the year 2000.

➢ **FIGURE 2 Fresh Kills landfill on Staten Island.**

➢ **FIGURE 15.5 Geology of the three classes of landfills. Note that a Class I fill offers a maximum protection to underground water and that a Class III fill offers no protection.**

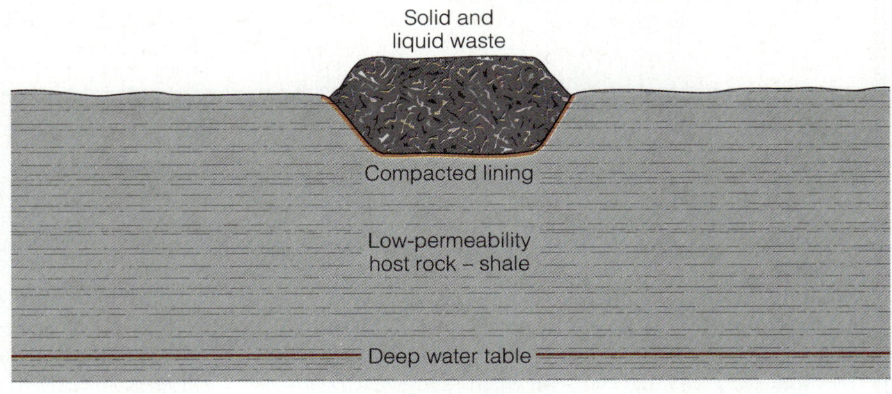

(a) Class I landfill

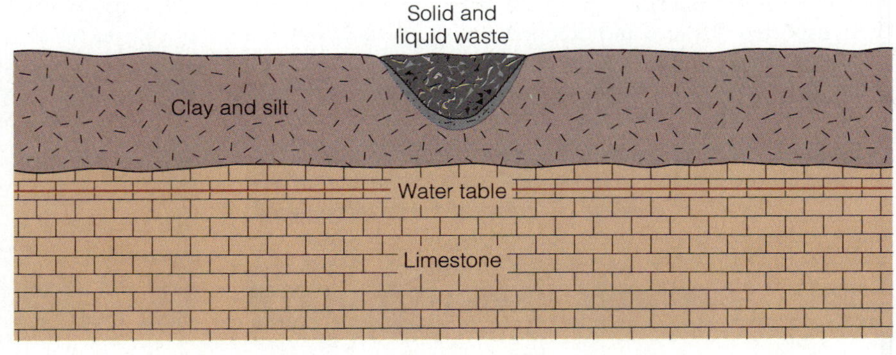

(b) Class II landfill

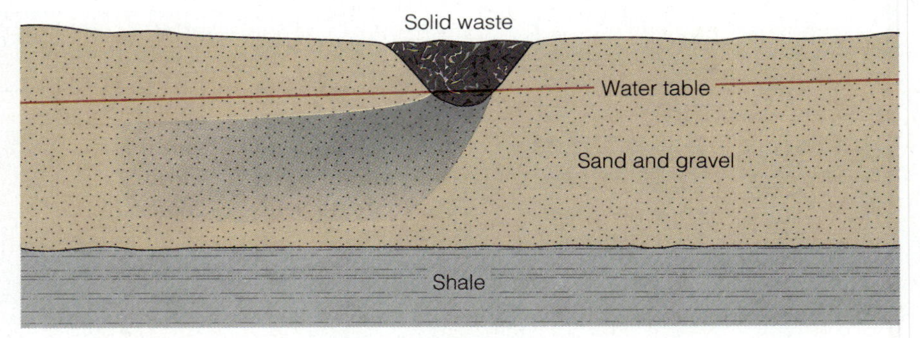

(c) Class III landfill

INCINERATION. Burning is the only proven way to significantly reduce the volume of garbage, as much as 90 percent. It can be burned to produce electricity (see the section on Resource Recovery from Municipal Waste) or simply burned and the heat and gases wasted to the atmosphere. It is estimated that there are 155 large modern incinerators operating in the United States. New York City is a leader in incinerating its trash, and about 720,000 tons per year, or 11 percent of all its waste, is burned. Wastes must be burned at very high temperatures, and incinerator exhausts are fitted with sophisticated scrubbers that remove dioxins and other toxic air pollutants (see Chapter 13). Incinerator ash presents another problem because it is in itself a hazardous waste, containing high ratios of heavy metals that are chemically active (see Case Study 15.2).

OCEAN DUMPING. Greek mythology relates how Heracles (Latin, *Hercules)* was given the task of cleaning King Augeas's stables, which contained 30 years' accumulation of filth from 3,000 head of cattle. Inasmuch as one cow produces roughly 18 wet tons of manure per year, Heracles was up to his ears in 1.6 million tons of you-know-what, about the amount of sewage sludge that

➢ **FIGURE 15.6 Oil-field brine being dumped in a Class I landfill. Other parts of this landfill have a lower classification because their bedrock is permeable.**

TABLE 15.4 Selected Seawater Pollutants from Human and Natural Sources

SEAWATER POLLUTANTS*	
	Organic
PCBs	polychlorinated biphenyls
PCDFs	polychlorinated dibenzofurans
PCDDs	polychlorinated dioxins
DDT	chlorinated hydrocarbon pesticide
	Inorganic
Cd	cadmium
Pb	lead
Co	cobalt
Hg	mercury
Zn	zinc

*Defined as pollutants originating in sludge, dredged materials from harbors or estuaries, and from sewage treatment and industrial wastewater that are known to be carcinogenic or to cause mutations, birth defects, nervous disorders, or liver, kidney, or lung disease.

New York City generates in 4 months. Because Heracles was to perform the cleanup in a day, he ingeniously diverted the courses of two rivers to make them flow through the stables and wash the filth into an estuary (see Chapter 11). Had an environmental impact statement been required, it would have noted that a huge mass of solid sludge would cover a large area of the wetlands and bury bottom-dwelling organisms. In addition, phosphates and nitrates in the effluent would promote explosive algal blooms at the expense of other organisms, and dissolved oxygen in the water would decrease to the point of mass mortality of swimming and bottom-dwelling organisms.

Historically, all coastal countries used the sea for waste disposal. This is not surprising, since the sea is convenient, about three times more ocean area exists than land area, and as waste matter disperses, sinks to the bottom, or is diluted, the ocean gives the visual impression that it is able to accommodate an unlimited amount of bad things. What existed then was the *assimilative capacity* approach; that is, the assumption that a body of seawater can hold a certain amount of material without adverse biological impact. This attitude is still held in many coastal areas, particularly in Third World countries, where floatables in trash present a highly unsightly appearance. No U.S. municipality has dumped garbage in the sea since New York City stopped the practice in 1934.

Ocean pollution is governed by the Marine Protection Reserve and Sanctuary Act of 1972, better known as the *Ocean Dumping Act*. This act requires anyone dumping waste into the ocean to have an EPA permit and to provide proof that the dumped material will not degrade the marine environment or endanger human health. Hazardous substances entering the sea from the land (Table 15.4) are contained in sewage, sewage sludge (the solid part of sewage), dredged materials, industrial effluents, and natural runoff. Sludge and other particulate matter can create a biological imbalance in three ways:

- they are consumed by marine life as a food source,
- they introduce heavy-metal and chlorinated organic compounds when they are attached to the particles, and
- they inhibit light in the water column.

In the late 1970s the EPA announced its intention to ban the dumping of all sludge that degrades the environment by 1981. However, the city of New York challenged enactment of the law and won. Because the city dumps so much sludge (about 5 million wet tons per year), the EPA required the city to move its dumpsite from 12 miles offshore to what is known as *Site 106*, 106 miles off the New Jersey coast in water 2.4 kilometers (1½ mi) deep (➢ Figure 15.7).

Ideally, a sewage outfall that is sited in deep water below the *thermocline*—the zone where water temperature decreases and density increases rapidly with depth—could accept untreated raw sewage. The thermocline acts as a floor for warm surface water and as a ceiling for cool, dense bottom water and prevents the two water layers from mixing. Most outfalls, however, are in water depths *above* the thermocline, and thus sewage effluent mixes with surface waters, which may then pollute shallow-water ecosystems and beaches. According to the EPA, the New York City sludge was supposed to disperse totally during its descent so that it would leave no measurable impact on the bottom fauna. In 1989 a research team from Woods Hole Oceanographic Institution visited Site 106 in the research submersible *DSV Alvin*. Samples taken by the team contained trace metals and bacteria indicative of human sewage. Ongoing studies will help us

(a)

(b)

➢ **FIGURE 15.10** **(*a*) Mission Canyon landfill near Santa Monica, California, as it was being finished. (*b*) The reclaimed landfill is now a golf course and expensive residential neighborhood.**

the gas (➢ Figure 15.11). One strategy for handling the large volumes of methane generated at large landfills is to extract it through perforated plastic pipes, remove the impurities (mostly CO_2), and use it as fuel—the so-called refuse-derived fuel, *RDF*. This can be accomplished either by piping the gas away via pipelines or by using it on the site to generate electricity. Because the gas may be as much as 50 percent CO_2, it is more expedient to use the fuel directly to generate electricity than to clean out the impurities and put it into pipelines. The Los Angeles County (California) Sanitation District currently extracts landfill methane and generates sufficient electricity to supply 10,000 homes at a completed fill at Palos Verdes and another 45,000 homes at a similar facility at Puente Hills.

➢ **FIGURE 15.11** **Methane-extraction well and drill rig. The extracted gas is conveyed to a central location for conversion to electricity.**

LEACHATE. The compositions of leachates in and leaving landfills are highly variable, but they must be considered dangerous to the environment until established otherwise (➢ Figure 15.12). A leachate may be such that a receiving body of water can assimilate it without any impairment in water quality. Some purification of a leachate occurs as it filters through clayey soils, where solid particles are absorbed on mineral surfaces. Some purification also occurs when contained organic pollutants are oxidized. Ground-water-quality measurements taken in the vicinity of a landfill, at the fill, and downstream from the fill are shown in Table 15.5. Note that the leachate has caused the deterioration of ground-water quality in the monitoring well. To prevent this deterioration, the leachate will have to be intercepted by wells or a barrier will have to be installed before the leachate reaches the water table. Under existing regulations, a landfill that is sited in a geologic setting where leachate threatens local water bodies is allowed to accept only inert (Group 3) wastes.

Resource Recovery from Municipal Waste

Resource recovery in waste management systems refers to the retrieval of something of value from urban trash. Recoverable resources in landfills include energy and reusable materials such as aluminum, glass, ferrous metals, and paper products. Perhaps the biggest benefit of recovery is the extension of landfills' lifetimes by the resulting reductions of trash volume. The New York State Department of Environmental Conservation estimates that optimal resource recovery could reduce the volume of solid waste by 90 percent and its weight by 75 percent.

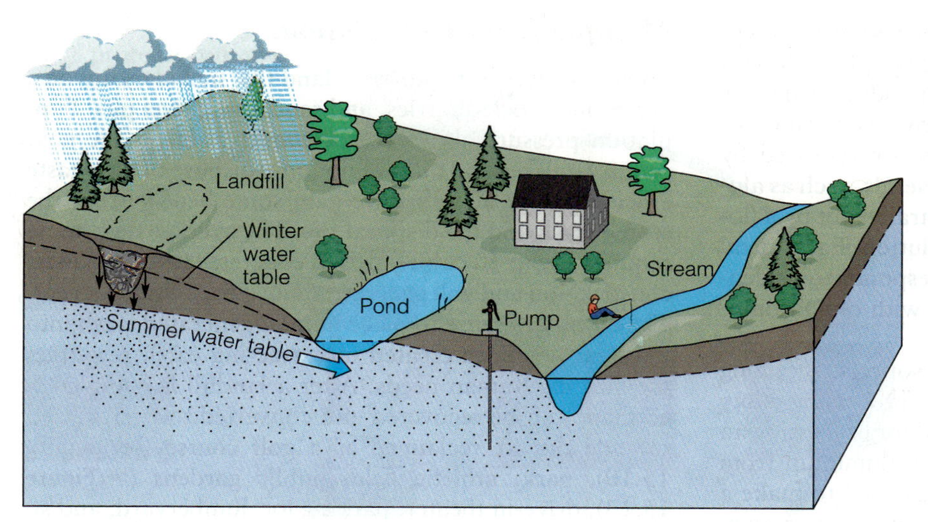

➢ **FIGURE 15.12 The flow of leachate from a landfill can contaminate underground water, ponds, and streams. See also Table 15.5.**

ENERGY RECOVERY. Energy is recovered from waste by burning and by fuel extraction. Mass burning of unprocessed solid waste yields heat that is used for space heating and cooling or that is converted into steam for generating electricity. Ferrous metals can be recovered from the ash before they are placed in a landfill as explained in Case Study 15.2. Methane (natural gas) can be extracted for space heating or for generating electrical energy as explained in the preceding subsection. Other refuse-derived fuels are obtained by separating combustibles from noncombustible materials and shredding them before burning.

LANDSPREADING. Incorporating organic wastes into the soil increases the soil's productivity. Sewage sludge and septic-tank wastes can be spread on soils to reclaim poor soils at abandoned mining areas, on deforested lands, and in some cases, on agricultural lands.

RECYCLING. An effective way to lengthen the life of landfills and our natural resources is to recycle municipal wastes. Recycling includes such practices as composting, making concrete from incineration ash, and returning beverage bottles for reuse—the bottles average 15 to 20 fillings before breaking. Many cities have instituted regulations that require residents to separate glass, metal, plastic, and paper products from other trash materials. Recycled paper, for example, is de-inked, shredded, pulped, and remade into newsprint, cardboard boxes, paper bags, and other products. Several kinds of plastic are now recycled by melting and then extruding to make plastic "lumber," trash cans, furniture frames, and the like. Many states now require plastic objects to be stamped with a code to indicate the type of plastic used, which facilitates reprocessing.

New recycling technology separates out heavier materials such as metals, glass, and plastics before mass burning of the lighter, "wet" garbage to produce energy.

TABLE 15.5 Ground-Water Quality Found in the Vicinity of a Landfill

WATER CHARACTERISTIC	MEASUREMENT SITE		
	Local Ground Water	*Leachate Fill**	*Monitoring Well***
Total dissolved solids, ppm	636	6,712	1,506
pH***	7.2	6.7	7.3
BOD, mg/L****	20	1,863	71
Hardness, ppm	570	4,960	820
Sodium content, ppm	30	806	316
Chloride content, ppm	18	1,710	248

*Quality in a saturated fill.

**Quality in a monitoring well about 150 feet downstream from the landfill at a depth of 11 feet in sandy material.

***A quantitative measurement of the acidity or basicity of a solution. A solution with a pH of 7 is neutral, less than 7 is acidic, and greater than 7 is basic.

****Biological oxygen demand: the amount of oxygen per unit volume of water required for total aerobic decomposition of organic matter by microorganisms. This is discussed later in the chapter in the section on wastewater treatment.

SOURCE: D. R. Brunner, and D. J. Keller, *Sanitary Landfill Design and Operation* (Washington, D.C.: U.S. Environmental Protection Agency, 1992).

➢ **FIGURE 15.17 Eagle Mountain Mine, site of the proposed Eagle Mountain landfill. As designed, the fill could accommodate up to 20,000 tons of trash per day, which would make it the largest-capacity landfill in the world. It would accept only nonhazardous waste and would have a recycling center.**

hazardous waste), which would be delivered in closed railroad cars. The tracks that served the mine are still in place, and the site is known to be free from faults and potential ground-water contamination. Far from any population centers, the site would accommodate half of all the household trash generated in southern California in the next 115 years! The open pit is in dense bedrock that would be lined with clay and high-density polyethylene plastic to protect ground water. The proposed site is acceptable geologically, and the only objections raised at public hearings regarding the proposal, from a technical point of view, seem to be entirely NIMBY, PIITBY, and NIMTOO (see Case Study 15.1). The Metropolitan Water District, which operates an open canal system 2 kilometers (1¼ mi) away that delivers millions of gallons of fresh water daily from the Colorado River to southern California, has no objection to the project.

Hazardous Waste Disposal

Hazardous wastes are primarily such industrial products as sludges, solvents, acids, pesticides, and PCBs (polychlorinated biphenyls, highly toxic organic fluids used in plastics and electrical insulation). The threats that these and other pollutants pose to ground water are discussed in Chapter 10. The United States produces the largest amount of hazardous waste and the second-largest amount per unit area (see Table 15.2). The EPA has estimated that U.S. industries generated 150 million metric tons of hazardous waste in 1981, most of it in liquid forms. This amounts to 40 billion gallons, which, if placed in 55-gallon drums end to end, would encircle the earth 16 times (➢ Figure 15.18). Examples of hazardous wastes generated by consumers include drain cleaners, paint products, used crankcase oil, and discarded fingernail polish (not trivial). The EPA estimates that 90 percent of our hazardous waste, coming from about 750,000 sources, is disposed of improperly, and that this could leave problems that will last for generations. Fortunately, knowledge of how to dispose of these substances with a high degree of safety exists.

Hazardous Waste Disposal Methods

SECURE LANDFILLS. Secure landfills are designed to totally isolate the received waste from the environment. Wastes are packaged and placed in an underground vault that is surrounded by a clay barrier, thick plastic liners, and a leachate removal system (➢ Figure 15.19). This is the *preferred* method of disposing of hazardous waste; unfortunately, few of the nation's existing 75,000 industrial landfills have plastic liners, clay barriers, or leachate drains. Many older fills will either have to be retrofitted with modern environmental-protection systems or shut down. Additional protection is provided by existing regulations that mandate ground-water monitoring at disposal sites. An ideal secure landfill for hazardous waste would be a site in a dry climate with a very low water table. In humid climates where water tables are high, alternative means of safe disposal must be found.

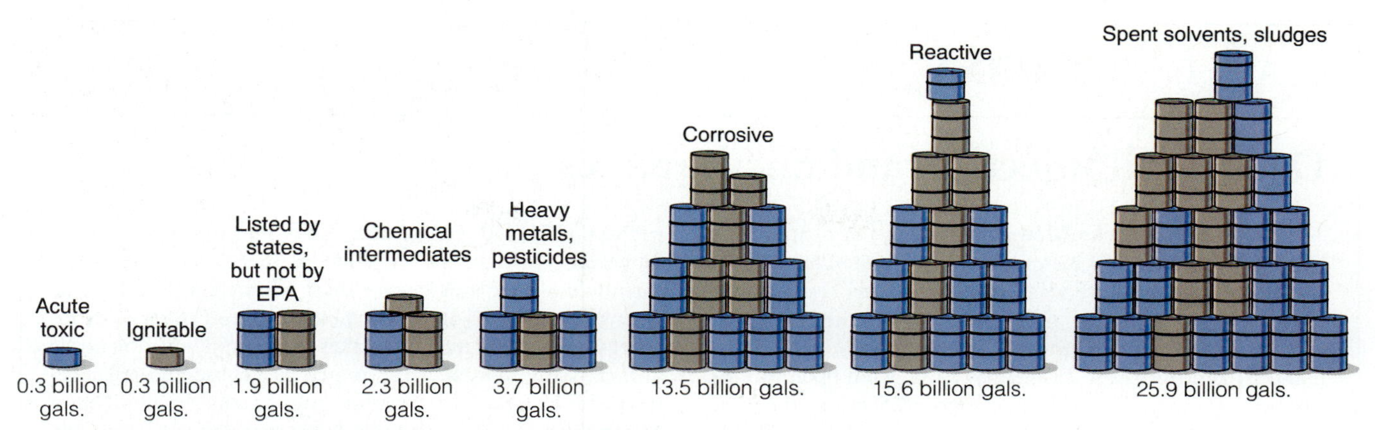

➢ **FIGURE 15.18 Breakdown by category of the 40 billion gallons of toxic waste U.S. industries generated in 1981. The total of all categories is greater than 40 billion due to overlap; for example, a fluid may be both corrosive and reactive.**

DEEP-WELL INJECTION. Deep-well injection into permeable rocks far below fresh-water aquifers is another method of isolating toxic and other hazardous liquid substances (➢ Figure 15.20). Before such a well is approved, thorough studies of subsurface geology, such as the location of faults, the state of rock stresses at depth, and the impact of fluid pressures on the receiving formations must be made (Case Study 15.3). These boreholes are lined with steel casing to prevent the hazardous fluids from leaking into fresh-water zones above the injection depth.

Hazardous waste disposal, both liquid and solid, is regulated by the Resource Conservation and Recovery Act of 1976 (RCRA). Exempt from the requirements of this act are businesses that generate less than 1 metric ton of waste per month. According to EPA estimates, about

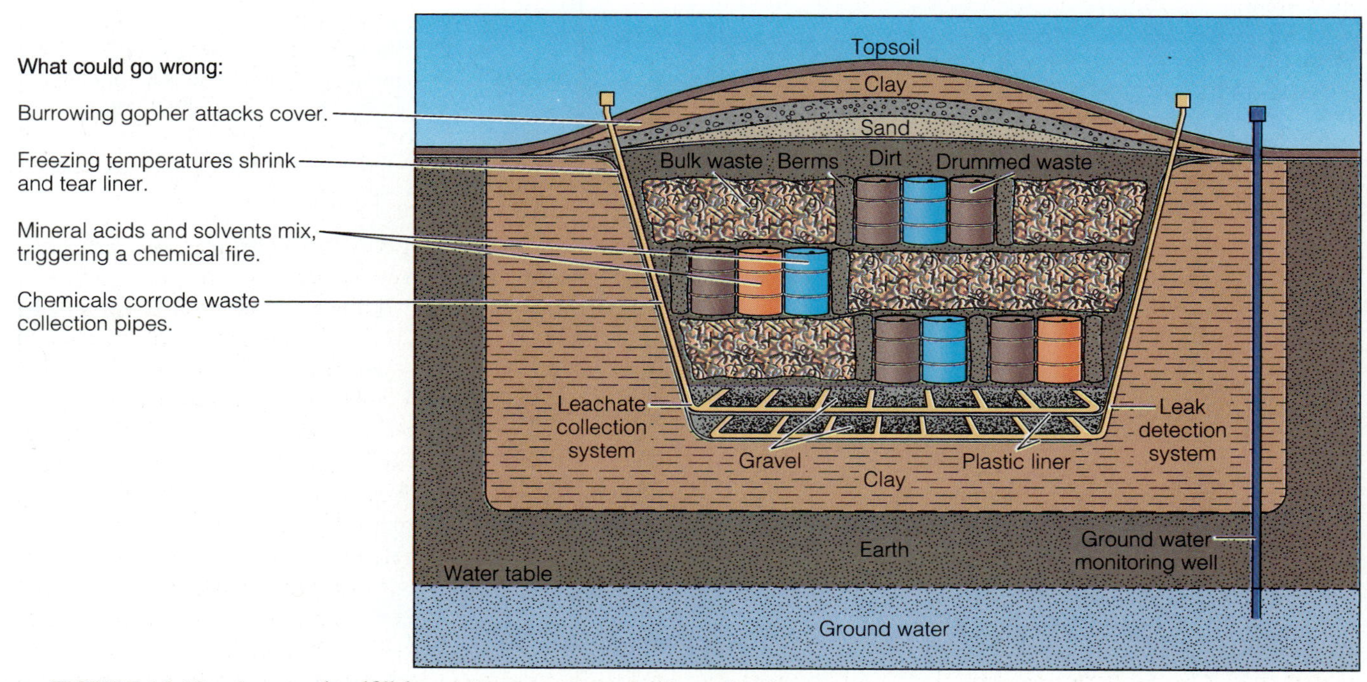

➢ **FIGURE 15.19 A secure landfill for toxic waste. Note the liners, leachate collection system, leak detection system, and monitoring well.**

CASE STUDY 15.3

Deep-Well Injection and Earthquakes

The U.S. Army's Rocky Mountain Arsenal, in a suburb of Denver, drilled a 3,764-meter (2⅓-mi)-deep well for disposing of chemical-warfare wastes in 1961. The Tertiary and Cretaceous sedimentary rocks into which the well was drilled were cased off to prevent any pollution of these formations, and the lower 23 meters (75 ft) were open to Precambrian granite. About 40 kilometers (25 mi) to the west of the arsenal is the frontal fault system of the Rocky Mountains (➢ Figure 1).

Injection began in 1962 at a rate of about 700 cubic meters per day (176,000 g/day). Soon after injection was started, numerous earthquakes were felt in the Denver area. The earthquake epicenters lay within 8 kilometers (5 mi) of the arsenal and plotted along a line northwest of the well. A local geologist attributed the quakes to the fluid injection at the arsenal. Although the army denied any cause-and-effect relationship, a graph of injection rates versus the number of earthquakes showed an almost direct correlation—especially during the quiet period when fluid injection was stopped (➢ Figure 2).

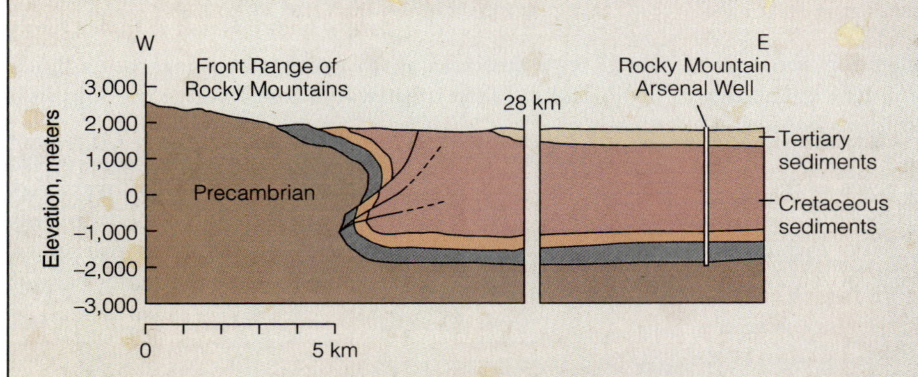

➢ **FIGURE 1 Relationship between the Rocky Mountain Arsenal and the Rocky Mountain frontal fault system.**

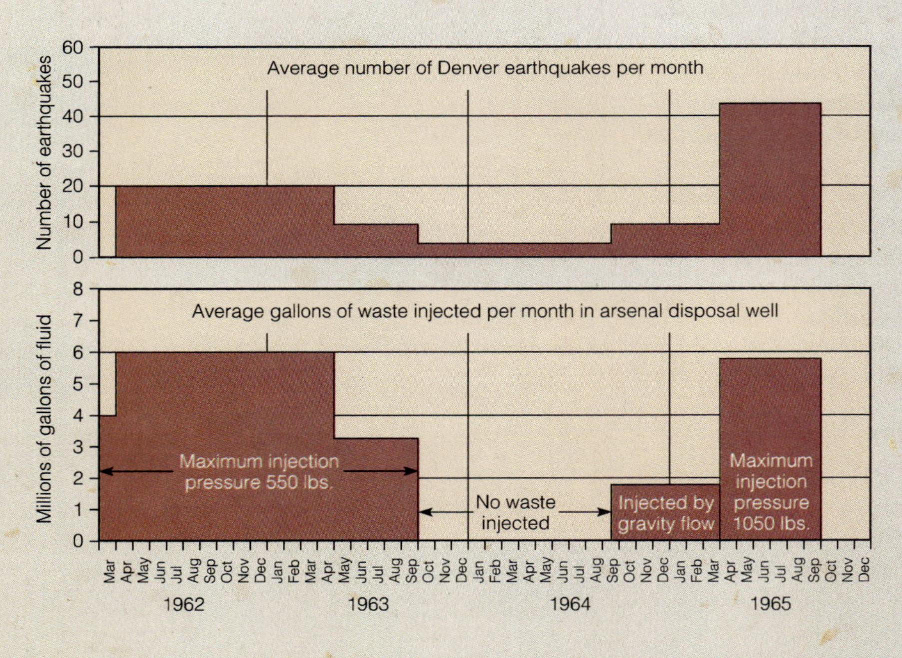

➢ **FIGURE 2 Correlation of fluid waste injection and the number of earthquakes per month in the Denver area, 1962–1965.**

The fractured granite into which the fluid was being injected was in a stressed condition, and the high fluid pressures at the bottom of the hole decreased the frictional resistance (effective stress, see Chapter 7) on the fracture surfaces. Earthquakes were generated along a fracture propagating northwest of the well as the rocks adjusted to relieve internal stresses.

Thus earthquakes *can* be "triggered" by human activities, and the Denver earthquakes initiated research into the possibility of modifying fault behavior by fluid injection; that is, relieving fault stresses by producing a large number of small earthquakes. Seismic studies conducted during a water-injection program at the nearby Rangely Oil Field in Colorado (see Chapter 13) confirmed that earthquakes *could* be turned on or off at will, so to speak, in areas of geologic stress. To date no group or government entity has assumed responsibility for triggering earthquakes by fluid injection, for obvious reasons. What would be the consequences if, instead of triggering many small earthquakes, a major earthquake was triggered, accompanied by loss of life and property damage? More research is needed in this area.

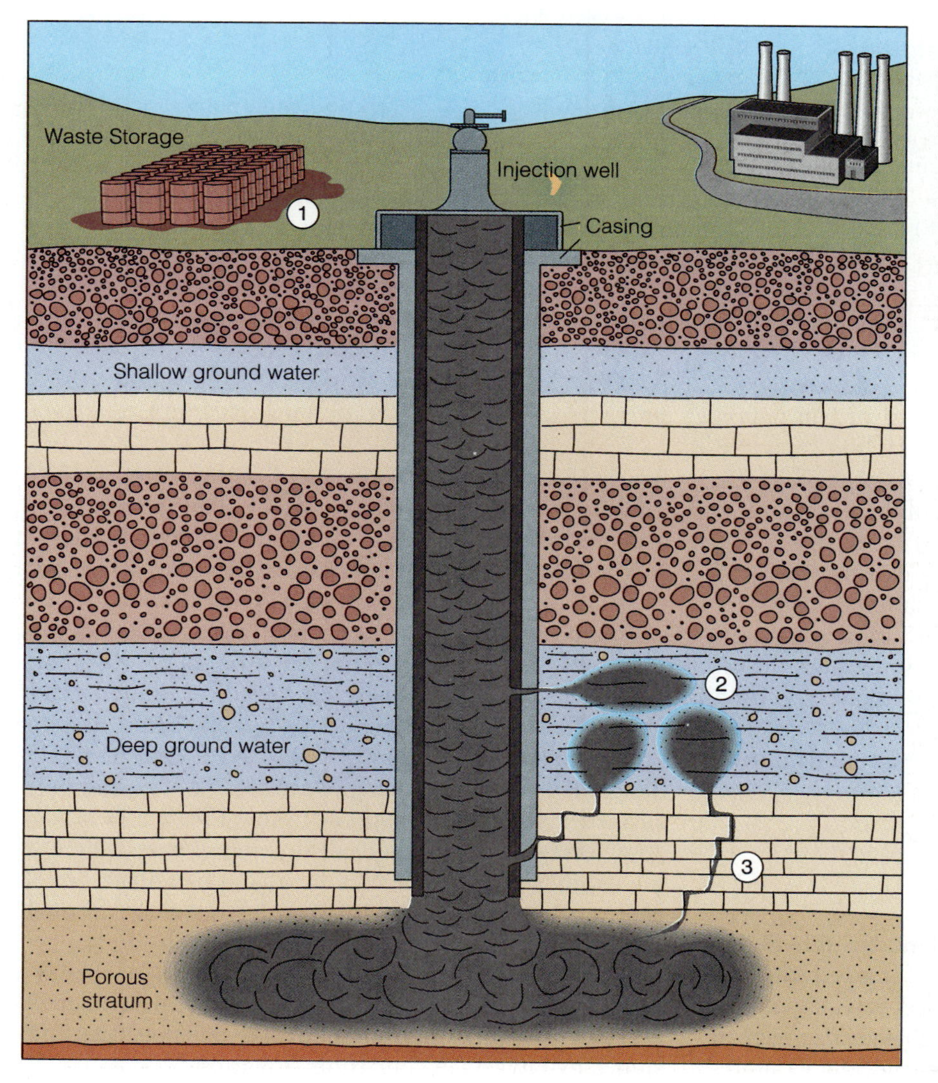

➢ **FIGURE 15.20 Deep-well injection of hazardous waste. This method presumes that wastes injected into porous-rock strata deep within the earth are isolated from the environment forever. What could go wrong? 1. Toxic spills may occur at the ground surface. 2. Corrosion of the casing may allow injected waste to leak into an aquifer. 3. Waste may migrate upward through rock fractures into the ground water.**

TABLE 15.6 Chronology of Selected Hazardous Substances in the United States

DATE	SUBSTANCE	NATURE OF HAZARD	EVENT
1700s	Phenols-polyaromatic hydrocarbons	Carcinogenic	
1860	Copper acetoarsenite (Paris Green)	Chronic exposure results in serious health problems or death	Used against the Colorado potato beetle
1900	Arsenic, sulfur, and petroleum oils, pyrethrum	Toxic	Commonly used pesticides
1920	Phenol, cresol hydrocarbons, napthalene	Produce disagreeable tastes in drinking water	Effects identified
1925	Ethylene, ethylene oxide, ethylene glycol (antifreeze)	Carcinogenic; causes liver disease, mutations, nervous disorders	Effects identified
1927	Polychlorinated biphenyls (PCBs)	Causes mutations, cancer, birth defects, and liver disease	Used in heat transfer for electrical transformers, motors, appliances
1932	Butylenes refined to produce methylethyl ketone (MEK), other halogenated solvents	Highly toxic	Identified as hazardous ground-water contaminants
1940s	Chlorinated hydrocarbons (chlordane, toxaphene, dieldrin, aldrin, DDT) and organophosphate (parathion)	Highly toxic or carcinogenic; persistent for years, low biodegradation, bioaccumulative in humans	Used as insecticides
1940s	Plastics of all types (polyvinyl chloride (PVC), styrene, polystyrene), urethanes	Indiscriminate disposal of chemical wastes	Synthetic organic polymer industry expands rapidly
1950s	Polyethylene and high-density polyethylene (HDPE)	Indiscriminate disposal of chemical wastes	Plastics expand into almost all sectors of manufacture of durable goods

table continues next page

700,000 of these small firms generate 90 percent of our hazardous waste. Consequently, most of these wastes end up in sanitary landfills (see Figure 15.6).

Discharge into sealed pits is the least expensive way to dispose of large amounts of water containing relatively small amounts of hazardous substances. If the pit is well sealed and evaporation of water equals or exceeds the input of contaminated water, the pit may receive and hold hazardous waste almost indefinitely. Problems can arise from leaky seals and overflow of holding ponds during heavy storms or floods.

Superfund

What about the thousands of old and mostly unrecorded sites where hazardous waste has been dumped improperly in the past? Hazardous substances have been disposed of indiscriminately since the eighteenth century in the United States (Table 15.6). **Superfund,** the Comprehensive Environmental Response, Compensation, and Liability Act (CERCLA), was passed in 1980 to rectify past and present abuses from toxic waste dumping. Managed by the EPA, it allows the agency to clean up a spill and then bill the responsible party for the cleanup. CERCLA also empowered the EPA to establish a National Priorities List (NPL) for cleaning up the worst of the 15,000 abandoned toxic or hazardous-waste sites, now known as *Superfund sites,* some of which date back many decades (Case Study 15.4). By 1993 1,236 NPL sites had been designated or were pending. Of those, 52 sites had been cleaned up, and work had begun on at least 500 others. New Jersey has the dubious distinction of harboring 6 of the worst 15 Superfund sites as ranked by impact on the environment—5 landfills and 1 industrial site. Although the cleanup costs are staggering, toxic dumps do not just "go away," and we must correct past abuses for the sake of future generations (➢ Figure 15.21). Unfortunately, 40 percent of the Superfund budget to date has been consumed by lawyers in courtroom litigation.

TABLE 15.6 Chronology of Selected Hazardous Substances in the United States *Continued*

DATE	SUBSTANCE	NATURE OF HAZARD	EVENT
1955	Halogenated hydrocarbons	Chemical warfare agents, pesticides, propellants, refrigerants. Some are ozone scavengers	Toxicity discovered by Van Oettingen of U.S. Public Health Service
1959	Dibromochloropropane (DBCP)	Toxic	Used against nematode worms
1968	All paints with white lead pigment	Human health hazard	Banned
1970	PCBs		Manufacturing ceased
1977	DBCP	Toxic	Banned
1980s	Cyanide, pesticides, acids	Toxic	Local aquifers at Swartz Creek, Michigan,* were found to be contaminated by illegal burial of drums containing these substances in 1985
1980s	Dioxin-containing oil	Carcinogenic, birth defects	Gravel roads at Times Beach, Missouri,* were surfaced with this oil. The EPA bought the town and relocated 2,200 residents at a cost of $33 million in 1985
1985	Paint thinner, pesticides, TV sets, chemical bleaches, batteries, fingernail polish	Various	These substances were found dumped with ordinary wastes at BKK landfill, Covina, California*
1990s	Cyanide, mercury, lead, sulfur, hydrocarbons, etc.	Various	Abandoned mines and oil wells on lands administered by Dept. of the Interior. Cleanup costs estimated at $50 billion in 1993
1990s	Radioactive and toxic wastes	Cancer causing, may be airborne	Cleanup of Department of Energy complexes. Estimated as $200 billion over 25 years in 1993

*Superfund site.

SOURCE: Pre-1980 data adapted from Allen Hatheway, "Pre-RCRA History in Industrial Waste Management in Southern California," in B. W. Pipkin and R. J. Proctor, eds., *Engineering Geology Practice in Southern California,* Association of Engineering Geologists Special Publication 4 (Belmont, Calif.: Star Publishing Company, 1992).

WASTEWATER TREATMENT

Sewage is required to be purified to some predetermined standard before it is delivered to a receiving stream or aquifer. An important measure of purity of sewage effluent is **biochemical oxygen demand (BOD),** the amount of oxygen per unit volume of water—milligrams of O_2 per liter of water—required for the total aerobic decomposition of organic matter by microorganisms. BOD is determined by measuring the amount of oxygen in a closed liter bottle of treated water at intervals until uptake ceases. The total oxygen used is the BOD. Water that is totally free of organic matter and microorganisms has a BOD of zero. As many as three levels of sewage treatment may be required to bring BOD, water clarity, and chemical quality to within acceptable limits for a particular receiving water body (Table 15.7):

- *Primary treatment* consists mostly of removing solids by gravity settling, screening, and then aerating the fluid portion to remove some organic material. There is considerable debris in raw sewage, including wood, rags, paper, and sediment. (A group of students once reported seeing a small bicycle on the screen during a field trip to a treatment facility!)
- *Secondary treatment* reduces the BOD of the wastewater to a specified level of purity. It introduces bacteria to digest organic matter and facilitates further sludge settling and clarification of the effluent. Secondary treatment requires the presence of microorganisms and a mechanism for replenishing oxygen in the wastewater. One method is to spray wastewater onto a trickling filter, where microbial films develop on the filter material. Another method uses *activated sludge,*

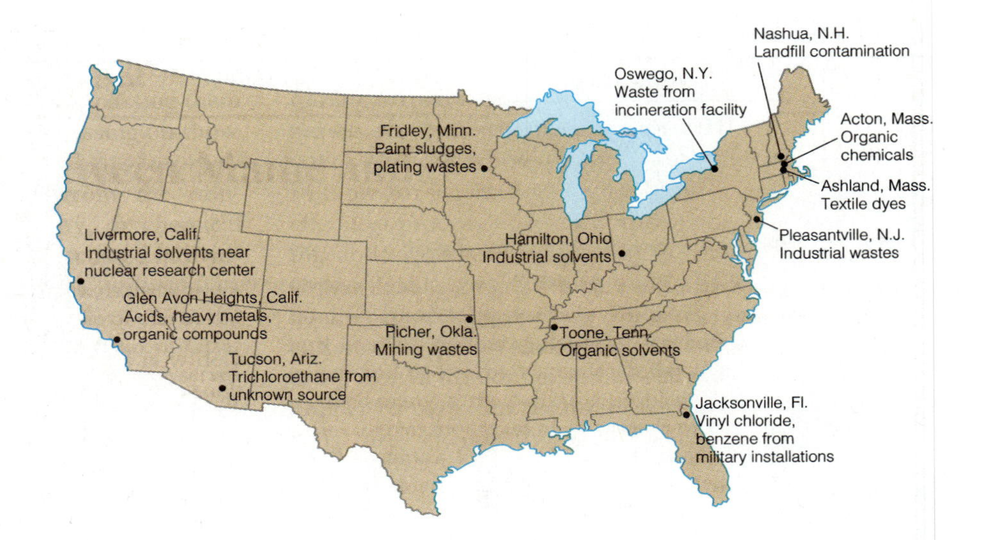

➢ **FIGURE 15.21 Only 13 of the thousands of identified U.S. cases of water contamination.**

whereby a microbial growth that consumes organic matter is suspended in the wastewater and settles to the floor of the reaction tank. Some of the settled sludge is recycled to the incoming wastewater to maintain the biomass required for decomposition. Regardless of the level of wastewater treatment, a huge volume of solids always remains that requires disposal. The alternative of dumping sludge at sea will soon be history. In landlocked areas and coastal areas where ocean dumping is prohibited, sludge must be dried and hauled to a landfill. Some is used commercially as a soil amendment for holding moisture and for fertilizer, although dried sludge has low nitrogen and little plant-nutrient value.

- *Tertiary treatment* removes nitrogen and phosphorus compounds, reducing their levels to below those that cause eutrophication of the receiving waters. **Eutrophication** (from the Greek word, "to thrive") is a water problem that arises from nutrient enrichments and their associated explosive algal blooms. Sewage effluent is high in ammonia (NH_3), which is converted to nitrate $(NO_3)^-$, a plant fertilizer, by nitrifying bacteria. Tertiary treatment is expensive, but it is essential where the discharge of partially treated sewage effluent would jeopardize the health of the receiving water body, such as at Lake Tahoe (Case Study 15.5). Tertiary-treated effluent is pure enough for human consumption.

TABLE 15.7 What's Left after Two-stage Sewage Treatment

WASTEWATER COMPONENT	REMAINING AFTER TREATMENT(%)	
	Primary	*Secondary*
Biological oxygen demand (BOD) Dissolved oxygen necessary to decompose organic matter in wastewater.	65	15
Nitrogen content Nitrogen becomes a plant nutrient when it combines with hydrogen or oxygen to form ammonia, nitrate, or nitrite.	80–85	70
Phosphorus content Phosphorus becomes a plant nutrient when it combines with oxygen to form phosphate.	80–85	70
Suspended solids Undissolved material suspended in water (sludge)	40	15

Data from T. M. Hawley, "Herculean labors to clean wastewater" *Oceanus* 33 (1990) no. 2, p. 73.

Private Sewage Disposal

Modern household sewage-disposal systems utilize a septic tank for separating solids from the effluent and leach lines for disseminating the liquids into the surrounding soil or rock (➢ Figure 15.22). A septic tank is constructed in such a way that all solid material stays within the tank and only liquid passes on into the leach field. A leach field consists of rows of tile pipe surrounded by a gravel filter through which the effluent percolates prior to percolating into the soil or underlying rock. Sewage is broken down by *anaerobic* bacteria in the septic tank and the first few feet of the leach lines. As the effluent percolates into the soil or bedrock, it is filtered and oxidized, and *aerobic* conditions are established. Studies have shown that the movement of effluent through a few feet of unsaturated fine-grained soil under aerobic conditions reduces bacterial counts to almost nil. Fine-grained soils have been found to be the best filters of harmful pathogens, whereas

CASE STUDY 15.4

Love Canal—No Love Lost Here

Love Canal, in the honeymoon city of Niagara Falls, New York, may sound very romantic, but it's not. Excavated by William T. Love in the 1890s but never finished, the canal is 1,000 meters long and 25 meters wide (3,300 ft × 80 ft). Unused and drained, Love Canal was bought by Hooker Chemical and Plastics Corporation in 1942 for use as a waste dump. In the next 11 years Hooker dumped 21,800 tons of toxic waste into the canal, which was believed to be impermeable and thus an ideal "grave" for hazardous substances. After the canal was filled and covered, Hooker Chemical sold the land to the Niagara Falls School Board for one dollar. A school was built on the landfill, and hundreds of homes and all the infrastructure needed to support a thriving suburban community were built nearby.

In the early 1970s, after several years of heavy rainfall, water leaked through the "impermeable" clay cap and into basements and yards in the Love Canal area, and also into the local sewer system. The residents of the new community experienced a high incidence of miscarriages, birth defects, liver cancer, and seizure-inducing diseases among children. Residents complained that toxic chemicals in the water were causing these adverse health effects.

In 1978 the New York State Health Commissioner requested the EPA's assistance in investigating the chemistry of fluids that were leaking into a few houses around the canal. The study revealed the presence of 82 toxic chemicals including benzene, chlorinated hydrocarbons, and dioxin, and the state commissioner declared Love Canal a health "emergency." Five days later President Carter declared it a federal disaster area, the first human-made environmental problem to be so designated in the United States. The recommendation that pregnant women and children under the age of two be evacuated from the area followed, and New York State appropriated $22 million to buy homes and repair the leaks. In all, a thousand families were relocated (➢ Figure 1).

The results of an EPA monitoring study were released to the public in 1982. An effective ground-water barrier and drain system were installed to intercept the toxic chemicals. The presence of relatively impermeable clays in the area suggests that future long-distance migration of pollutants from Love Canal is unlikely. Because the shallow and deep aquifers in the area are not hydraulically connected, there is little danger of contaminants migrating into deeper ground-water zones.

Love Canal was a precedent-setting case, because it caused the federal government to create Superfund under CERCLA, a law that provides federal money for toxic cleanup without long appeals. It also permits governmental entities to sue polluters for the costs of cleanup and relocation of victims. Although Love Canal has been essentially contained, some chemicals still infect a nearby stream and schoolyard, and maintenance costs amount to a half-million dollars annually. The EPA no longer ranks Love Canal high among the nation's most dangerous waste sites, and the New York State Health Commissioner has announced that recent federal–state studies found four of the seven polluted Love Canal areas to be habitable. In spite of the commissioner's statement, there has been no stampede to reinhabit the area.

➢ FIGURE 1 Love Canal in Niagara Falls, New York, where the honeymoon ended and the nightmares began.

CASE STUDY 15.5

The Cinder Cone Disposal Site—Getting Your Sewage Together

Prior to 1965 sewage disposal in the north Lake Tahoe area was not much of a problem. Most households used cesspools or septic tanks to dispose of their waste. With the area's population growth in the 1960s, however, it became clear that too much phosphate and nitrate were seeping from private septic-system leach lines into the lake. The nutrients in the effluent were stimulating algal growth, which in turn was clogging the shore zone and causing problems. It was decided to build a sewage disposal site outside the Tahoe basin to handle the 2.5 million gallons of sewage generated each day in the north Lake Tahoe area. This was to be an interim solution until a permanent tertiary treatment facility could be built to serve the area.

Geologic studies indicated that a volcanic cinder cone near the Truckee River northwest of Tahoe City would be an ideal temporary disposal site (➢ Figure 1). It showed no evidence of recent activity, and its clinkery cinders and aa-type lava flows would be good percolation media for sewage effluent (➢ Figure 2). Furthermore, springs on the cone's flanks would allow chemical monitoring of the sewage effluent to determine its purity after percolation through the volcano. An interim primary treatment plant was built in Tahoe City to separate the solid waste from sewage. The sludge from that plant was dried and trucked to a disposal site some 30 kilometers (19 mi) away, and the effluent was collected in ponds and pumped to several miles of percolation trenches in the central crater (➢ Figure 3). Winter freezing was not a problem, and chemical analyses indicated that the water emerging from the cone was in a relatively pure state.

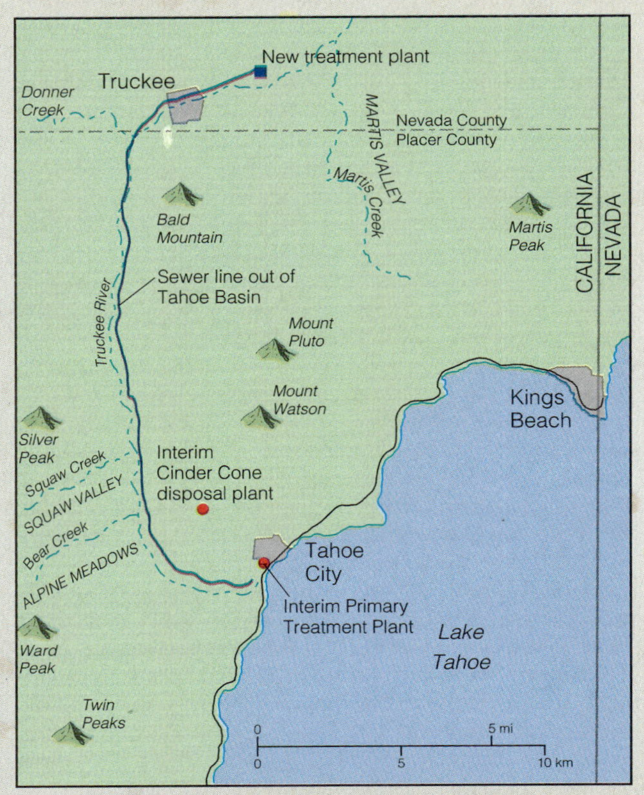

➢ **FIGURE 1 Temporary Cinder Cone disposal site and new water treatment plant for Tahoe City, California.**

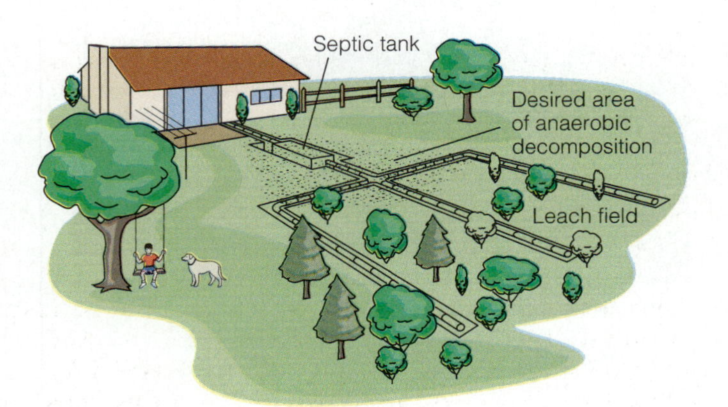

➢ **FIGURE 15.22 A septic tank near the house drains into a leach field of drain tiles that are laid in trenches and covered with filter gravel. The excavation is then covered with soil and landscaped.**

coarse-grained soils allow pathogenic organisms and viruses to disperse over greater distances.

Under the best of conditions, leach lines have a life expectancy of 10–12 years. Factors that reduce infiltration rates and thus limit the life of a septic system are dispersion of clay particles by transfer of sodium ions (Na^+) in sewage to clay particles in the soil, plugging by solids, physical breakdown of the soil by wetting and drying, and biological "plugging." This plugging occurs when the leach system becomes overloaded with sewage and anaerobic conditions invade the leach lines, resulting in a growth of organisms that plug the filter materials.

After passing through the leach field, the effluent is still foul, smelly, and highly toxic—a harmful, pathogen-carrying fluid. It must be kept underground until oxidation and bacteria can purify it, making it indistinguishable from local ground water. The geologic condition

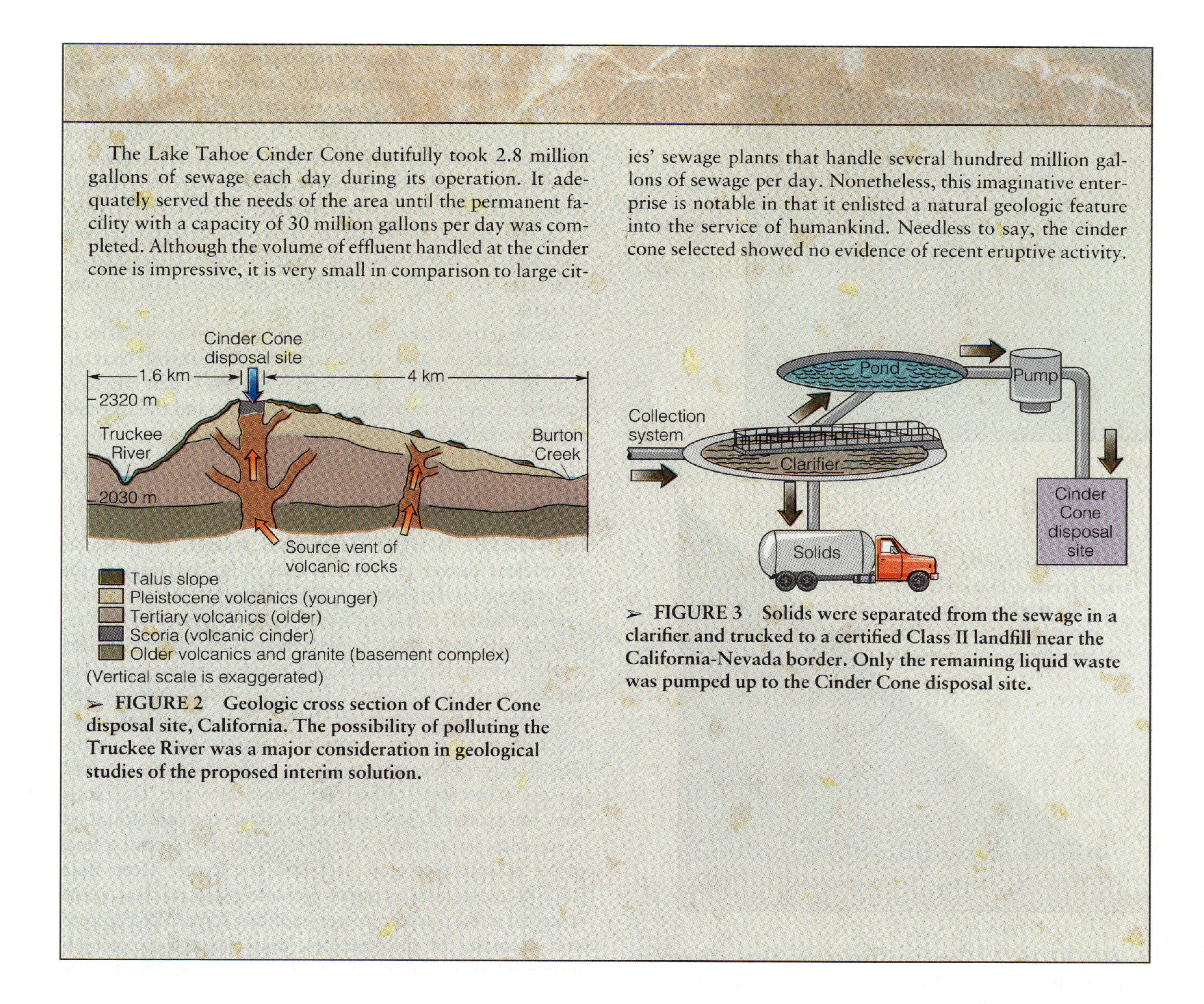

The Lake Tahoe Cinder Cone dutifully took 2.8 million gallons of sewage each day during its operation. It adequately served the needs of the area until the permanent facility with a capacity of 30 million gallons per day was completed. Although the volume of effluent handled at the cinder cone is impressive, it is very small in comparison to large cities' sewage plants that handle several hundred million gallons of sewage per day. Nonetheless, this imaginative enterprise is notable in that it enlisted a natural geologic feature into the service of humankind. Needless to say, the cinder cone selected showed no evidence of recent eruptive activity.

➢ FIGURE 2 Geologic cross section of Cinder Cone disposal site, California. The possibility of polluting the Truckee River was a major consideration in geological studies of the proposed interim solution.

➢ FIGURE 3 Solids were separated from the sewage in a clarifier and trucked to a certified Class II landfill near the California-Nevada border. Only the remaining liquid waste was pumped up to the Cinder Cone disposal site.

that most commonly causes a private sewage-disposal system to fail is a high water table that intersects or is immediately beneath the leach lines (➢ Figure 15.23a). This condition initiates the anaerobic growth that plugs filter materials and tile drains. Water tables may rise due to prolonged heavy precipitation or due to the addition of a number of septic tanks in a small area. One way to avoid exceeding the soil's ability to accept effluent is to design leach fields on the basis of percolation tests performed during wet periods. If necessary, a system's lifetime can be extended by constructing multiple leach fields whose use can be rotated regularly by switching a valve. Another problem is the improper placement of leach lines. This results in effluent seepage on hill slopes or stream valley walls (➢ Figure 15.23b; see also Chapter 10). The proper horizontal distance of leach lines from a hill face is a function of *vertical permeability* and thus is best determined by a qualified professional. Although typical plumbing codes specify a minimum of 5 meters (16 ft) horizontal distance from a hill face, this may not be sufficient if the vertical permeability is low and the hill slope is steep.

Nitrates in sewage effluent pose a eutrophication problem if households are built near a lake or pond. Eutrophication is a natural process in the life cycle of a lake, but cultural (human-caused) eutrophication accelerates the process and can convert an attractive, "real-estate" lake into a weed-choked bog in a short time. Nutrients from sewage systems and fertilized lawns must not be allowed to flow unchecked into bodies of water. Solving this problem requires the help of such experts in water flow and quality as aquatic biologists and hydrogeologists.

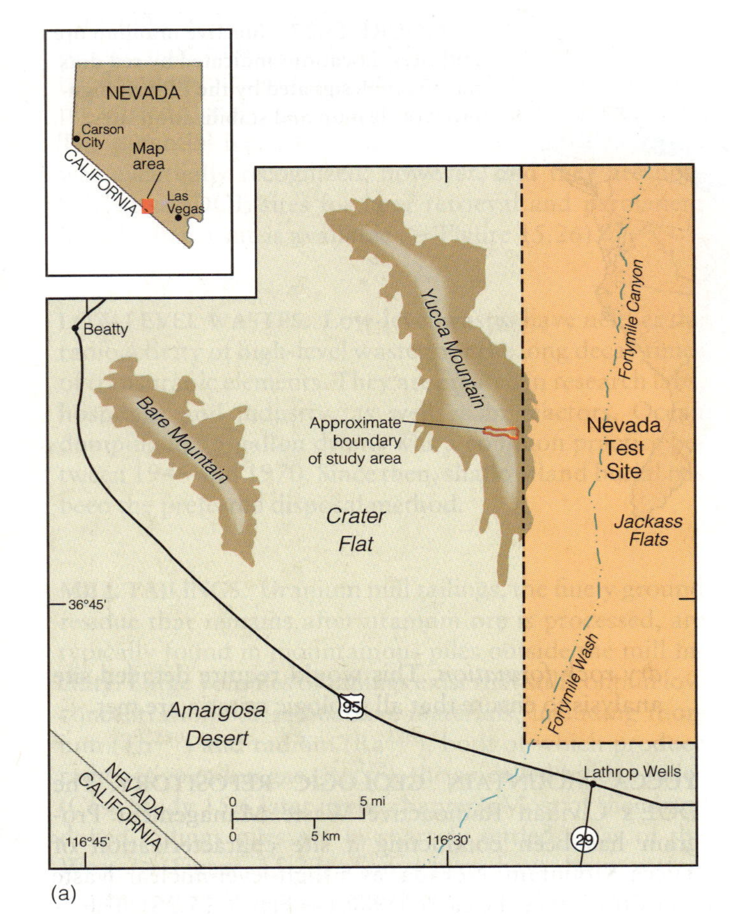

(a)

(b)

➢ **FIGURE 15.28** **(*a*) Location of proposed Yucca Mountain nuclear repository. (*b*) Aerial view of the Yucca Mountain study area. Structures are at the site of the proposed repository.**

activity, and have conducted countless, rigorous geophysical studies of such factors as seismic and volcanic potential. Preliminary studies indicate that Yucca Mountain could be an acceptable geologic repository, but until it receives final certification, radioactive wastes are piling up in temporary storage facilities at an alarming rate.

Completion of the Yucca Mountain repository was originally scheduled for 1998, but the year has been extended twice to date, first to 2003 and then to 2010. The legal stumbling blocks placed in the way of the project by the State of Nevada are classic examples of NIMBY (see Case Study 15.1). When and if the repository is ever built and filled, the DOE will be required to monitor it for at least 50 years from the opening date. This will allow nuclear scientists to monitor heat flow, rock conditions, and leakage and to retrieve spent fuel if it is deemed necessary. Eventually all openings will be backfilled with the removed rock and soil and then sealed. Buildings will be removed, and the site surface will be returned to its original condition.

Health and Safety Standards for Waste

There is general agreement that the Yucca Mountain repository (or any geologic repository for that matter) will leak over the course of the next 10,000 years; that is, that it will not contain 100 percent of the radionuclides. Nuclear waste emits three types of radiation:

1. Alpha (α) particles, the most energetic but least penetrating type of radiation (they can be stopped by a piece of paper). If α emitters are inhaled or ingested, alpha-radiation can attack lung and body tissue.
2. Beta (β^-) radiation, penetrating but most harmful when a β^- emitter is inhaled or ingested. For example, strontium-90 is a β^- emitter that behaves exactly like calcium (builds bone) in the food chain.
3. Gamma (γ) rays, nuclear X rays that can penetrate deeply into tissue and damage human organs.

Nuclear waste policy limits the radioactivity that may be emitted from a repository, as measured by cancer deaths directly attributable to leaking radiation. The mortality limit is 10 deaths per 100 years, and for this reason the maximum allowable radiation dose per year per individual in the proximity of a repository is set at 25 millirems (mrem). The *rem* is the unit used for measuring the amounts of ionizing radiation that affect human tissue, and a *millirem* is 0.001 rem. Doses on the order of 10 rems can cause weakness, redden the skin, and reduce blood-cell counts. A dose of 500 rems would kill half of the people exposed. Rem units take into account the type of radiation—alpha, beta, or gamma—and the amount of radioactivity deposited in body tissue. ➢ Figure 15.30 shows how the average annual dose of 360 mrems is partitioned for U.S. citizens and common sources of radiation. (Case Study 15.6 explains the hazards of high levels

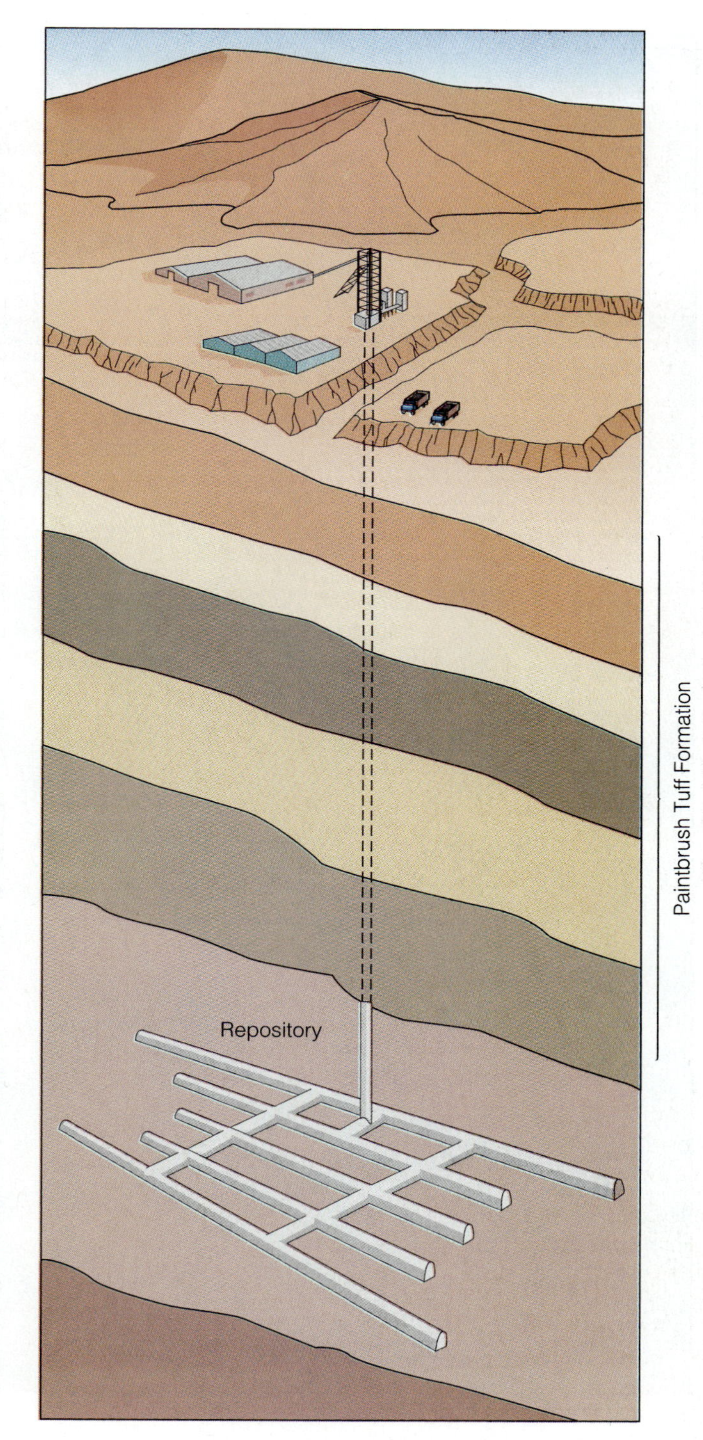

➢ **FIGURE 15.29** **(*a*) Paintbrush Tuff Formation at Yucca Mountain. Rock layers called *members* are distinguishable beds within the formation. The proposed position of the repository is shown. (*b*) Artist's conception of a typical repository.**

of radon gas in the home and methods for testing and mitigating them.) It is estimated that a person living within 5 kilometers (3 mi) of the Yucca Mountain repository would receive less than 1 mrem per year as a result of storing nuclear waste there (currently no one lives that close to the site). The 25-mrem standard for maximum annual dosage is equal to the radiation a person would receive in 2 or 3 medical X rays.

Another standard is that the travel time for underground water flowing from the repository to areas where it is consumed must be more than 1,000 years. Calculations based upon the permeability of the rocks near Yucca Mountain indicate that the ground-water flow rate there is 3,400 to 8,300 years per mile. Since this area receives only 15 centimeters (6 in) of rainfall a year, heavy infiltration through the repository to the water table should not be a problem. Contamination of water below the zone of saturation is most likely if the waste is in liquid form and the containers leak. For this reason, most of the waste would be stored in solid form. Forecasting the climate 10,000 years from now is a problem. If there is a significant change, say another ice age, pluvial (wet) climate conditions could exist in the desert that would increase the amount of water infiltrating downward through the repository and into the saturated zone. A rise in the water table would decrease the time in which any leaking radionuclides would spread into the environment.

The final repository for nuclear-weapons waste now stored at facilities around the country is the 10,000-acre Waste Isolation Plant (WIP) in New Mexico. To expand the use of this facility would simply require that the Department of Energy withdraw the land from the public domain. Federal legislators have objected to this, however, because storage plans are not subject to EPA review.

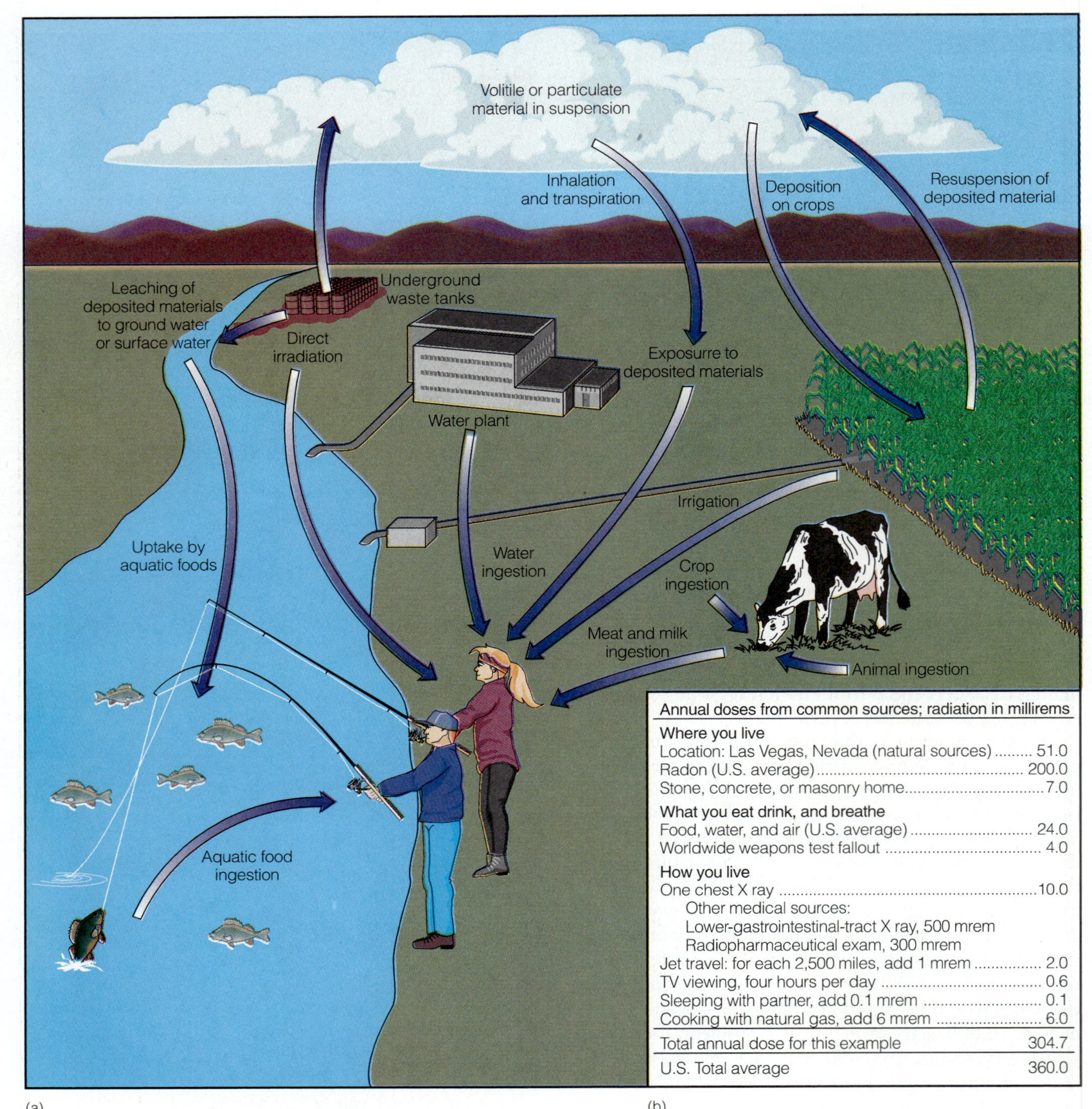

Annual doses from common sources; radiation in millirems	
Where you live	
Location: Las Vegas, Nevada (natural sources)	51.0
Radon (U.S. average)	200.0
Stone, concrete, or masonry home	7.0
What you eat drink, and breathe	
Food, water, and air (U.S. average)	24.0
Worldwide weapons test fallout	4.0
How you live	
One chest X ray	10.0
Other medical sources:	
Lower-gastrointestinal-tract X ray, 500 mrem	
Radiopharmaceutical exam, 300 mrem	
Jet travel: for each 2,500 miles, add 1 mrem	2.0
TV viewing, four hours per day	0.6
Sleeping with partner, add 0.1 mrem	0.1
Cooking with natural gas, add 6 mrem	6.0
Total annual dose for this example	304.7
U.S. Total average	360.0

➢ **FIGURE 15.30** (*a*) **Pathways through which humans are subjected to radiation.** (*b*) **Whole-body radiation dose received annually by the average U.S. citizen. Persons who work in medical research or certain industrial activities or who live in high-radon areas may exceed the average by considerable amounts.**

CASE STUDY 15.6

Radon and Indoor Air Pollution

Radon (Rn^{222}) is a radioactive gas formed by the natural disintegration of uranium as it transmutes step-by-radioactive-step to form stable lead. The immediate parent element of radon is radium (Ra^{226}), which emits an alpha particle (He^4) to form Rn^{222}. $Radon^{222}$ has a half-life of only 3.8 days, but it in turn breaks down into other elements that emit dangerous radionuclides. *Emanation* is sometimes used to describe the behavior of this element, and concern has been growing about the health hazard posed by radon emanating from rocks and soils and seeping into homes. Radon was first suggested as a danger to miners by Georgius Agricola (1494–1555). Known as "the father of mineralogy," Agricola proposed that the lung disease contracted by miners might have something to do with emanations from inside the mines. The EPA estimates that between 5,000 and 20,000 people die every year of lung cancer because they have inhaled radon and its decay products.

Radon is only a problem in areas underlain by rocks whose uranium concentrations are greater than 10 ppm—which is 3 times the average amount found in granite. For comparison, uranium ore contains more than 1,000 ppm of uranium. As radon disintegrates in the top few meters of rock and soil, it seeps upward into the atmosphere, or into homes through cracks in concrete slabs or around openings in pipes. The measure of radon activity is picocuries per liter (pCi/L)—the number of nuclear decays per minute per liter of air. One pCi per liter represents 2.2 potentially cell-damaging disintegrations per minute. Inasmuch as the average human inhales between 7,000 and 12,000 liters (2,000–3,000 gallons) of air per day, high radon concentrations in household air can be a significant health hazard. In addition, radon can access the human body through well water if it is injested within a few weeks of withdrawal from underground. Average uranium concentrations of some common rocks are

black shale	8.0 ppm
granite	3.0 ppm
schist	2.5 ppm
limestone	2.5 ppm
basalt	1.0 ppm

Bear in mind that these figures do not indicate the *range* of uranium content in the rocks, which for black shales and granites can be quite great.

Cancer due to bombardment of lung tissue by breathing Rn^{222} or its decay products is a major concern of health scientists. The EPA has established 4 pCi/L as the maximum allowable indoor radon level. This is equivalent to smoking a half-pack of cigarettes per day or having 300 chest X rays in a year. ➢ Figure 1 vividly illustrates the dangers of inhaling indoor radon (and of smoking). Cancer rarely appears before 5–7 years of exposure, and it is rare in exposed individuals under the age of 40.

pCi/L	Comparable Exposure Level	Comparable Risk *
200	1,000 times average outdoor level	More than 75 times nonsmoker risk of dying from lung cancer
		4-pack/day smoker
100	100 times average indoor level	10,000 chest X rays per year
40		30 times nonsmoker risk of dying from lung cancer
20	100 times average outdoor level	2-pack/day smoker
10	10 times average indoor level	1-pack/day smoker
4		3 times nonsmoker risk of dying from lung cancer
2	10 times average outdoor level	200 chest X rays per year
1	Average indoor level	Nonsmoker risk of dying from lung cancer
0.2	Average outdoor level	

* Based on lifetime exposure

➢ FIGURE 1 Lung cancer risk due to radon exposure, smoking, and X-ray exposure. A few houses have been found with radon radiation levels greater than 100 pCi/L, and one in Pennsylvania had a level of 2,000 pCi/L.

Widespread interest in radon pollution first appeared in the 1980s with the discovery of high radon levels in houses

continued next page

CASE STUDY 15.6 *Continued*

in Pennsylvania, New Jersey, and New York. The discovery was made accidentally when a nuclear-plant worker set off the radiation monitoring alarm when he arrived at work one day. Since his radioactivity was at a safe level, he did not worry about it. Returning to work after a weekend at home, he once again triggered the alarm. An investigation of his home revealed very high levels of radon. This was the first time radon was recognized as a public health hazard.

There are currently two types of radon "detectors." One is a charcoal canister that is placed in the household air for a week and then returned to the manufacturer for analysis. The other is a plastic-film alpha-track detector that is placed in the home for 2–4 weeks (➢ Figure 2). It is recommended that the less-expensive canister type be used first and, if a high radon level is indicated, to make a follow-up measurement with a plastic-film detector. Any reading above 4 pCi/L requires follow-up measurements. If the reading is between 20 and 200 pCi/L, one should consider measures to reduce radon levels in the home (➢ Figure 3). Common methods are increasing the natural ventilation below the house in the basement or crawlspace, using forced air to circulate subfloor gases, and intercepting radon before it enters the house by installing gravel-packed plastic pipes below floor slabs. For houses with marginal radon pollution, sealing all openings from beneath the house that go through floor or walls, is usually effective.

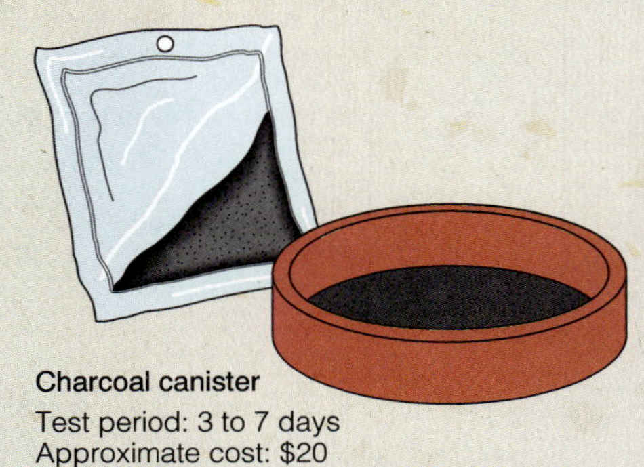

➢ **FIGURE 2 Canister and plastic-film radon detectors.**

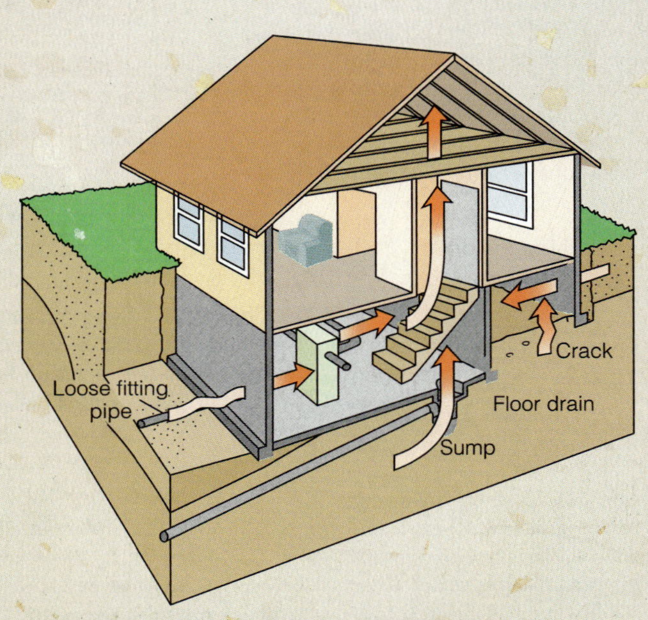

➢ **FIGURE 3 Routes of radon entry into a home. Plugging radon leaks is the most common method of reducing residential radon accumulations.**

Summary

Waste Disposal

METHODS: Isolation, attenuation, and recycling.

Municipal Waste Disposal

SANITARY LANDFILL: Trash is covered with clean soil each day.

OCEAN DUMPING: A favorite location in the past for disposal of sewage, sludge, dredgings, and canistered nuclear wastes. Now regulated by Marine Protection Reserve and Sanctuary Act (Ocean Dumping Act).

PROBLEMS: Leachate, settlement, gas generation, visual blight.

MULTIPLE LAND USE CONCEPT: A mine or quarry becomes a landfill, which when full may be converted to a park or other recreational area. Although some shallow landfills have been built upon, settlement is always a problem.

Hazardous Waste Disposal

SECURE LANDFILLS: Lined with plastic and impermeable clay seal. Built with leachate drains.

DEEP-WELL INJECTION: Into dry permeable rock body sealed from aquifers above and below. Has triggered earthquakes in Denver area.

SUPERFUND: Comprehensive Environmental Response, Compensation, and Liability Act (CERCLA) created in 1980 to correct and stop abuses from irresponsible toxic-waste dumping.

Wastewater Treatment

PURPOSE: To clarify and purify water. The measure of purification is biological oxygen demand (BOD).

1. *Primary*—Solids are separated by settling and some aeration to remove organic materials.
2. *Secondary*—BOD is reduced by bacterial action and aeration.
3. *Tertiary*—Inorganic nutrients such as nitrates and phosphates are removed.

Nuclear Waste Disposal

TYPES OF WASTES:

1. *High-Level*—No permanent storage has been found for this dangerous material, mostly spent fuel from nuclear power plants. This is a major health hazard at present.
2. *Transuranic*—Low radioactivity but long half-lives requires isolation.
3. *Low-level*—Generated by research labs, hospitals, industry, uranium mill tailings, as well as reactors. Shallow land burial is preferred disposal method. Mill tailings present a problem because they are subject to wind and water erosion, and have been used in building materials in the Western U.S.

ISOLATION CONSIDERATIONS: Sequestering nuclear waste must provide the following safeguards: isolation for 250,000 years; safe from sabotage and natural disasters; noncontamination of natural resources; and safe transportation and handling.

GEOLOGICAL REPOSITORY CONDITIONS: Geologically stable, dry, low water table, low permeability, relatively free from geologic hazards. The only site that meets these criteria at present is the Yucca Mountain, Nevada, site.

Key Terms

attenuation (waste disposal)
biological oxygen demand (BOD)
coking
composting
EPA
eutrophication
geological repository
high-level waste
isolation (waste disposal)
leachate
low-level waste
NEPA
NWPA
RCRA
recycling
sanitary landfill
Superfund
transuranic element (TRU)

Study Questions

1. Explain the federal legislation known as NEPA and RCRA and describe their impact on creating environmentally safe landfills.
2. What types of radiation are emitted by nuclear waste, and what hazards do they pose to the health of all organisms?
3. What is leachate, and what methods are used to prevent landfill leachates from entering bodies of water?
4. Explain the "cell" method of placing trash in a landfill.
5. How much solid waste does the average American generate each day in kilograms? In pounds?
6. Explain what Rn^{222} is and how it is formed. What environmental hazard does it present?
7. What problems are connected with operating a sanitary landfill?
8. What resources are found in a landfill, and how can they be recovered? What can individuals do to assist in recovering and extending the lives of certain limited natural resources?
9. By what two common methods are hazardous wastes sequestered from the environment? Explain the steps that are taken to keep these wastes from polluting underground water.
10. What is eutrophication, and what can be done in sewage treatment to minimize this threat to lakes and rivers?

average crustal abundance The percentage of a particular element in the composition of the earth's crust.

a-zone The upper soil horizon (topsoil) capped by a layer of organic matter. It is the zone of leaching of soluble soil salts.

back slope The gentler slope of a dune, scarp, or fault block.

ball mill A large rotating drum containing steel balls that grinds ore to a fine powder. Batteries of ball mills are used in modern milling operations.

banded-iron ore An iron deposit consisting essentially of iron oxides and chert occurring in prominent layers or bands of brown or red and black.

barchan dune An isolated, moving, crescent-shaped sand dune lying transverse to the direction of the prevailing wind. A gently sloping, convex side faces the wind, and the horns of the crescent point downwind.

barrier beach (barrier island) A narrow sand ridge rising slightly above hightide level that is oriented generally parallel with the coast and separated from the coast by a lagoon or a tidal marsh.

basalt A dark-colored, fine-grained volcanic rock that is commonly called *lava*. Basaltic lavas are quite fluid.

base level The level below which a stream cannot erode its bed. The ultimate base level is sea level.

base shear Due to a rapid horizontal ground acceleration during an earthquake. It can cause a building to part from its foundation; thus the term.

batholith A large, discordant intrusive body greater than 100 km^2 (40 mi^2) in surface area with no known bottom. Batholiths are granitic in composition.

bauxite An off-white, grayish, brown, or reddish brown rock composed of a mixture of various hydrous aluminum oxides and aluminum hydroxides and that rock contains such impurities as free silica, silt, iron oxides, and especially clay minerals; a highly aluminous laterite. The principal ore of aluminum.

beach cusps A series of regularly spaced points of beach sand or gravel separated by crescent-shaped troughs.

bedding *See* stratification.

bedding planes The divisions or surfaces that separate successive layers of sedimentary rocks from the layers above and below them.

bench placer Remnant of an ancient alluvial placer deposit that was concentrated beneath a former stream and that is now preserved as a bench on a hillside.

Benioff zone A zone of earthquake epicenters beneath oceanic trenches that extends from near the surface to a depth of 700 km (430 mi). The foci of such earthquakes were plotted in the 1940s and 1950s by Hugo Benioff, for whom they are named. They are thought to indicate active subduction.

bentonite A soft, plastic, light-colored clay that swells to many times its volume when placed in water. Named for outcrops near Fort Benton, Wyoming, bentonite is formed by chemical alteration of volcanic ash.

beta decay The radioactive emission of a nuclear electron (beta ray, β^-) that converts a neutron to a proton. The element is changed to the element of next-highest atomic number; for example, ${}_6C^{14} \rightarrow {}_7N^{14} + \beta^-$.

big bang The hypothesis that the currently observed expansion of the universe may be traced back to an exploding compact cosmic fireball.

biological oxygen demand (B.O.D.) Indicator of organic pollutants (microbial) in an effluent measured as the amount of oxygen required to support them. The greater the B.O.D. (ppm or mg/liter) the greater the pollution and less oxygen available for higher aquatic organisms.

biomass fuels Fuels that are manufactured from plant materials; e.g., grain alcohol (ethanol) is made from corn.

bioremediation The use of microbes to clean hydrocarbon residues from contaminated soils.

black smoker A submarine hydrothermal vent associated with a mid-oceanic spreading center that has emitted or is emitting a "smoke" of metallic sulfide particles.

block (volcanology) A mass of cold country rock greater than 32 mm (1¼ in) in size that is ripped from the vent and ejected from a volcano.

block glide (translational slide) A landslide in which movement occurs along a well-defined plane surface or surfaces, such as bedding planes, foliation planes, or faults.

bomb (volcanology) A molten mass greater than 32 mm (1¼ in) in diameter that is ejected from a volcano and cools sufficiently before striking the ground to retain a rounded, streamlined shape.

breakwater An offshore rock or concrete structure, attached at one end or parallel to the shoreline, intended to create quiet water for boat anchorage purposes.

breeder reactor A reactor that produces more fissile (splittable) nuclei than it consumes.

British thermal unit (Btu) The amount of heat, in calories, needed to raise the temperature of 1 pound of water 1° F.

caldera Large crater caused by explosive eruption and/or subsidence of the cone into the magma chamber, usually more than 1.5 km (1 mi) in diameter.

caliche In arid regions, a sediment or soil that is cemented by calcium carbonate ($CaCO_3$).

calving The breaking away of a block of ice, usually from the front of a tidewater glacier, that produces an iceberg.

caprock An impermeable stratum that caps an oil reservoir and prevents oil and gas from escaping to the ground surface.

capillary fringe The moist zone in an aquifer above the water table. The water is held in it by capillary action.

capillary water Water that is held in tiny openings in rock or soil by capillary forces.

carrying capacity The number of creatures a given tract of land can adequately support with food, water, and other necessities of life.

central-core meltdown The melting of an atomic-reactor fuel containment vessel as a result of runaway atomic fission that generates extreme heat.

chain reaction A self-sustaining nuclear reaction that occurs when atomic nuclei undergo fission. Free neutrons are released that split other nuclei, causing more fissions and the release of more neutrons, which split other nuclei, and so on.

chlorofluorocarbons (CFCs) A group of compounds that, on escape to the atmosphere, break down, releasing chlorine atoms that destroy ozone molecules; e.g., Freon, which is widely used in refrigeration, air-conditioning, and formerly was a propellant in aerosol-spray cans.

cinder cone A small, straight-sided volcanic cone consisting mostly of pyroclastic material (usually basaltic).

cinders Glassy and vesicular (containing gas holes) volcanic fragments, only a few centimeters across, that are thrown from a volcano.

clastic Describing rock or sediment composed primarily of detritus of preexisting rocks or minerals.

clay minerals Hydrous aluminum silicates that have a layered atomic structure. They are very fine grained and become plastic (moldable) when wet. Most belong to one of three clay groups: kaolinite, illite, and smectite, the last of which are expansive when they absorb water between layers.

claypan A dense, relatively impervious subsoil layer that owes its character to a high clay-mineral content due to concentration by downward-percolating waters.

cleavage The breaking of minerals along certain crystallographic planes of weakness that reflects their internal structure.

coastal desert A desert on the western edge of continents in tropical latitudes; i.e., near the Tropic of Capricon or Cancer. The daily and annual temperature fluctuations are much less than in inland tropical deserts.

columnar jointing Vertically oriented polygonal columns in a solid lava flow or shallow intrusion formed by contraction upon cooling. They are usually formed in basaltic lavas.

compaction The reduction in bulk volume of fine-grained sediments due to increased overburden weight or to tighter packing of grains. Clay is more compressible than is sand and therefore compacts more.

composite cone A large cone or volcanic structure that is slightly concave upward and consists of alternating layers of ash and lava.

concentrate Enriched ore, often obtained by flotation, produced at a mill.

concentration factor The enrichment of a deposit of an element, expressed as the ratio of the element's abundance in the deposit to its average crustal abundance.

cone of depression The cone shape formed by the water table around a pumping well when pumping exceeds the flow of water to the well.

confined aquifer A water-bearing formation bounded above and below by impermeable beds or beds of distinctly lower permeability.

conservative boundary A tectonic plate boundary where plates simply slide past one another, neither creating nor destroying crust. Such boundaries are marked by *transform faults,* such as the San Andreas fault in California.

contact metamorphism A change in the mineral composition of rocks in contact with an invading magma from which fluid constituents are carried out to combine with some of the country-rock constituents to form a new suite of minerals.

continental divide A drainage divide that separates streams flowing toward opposite sides of a continent.

continental glacier A glacier of considerable thickness that completely covers a large part of a continent or an area of at least 50,000 km^2 (20,000 mi^2), obscuring the relief of the underlying surface.

convection The circulatory movement within a body of nonuniformly heated fluid. Warmer material rises in the center, and cooler material sinks at the outer boundaries. It has been proposed as the mechanism of plate motion in plate tectonic theory.

convergent boundary Tectonic boundary where two plates collide. Where a tectonic plate sinks beneath another plate and is destroyed, this boundary is known as a *subduction zone.* Subduction zones are marked by deep-focus seismic activity, strong earthquakes, and violent volcanic eruptions.

core The central interior of the earth beginning at a depth of 2,900 km (1,800 mi). It probably consists of an iron–nickel alloy, with an outer liquid part and a solid interior.

Coriolis effect A condition peculiar to motion on a rotating body such as the earth. It appears to deflect moving objects in the Northern Hemisphere to the right and objects in the Southern Hemisphere to the left, to an observer on the rotating body.

craton A part of the earth's continental crust that has attained stability—that has not been deformed for a long period of time. The term is restricted to continents and includes their most stable areas, the continental shields.

creep (soil) The imperceptibly slow downslope movement of rock and soil particles by gravity.

cross-bedding Arrangement of strata, greater than 1 cm thick, inclined at an angle to the main stratification.

crust The outermost layer of the earth consisting of granitic continental crust and basaltic oceanic crust; that part of the earth above the Mohorovičić discontinuity.

crystal settling In a magma, the sinking of crystals because of their greater density, sometimes aided by magmatic convection. (*See* gravity separation).

cubic feet per second (cfs) Unit of water flow or stream discharge. One cubic foot per second is 7.48 gallons passing a given point in one second.

cutoff A new channel that is formed when a stream cuts through a very tight meander. (*See* oxbow lake).

cyanidation The process of using cyanide to dissolve and recover gold and silver from ore.

c-zone Soil zone of weathered parent material leading downward to fresh bedrock.

Darcy's Law A derived formula for the flow of fluids in a porous medium.

debris avalanche A sudden, rapid movement of a water–soil–rock mixture down a steep slope.

debris flow A moving mass of a water, soil, and rock mixture. More than half of the soil and rock particles are coarser than sand, and the mass has the consistency of wet concrete.

deflation The lifting and removal of loose, dry, fine-grained particles by the action of wind.

delta The nearly flat land formed where a stream empties into a body of standing water, such as a lake or the ocean.

desert An arid region of low rainfall, usually less than 25 cm (10 in) annually, and of high evaporation or extreme cold. Deserts are generally unsuited for human occupation under natural conditions.

desert pavement An interlocking cobble mosaic that remains on the earth surface in a dry region after winds have removed all fine-grained sediments.

desertification The process by which semiarid grasslands are converted to desert.

detrital *See* clastic.

dilatancy–diffusion model A model for explaining certain earthquake precursory phenomena, for example changes in P-wave velocities.

discharge (*Q*) The volume of water, usually expressed in cfs, that passes a given point within a given period of time.

disseminated deposit A mineral deposit, especially of a metal, in which the minerals occur as scattered particles in the rock, but in sufficient quantity to make the deposit a commercially worthwhile ore.

divergent boundary Found at mid-ocean ridge or spreading center where new crust is created as plates move apart. For this reason they are sometime referred to as *constructive* boundaries and are the sites of weak shallow-focus earthquakes and volcanic action.

divining rod A forked wooden stick or similar object purportedly useful for locating a body of ground water or a mineral deposit.

doldrums The area of warm rising air, calm winds, and heavy precipitation between 5° north latitude and 5° south latitude. It is where the northeast and southeast tradewinds meet.

doubling time The number of years required for a population to double in size.

drainage basin The tract of land that contributes water to a particular stream, lake, reservoir, or other body of surface water.

drainage divide The boundary between two drainage basins (*see* continental divide).

drawdown The lowering of the water table immediately adjacent to a pumping well.

dredging The excavation of earth material from the bottom of a body of water by a floating barge or raft equipped to scoop up, discharge by conveyors, and process or transport materials.

drift A general term for all rock material that is transported and deposited by a glacier or by running water emanating from a glacier.

dripstone A general term for calcite or other minerals deposited in caves by dripping water.

dune A mound, bank, ridge, or hill of loose, windblown granular material (generally sand), either bare or covered with vegetation. It is capable of movement, but maintains its characteristic shape.

Dust Veil Index (DVI) A measure of the impact of a volcanic eruption on incoming solar radiation and climate.

eccentricity The deviation of the earth's path around the sun from a perfect circle.

effective stress The average normal force per unit area (stress) that is transmitted directly across grain-to-grain boundaries in a sediment or rock mass.

effluent stream A stream that receives water from the zone of saturation.

elastic Said of a body in which strains are totally recoverable, as in a rubber band. (*Contrast with* plastic).

Elastic Rebound Theory The theory that movement along a fault and the resulting seismicity is the result of an abrupt release of stored elastic strain energy between two rock masses on either side of the fault.

element A substance that cannot be separated into different substances by usual chemical means.

end moraine (thermal moraine) A moraine that is produced at the front of an actively flowing glacier; a moraine that has been deposited at the lower end of a glacier.

ENSO Abbreviation for the periodic meteorological event known as El Niño–Southern Oscillation.

environmental geology The relationship between humans and their geological environment.

eolian Of, produced by, or carried by wind.

epicenter The point at the surface of earth directly above the focus of an earthquake.

equatorial low The belt of low-pressure air masses near the equator. (*See* doldrums.)

erosion The weathering and transportation of the materials of the earth's surface.

erratic A rock fragment that has been carried by glacial ice or floating ice and deposited when the ice melted at some distance from the outcrop from which it was derived.

estuary A semienclosed coastal body where outflowing river water meets tidal sea water.

eutrophication The increase in nitrogen, phosphorous and other plant nutrients in the aging of an aquatic system. Blooms of algae develop preventing light penetration, and causing reduction of oxygen needed in a healthy system.

evaporite A nonclastic sedimentary rock composed primarily of minerals produced when a saline solution becomes concentrated by evaporation of the water; especially a deposit of salt that precipitated from a restricted or enclosed body of seawater or from the water of a salt lake.

evapotranspiration That portion of precipitation that is returned to the air through evaporation and transpiration, the latter being the escape of water from the leaves of plants.

exfoliation The process whereby slabs of rock bounded by sheet joints peel off the host rock, usually a granite or sandstone.

exfoliation dome A large rounded dome resulting from exfoliation; e.g., Half Dome in Yosemite National Park.

exotic terranes Fault-bounded bodies of rock that have been transported some distance from their place of origin and that are unrelated to adjacent rock bodies or terranes.

expansive soils Clayey soils that expand when they absorb water and shrink when they dry out. They can generate uplift pressures sufficient to break foundations and crack walls.

extrusive rocks Igneous rocks that formed at or near the surface of the earth. Because they cooled rapidly, they generally have an *aphanitic* texture.

fault creep (tectonic) The gradual slip or motion along a fault without an earthquake.

fault Fracture in the earth along which there has been displacement.

felsic Descriptive of magma or rock with abundant light-colored minerals and a high silica content. The term is derived from *fel*dspar + *si*lica + *c*.

fetch The unobstructed stretch of sea over which the wind blows to create wind waves.

firn A transitional material between snow and glacial ice, being older and denser than snow, but not yet transformed into glacial ice. Snow becomes firn after surviving one summer melt season; firn becomes ice when its permeability to liquid water becomes zero.

first-motion studies Seismological study of an earthquake that determines areas of compression and dilatation adjacent to the fault plane and study indicates the direction of motion on the causative fault.

fissile isotopes Isotopes with nuclei that are capable of being split into other elements (*see* fission).

fission The rupture of the nucleus of an element into lighter elements (fission products) and free neutrons spontaneously or by absorption of a neutron.

fjord A narrow, steep-walled inlet of the sea between cliffs or steep slopes, formed by the passage of a glacier. A drowned glacial valley.

flood frequency The average time interval between floods that are equal to or greater than a specified discharge.

floodplain The portion of a river valley adjacent to the channel that is built of sediments deposited during times when the river overflows its banks at flood stage.

floodway Floodplain area under federal regulation for flood insurance purposes.

flotation concentration The process of concentrating minerals with distinct non-wettable properties by floating them in liquids containing soapy frothing agents such as pine oil.

fluvial Pertaining to rivers.

focus (hypocenter) Point within the earth where an earthquake originates.

foliation The planar or wavy structure that results from the flattened growth of minerals in a metamorphic rock.

foreshore slope The zone of a beach that is regularly covered and uncovered by the tide.

fossil fuels Coal, oil, natural gas, and all other solid or liquid hydrocarbon fuels.

fracture Break in a rock caused by tensional, compressional or shearing forces (*see also* joints).

frost wedging The opening of joints and cracks by the freezing and thawing of water.

fusion The combination of two nuclei to form a single heavier nucleus accompanied by a loss of some mass that is converted tó heat.

Gaia The controversial hypothesis that proposes that life has a controlling influence on the oceans and atmosphere. It states that the earth is a giant self-regulating body with close connections between living and non-living components.

geologic agents All geologic processes—e.g., wind, running water, glaciers, waves, mass wasting—that erode, move, and deposit earth materials.

geothermal system A subsurface geologic area of circulating hot water and steam; usually associated with cooling magma or hot rocks.

gigawatt (GW) A thousand million (a billion) watts; 1,000 megawatts.

glacial outwash Deposits of stratified sand, gravel, and silt that have been removed from a glacier by meltwater streams.

glacial polish A smooth surface produced on bedrock by glacial abrasion.

glacial striation One of a series of long, delicate, finely cut, usually straight and parallel furrows or lines inscribed on a bedrock surface by the rasping action of rock fragments entrained at the base or sides of a moving glacier.

glacier A large mass of ice formed, at least in part, on land by the compaction and recrystallation of snow. It moves slowly downslope by creep or outward in all directions due to the stress of its weight, and it survives from year to year.

glaciofluvial Pertaining to meltwater streams that flow from wasting glacier ice and especially to the deposits and landforms such as outwash plains that are produced by such streams.

Gondwana, Gondwanaland The southern part of the Permo-Triassic drift landmass of Pangaea.

graded stream A stream that is in equilibrium, showing a balance of erosion, transporting capacity, and material supplied to it. Graded streams have a smooth profile.

gradient (stream) The slope of a streambed usually expressed as the amount of drop per horizontal distance in meters per kilometer (m/km) or in feet per mile (ft/mi).

gravity separation *See* crystal settling.

gravity stamp mill A battery of heavy iron weights that are alternately lifted and dropped to pulverize ore as part of the milling process for recovering gold, silver, or copper.

greenhouse warming The warming of the atmosphere by an increase in the amount of certain gases in the atmosphere, primarily CO_2 and CH_4, that increase the retention of heat that has been radiated by the earth.

groin A structure of rock, wood, or concrete built roughly perpendicular to a beach to trap sand.

ground water That part of subsurface water that is in the zone of saturation (below the water table).

gullying The cutting of channels into the landscape by running water. When extreme, it renders farmland useless.

half life The period of time during which one-half of a given number of atoms of a radioactive element or isotope will disintegrate.

hardness scale A standard of ten minerals by which the hardness of other minerals may be rated on a scale of 1 (softest) to 10 (hardest). Also called Mohs Scale after Friedrich Mohs, the German mineralogist who devised the scale in 1812.

hard rock (*a*) A term loosely used for an igneous or metamorphic rock, as distinguished from a sedimentary rock. (*b*) In mining, a rock that requires drilling and blasting for economical removal.

hardpan Impervious layer just below the land surface produced by calcium carbonate ($CaCo_3$) in the B horizon (*see* claypan).

horse latitudes Oceanic regions in the *subtropical highs* about 30° north and south of the equator that are characterized by light winds or calms, heat, and dryness.

hot spot A point (or area) on the lithosphere over a plume of lava rising from the mantle.

hydration A process whereby anhydrous minerals combine with water.

hydraulic conductivity A ground-water unit expressed as the volume of water that will move in a unit of time through a unit of area measured perpendicular to the flow direction. Commonly called *permeability,* it is expressed as the particular aquifer's $m^3/day/m^2$ (or m/day) or $ft^3/day/ft^2$ (or ft/day).

hydraulic gradient The slope or vertical change (ft/ft or m/m) in water-pressure head with horizontal distance in an aquifer.

hydraulic mining, hydraulicking A mining technique by which high-pressure jets of water are used to dislodge unconsolidated rock or sediment so that it can be processed.

hydrocarbon One of the many chemical compounds solely of hydrogen and carbon atoms; may be solid, liquid, or gaseous.

hydrocompaction The compaction of dry, low-density soils due to the heavy application of water.

hydrogeology The study of the geology and production of subsurface waters and the related geologic aspects of surface waters.

hydrograph A graph of the stage (height), discharge, velocity, or some other characteristic of a body of water over time.

hydrologic cycle (water cycle) The constant circulation of water from the sea to the atmosphere, to the land, and eventually back to the sea. The cycle is driven by solar energy.

hydrology The study of liquid and solid water on, under, and above the earth's surface, including economic and environmental aspects.

hydrolysis The chemical reaction between hydrogen ions in water with a mineral, commonly a silicate, usually forming clay minerals.

hydrothermal Of or pertaining to heated water, the action of heated water, or a product of the action of heated water—such as a mineral deposit that precipitated from a hot aqueous solution.

hypocenter *See* focus.

ice cap A dome-shaped or platelike cover of perennial ice and snow that covers the entire summit area of a mountain mass, spreading outward in all directions due to its weight, with an area of less than 50,000 km^2 (20,000 mi^2).

ice field An extensive mass of land ice covering all but the highest peaks and ridges of a mountain region and consisting of many interconnected valley and other types of glaciers.

igneous rocks Rocks that crystallize from molten material at the surface of the earth (volcanic) or within the earth (plutonic).

inactive sea cliff A cliff whose erosion is dominated by subaerial processes rather than wave action.

indicated reserves The hydrocarbons that can only be recovered by secondary recovery techniques such as fluid injection. These reserves are in addition to the reserves directly recoverable from a known reservoir.

infiltration The flow of fluid within a solid substance through pores or small openings.

influent stream A stream or reach of a stream that contributes water to the zone of saturation.

intensity scale An earthquake rating scale (I–XII) based upon subjective reports of human reactions to ground shaking and upon the damage caused by an earthquake.

intrusive rocks Igneous rocks that have "intruded" into the crust, hence they are slowly cooled and generally have phaneritic texture.

ion An atom with a positive or negative electrical charge because it has gained or lost one or more electrons.

isoseismals Lines on a map that enclose areas of equal earthquake shaking based upon an intensity scale.

isostasy The condition of equilibrium, comparable to floating, of large units of lithosphere above the asthenosphere. Isostatic compensation demands that as mountain ranges are eroded, they rise—just as a block of wood floating in water would rise if a part of it above the waterline were removed.

isostatic rebound Uplift of the crust of the earth that results from unloading such as results from the melting of ice sheets.

isotope One of two or more forms of an element that have the same atomic number but different atomic masses.

J-curve The shape of a graphing of geometric (exponential) growth. Growth is slow at first, and the rate of growth increases with time.

jetties Structures built perpendicular to the shoreline to improve harbor inlets or river outlets.

joint Separation or parting in a rock that has not been displaced. Joints usually occur in groups ("sets"), the members of a set having a common orientation.

karst topography, karst terrane Area underlain by soluble limestone or dolomite and riddled with caves, caverns, sinkholes, lakes, and disappearing streams.

kerogen A carbonaceous residue in sediments that has survived bacterial metabolism. It consists of large molecules from which hydrocarbons and other compounds are released on heating.

kettle A bowl-shaped depression without surface drainage in glacial-drift deposits, often containing a lake, believed to have formed by the melting of a large detached block of stagnant ice left behind by a retreating glacier.

knob-and-kettle topography An undulating morainal landscape in which a disordered assemblage of knobs and ridges of glacial drift are interspersed with irregular depressions or kettles.

lahar A debris flow or mudflow consisting of volcanic material.

landslide The downslope movement of rock and/or soil as a semicoherent mass on a discrete slide surface or plane (*see also* mass wasting).

lapilli (cinders) Fragments only a few centimeters in size that are ejected from a volcano.

lateral moraine A low, ridge-like moraine carried on, or deposited at or near, the side of a mountain glacier. It is composed chiefly of rock fragments that have been loosened from valley walls by glacial abrasion and that have fallen onto the ice from bordering slopes.

lateral spreading The horizontal movement on nearly level slopes of soil and mineral particles due to liquefaction of quick clays.

laterite A highly weathered brick-red soil characteristic of tropical and subtropical rainy climates. Laterites are rich in oxides or iron and aluminum and have some clay minerals and silica. Bauxite is an Al-rich deposit of a similar origin.

Laurasia The northern part of the Permo-Triassic landmass of Pangaea. (*See* Gondwana)

lava dome (volcanic dome) Steep-sided protrusion of viscous, glassy lava, sometimes within the crater of a larger volcano (e.g. Mount St. Helens) and sometimes free-standing (e.g. Mono Craters).

lava plateau A broad elevated tableland, thousands of square kilometers in extent, underlain by a thick sequence of lava flows.

leaching The selective removal or dissolving of soluble materials from a rock or ore body by natural action of percolating water.

levees (natural) Embankments of sand or silt built by a stream along both banks of its channel. They are deposited during floods, when waters overflowing the stream banks are forced to deposit sediment.

liquefaction *See* spontaneous liquefaction.

lithification The conversion of sediment into solid rock through processes of compaction, cementation, and crystallization.

lithify To change into stone as in the transformation of loose sand to sandstone.

lithosphere The rigid outer layer from the earth's surface down to the asthenosphere. It includes the crust and part of the upper mantle and is of the order of 100 km (60 mi) thick. The "plates" in plate tectonic theory.

littoral drift Sediment (sand, gravel, silt) that is moved parallel to the shore by longshore currents.

loam A rich, permeable soil composed of a mixture of clay, silt, sand, and organic matter.

lode A mineral deposit consisting of a zone of veins in consolidated rock, as opposed to a *placer deposit.*

loess A blanket deposit of buff-colored calcareous silt that is porous and shows little or no stratification. It covers wide areas in Europe, eastern China, and the Mississippi valley. It is generally considered to be windblown dust of Pleistocene age.

longitudinal wave (P-wave) A type of seismic wave involving particle motion that is alternating expansion and compression in the direction of wave propagation. Also known as *compressional waves,* they resemble sound waves in their motion.

longshore current (littoral current) The current adjacent and parallel to a shoreline that is generated by waves striking the shoreline at an angle.

luster The manner in which a mineral reflects light using such terms as metallic, resinous, silky, and glassy.

L-wave Seismic waves that travel at the surface of the earth. Surface waves are generated by the unreflected energy of P- and S-waves striking the earth's surface. They are the most damaging of earthquake waves.

M-discontinuity (Moho) The contact between the earth's crust and the mantle. There is a sharp increase in earthquake P-wave velocity across this boundary.

maar A low-relief, broad volcanic crater formed by multiple shallow explosive eruptions. Maars commonly contain a lake and are surrounded by a low rampart or ring of ejected material. They typically form from *phreatic eruptions.*

mafic Descriptive of a magma or rock rich in iron and magnesium. The mnemonic term is derived from *ma*gnesium + *fe*ric + *ic.*

magma Molten material within the earth that is capable of intrusion or extrusion and from which igneous rocks form.

magnitude (earthquake) A measure of the strength or the strain energy released by an earthquake at its source. (*See* moment magnitude, Richter magnitude.)

manganese nodule A small irregular black to brown, laminated concretionary mass consisting primarily of manganese minerals with some iron oxides and traces of other metallic minerals, abundant on the floors of the world's oceans.

mantle The zone of the earth below the crust and above the core, generally thought to be composed of rocky material.

mass wasting A general term for all downslope movements of soil and rock material under the direct influence of gravity.

massive sulfide deposit Usually volcanogenic, often rich in zinc and sometimes in lead.

meander One of a series of sinuous curves in the course of a stream.

medial moraine (*a*) An elongate moraine carried on or in the middle of a glacier and parallel to its sides, usually formed by the merging of adjacent and inner lateral moraines below the junction of two coalescing valley glaciers. (*b*) An irregular ridge that is left behind in the middle of a glacial valley after the glacier has disappeared.

megawatt (MW) A million watts; 1,000 kilowatts.

meltdown *See* central-core meltdown.

metamorphic rocks Preexisting rocks that have been altered by heat, pressure, or chemically active fluids.

mid-latitude desert A desert area occurring within latitudes 30°–50° north and south of the equator in the deep interior of a continent, usually on the lee side of a mountain range that blocks the path of prevailing winds, and commonly characterized by a cold, dry climate (*see* rain-shadow desert).

mineral Naturally occurring crystalline substance with well-defined physical properties and a definite range of chemical composition.

mineral deposit A localized concentration of naturally occurring mineral material (e.g., a metallic ore or a nonmetallic mineral), usually of economic value, without regard to its mode of origin.

mineral reserves Known mineral deposits that are recoverable under present conditions but as yet undeveloped. The term excludes *potential ore.*

mineral resources The valuable minerals of an area that are presently legally recoverable or that may be so in the future; include both the known ore bodies *(mineral reserves)* and the *potential ores* of a region.

miscible Soluble; capable of mixing.

Modified Mercalli Scale Earthquake intensity scale from I (not felt) to XII (total destruction) based upon damage and reports of human reactions.

Mohorovičić discontinuity *See* M-discontinuity.

moment magnitude (earthquakes) A scale of seismic energy released by an earthquake based on the product of the rock rigidity along the fault, the area of rupture on the fault plane, and the amount of slip.

moraine A mound, ridge, hill, or other distinct accumulation of unsorted, unstratified glacial drift, predominantly till, deposited chiefly by direct action of glacial ice in a variety of landforms.

mother lode (*a*) A main mineralized unit that may not be economically valuable in itself but to which workable deposits are related; e.g., the Mother Lode of California. (*b*) An ore deposit from which a placer is derived; the *mother rock* of a placer.

neutron An electronically neutral particle (zero charge) of an atomic nucleus with an atomic mass of approximately 1.

nucleus The positively charged central core of an atom containing protons and neutrons that provide its mass.

nuée ardente *See* ash flow.

NWPA Nuclear Waste Policy Act of 1982 mandating selection of a repository for nuclear waste.

obsidian Dark-colored volcanic glass, high in SiO_2, that extrudes as a viscous mass, usually in volcanic domes.

open-pit mining Mining from open excavations, most commonly for low-grade copper and iron deposits, and coal.

ore deposit The same as a *mineral reserve* except that it refers only to a metal-bearing deposit.

ore mineral The part of an ore, usually metallic, that is economically desirable.

oxbow lake A crescent-shaped lake formed along a stream course when a tight meander is cut off and abandoned.

pahoehoe Basaltic lava typified by smooth, billowy, or ropy surfaces.

paleomagnetism Study of the remanent magnetism in rocks, which indicates the strength and direction of the earth's magnetic field in ancient landmasses.

paleoseismicity The rock record of past earthquake events in displaced beds and liquefaction features in trenches or natural outcrops.

Pangaea A supercontinent that existed from about 300–200 Ma, which included all the continents we know today.

paystreak A miner's colloquial term for an ore deposit.

pedalfer Soil of humid regions characterized by an organic-rich A horizon and clays and iron oxides in the B horizon.

pedocal Soil of arid or semiarid regions that is rich in calcium carbonate.

pedologist (Greek *pedo,* "soil," and *logos,* "knowledge.") A person who studies soils.

pedology The study of soils.

pegmatite An exceptionally coarse-grained igneous rock with interlocking crystals, usually found as irregular dikes, lenses, or veins, especially at the margins of batholiths.

perched water table The upper surface of a body of ground water held up by a discontinuous impermeable layer above the static water table.

permafrost Permanently frozen ground, with or without water, occurring in arctic, subarctic,and alpine regions.

permafrost table The upper limit of permafrost.

permeability The degree of ease with which fluids flow through a porous medium (*see* hydraulic conductivity).

phaneritic texture A coarse-grained igneous-rock texture resulting from slow cooling. Phanerites are coarse-grained igneous rocks.

photochemical smog Atmospheric haze that forms when automobile-exhaust emissions are activated by ultraviolet radiation from the sun to produce highly reactive oxidants known to be health hazards.

phreatic eruption Volcanic eruption, mostly steam, caused by the interaction of hot magma with underground water, lakes, or seawater. Where significant amounts of new (magmatic) material are ejected in addition to steam, the eruptions are described as *phreatomagmatic.*

piedmont glacier A thick continuous sheet of ice at the base of a mountain range formed by the spreading and coalescing of valley glaciers from higher elevations.

piping Subsurface erosion in sandy materials caused by the percolation of water under pressure. A somewhat similar effect is obtained by placing a garden-hose nozzle in a sandbox and jetting an opening through the sand.

placer deposit (pronounced "plass-er") A surficial mineral deposit formed by settling from streams of mineral particles from weathered debris. (*See also* lode.)

planetesimals Space debris, up to a few hundred kilometers in diameter, that first condensed from the solar nebula. Their accretion is believed to have formed the planets in our solar system.

plastic Capable of being deformed continuously and permanently in any direction without rupture.

playa A normally dry lakebed in an arid region, usually with a dry, mud-cracked or salt-encrusted surface. (Spanish, beach.)

Pleistocene The geologic Epoch of the Quaternary Period known as the Ice Age, falling by definition between 1.6 million and 12,000 years ago. Extensive glaciation is known as far back as the Pliocene Epoch 2.5 million years ago. The exact boundaries shift in time as new data are revealed.

Plinian eruption (volcanology) An explosive eruption in which a steady, turbulent, nearly vertical column of ash and steam is released from a vent at high velocity.

plutonic Pertaining to rock formed by any process at depth, usually with a phaneritic texture.

point bar A deposit of sand and gravel found on the inside curve of a *meander.*

polar desert The bitterly cold, arid region centered at the south pole that receives very little precipitation.

polar high The region of cold, high-density air that exists in the polar regions.

polarity The magnetic positive (north) or negative (south) character of a magnetic pole.

population dynamics The study of how populations grow.

population growth rate The number of live births less deaths per 1,000 people, usually expressed as a percentage; a growth rate of 40 would represent 4.0% growth.

pore pressure The stress exerted by the fluids that fill the voids between particles of rock or soil.

porosity A material's ability to contain fluid. The ratio of the volume of pore spaces in a rock or sediment to its total volume, usually expressed as a percentage.

porphyry copper A copper deposit, usually of low grade, in which the copper-bearing minerals occur in disseminated grains and/or in veinlets through a large volume of rock.

potable (water) Drinkable without ill effects.

potential ore As yet undiscovered mineral deposits and also known mineral deposits for which recovery is not yet economically feasible.

potentiometric surface The hydrostatic head of ground water and the level to which the water will rise in a well. The water table is the potentiometric surface in an unconfined aquifer. Also called *piezometric surface.*

precession The slow gyration of the Earth's axis, analogous to that of a spinning top, which causes the slow change in the orientation of the earth's axis relative to the sun that accounts for the reversal of the seasons of winter and summer in the Northern and Southern Hemispheres every 11,500 years.

precipitation The separation of a solid substance from a solution by a chemical reaction.

precursors (of earthquakes) Observable phenomena that occur before an earthquake and indicate that an event is soon to occur.

pregnant pond A catchment basin that holds a gold- and silver-bearing cyanide solution.

proglacial lake An ice-marginal lake formed just beyond the frontal moraine of a retreating glacier and generally in direct contact with the ice.

proton A particle in an atomic nucleus with a positive electrical charge and an atomic mass of approximately one.

protostar A star in the process of formation that has entered the slow gravitational contraction phase.

pumice Frothy-appearing rock composed of natural glass ejected from high-silica, gas-charged magmas. It is the only rock that floats.

P-wave *See* longitudinal wave.

pyroclastic ("fire broken") Descriptive of the fragmental material, ash, cinders, blocks, and bombs ejected from a volcano. (*See also* tephra.)

quad A quadrillion British thermal units (10^{15} Btu).

quick clay (sensitive clay) A clay possessing a "house-of-cards" sedimentary structure that collapses when it is disturbed by an earthquake or other shock.

radioactivity The property possessed by certain elements that spontaneously emit energy as the nucleus of the atom changes its proton-to-neutron ratio.

radiometric dating *See* absolute age dating.

rain-shadow desert A desert on the lee side of a mountain or a mountain range (*see* mid-latitude desert).

rank (coal) A coal's carbon content depending upon its degree of metamorphism.

Rapakivi An orbicular texture in granites consisting of rounded crystals of potassium-containing feldspar (orthoclase) a few centimeters in diameter that are surrounded in concentric layers by sodium-containing plagioclase (albite).

recessional moraine One of a series of nested end moraines that record the stepwise retreat of a glacier at the end of an ice age.

recrystallization The formation in the solid state of new mineral grains in a metamorphic rock. The new grains are generally larger than the original mineral grains.

recurrence interval The return period of an event, such as a flood or earthquake, of a given magnitude. For flooding, it is the average interval of time within which a given flood will be equalled or exceeded by the annual maximum discharge.

regional metamorphism A general term for metamorphism that affects an extensive area.

regolith Unconsolidated rock and mineral fragments at the surface of the earth.

regulatory flood plain The part of a floodplain that is subject to federal government regulation for insurance purposes.

relative age dating The chronologic ordering of geologic strata, features, fossils, or events with respect to the geological time scale without reference to their absolute age.

replacement ore body An ore deposit formed by capillary solution and deposition by which a new mineral of partly or wholly differing chemical composition may grow in the body of an old mineral or mineral aggregate.

replacement reproduction The birth rate at which just enough offsprings are born to replace their parents.

reserves Those portions of an identified resource that can be recovered economically.

reservoir rock A permeable, porous geologic formation that will yield oil or natural gas.

residence time The average length of time a substance (atom, ion, molecule) remains in a given reservoir.

residual soil Soil formed in place by decomposition of the rocks upon which it lies.

resonance The tendency of a system (a structure) to vibrate with maximum amplitude when the frequency of the applied force (seismic waves) is the same as the vibrating body's natural frequency.

revetment A rock or concrete structure built landward of a beach to protect coastal property.

Richter Magnitude Scale (earthquakes) The logarithm of the maximum trace amplitude of a particular seismic wave on a seismogram, corrected for distance to epicenter and type of seismometer. A measure of the energy released by an earthquake.

rift zone (volcanology) A linear zone of weakness on the flank of a shield volcano that is the site of frequent flank eruptions.

rill erosion The carving of small channels, up to 25 cm (10 in) deep, in soil by running water.

riparian (Latin *riparius,* "of a river bank") Pertaining to rivers, lakes, tidewater areas, and other bodies of surface water. Riparian laws govern the use of surface water.

roche moutonnée A small elongate knob sculptured by a glacier such that its long axis is oriented in the direction of ice movement and characterized by an upstream side that is gently inclined, smoothly rounded, and striated and a downstream side that is steep and jagged.

rock cycle The cycle or sequence of events involving the formation, alteration, destruction, and reformation of rocks as a result of processes such as erosion, transportation, lithification, magmatism, and metamorphism.

rocks Aggregates of minerals or rock fragments.

rotational landslide (slump) A slope failure in which sliding occurs on a well-defined, concave-upward, curved surface, producing a backward rotation of the slide mass.

runoff That part of precipitation falling on land that runs into surface streams.

runup The elevation to which a wind wave or tsunami advances onto the land.

salt dome A column or plug of rock salt that rises from depth because of its low density and pierces overlying sediments.

sea-floor spreading A hypothesis that oceanic crust forms by convective upwelling of lava at mid-ocean ridges and moves laterally to trenches where it is destroyed.

seawall A rock or concrete structure built landward of a beach to protect the land from wave action. (*See also* revetment.)

secondary enrichment Processes of near-surface mineral deposition, in which oxidation produces acidic solutions that leach metals, carry them downward, and reprecipitate them, thus enriching sulfide minerals already present.

sedimentary rock A layer of rock deposited from water, ice, or air and subsequently lithified to form a coherent rock.

seif (longitudinal) dune A very large, sharp-crested, tapering chain of sand dunes; commonly found in the Arabian Desert.

seismic gap The segment along an active fault that has seismically been quiet relative to segments at either end. It is the part of the fault most likely to rupture and generate the next earthquake. (*See also* asperity.)

seismogram The recording from a seismograph.

seismograph An instrument used to measure and record seismic waves from earthquakes.

sensitive clay *See* quick clay.

sheet erosion The removal of thin layers of surface rock or soil from an area of gently sloping land by broad continuous sheets of running water (sheet flow), rather than by channelized streams.

sheet joints (*See* joint) Cracks more or less parallel to the ground surface that result from expansion due to deep erosion and the unloading of overburden pressure.

shield volcano (cone) The largest type of volcanoes, composed of piles of lava in a convex-upward slope. Found mainly in Hawaii and Iceland.

SIGMET Significant meteorological observation warning of a dangerous atmospheric condition for all aircraft.

sinkhole Circular depression formed by the collapse of a shallow cavern in limestone.

slant-drilling (whipstocking) Purposely deflecting a drill hole from the vertical in order to tap a reservoir not directly below the drill site. The deflecting tool is a whipstock

slip face The steeper, lee side of a dune, standing at or near the angle of repose of loose sand and advancing downwind by a succession of slides whenever that angle is exceeded.

slump *See* rotational landslide.

smelter An industrial plant that mechanically and chemically produces metals from their ores.

soil (*a*) Loose material at the surface of the earth that supports the growth of plants (pedological definition). (*b*) All loose surficial earth material resting on coherent bedrock (engineering definition).

soil horizon A layer of soil that is distinguishable from adjacent layers by properties such as color, texture, structure, or chemical composition.

soil profile A vertical section of soil that exposes all of its horizons or zones.

solution (weathering) The dissolving of rocks and minerals by natural acids or water.

source rock The geologic formation in which oil and/or gas originate.

specific yield The ratio of the volume of water drained from an aquifer by gravity to the total volume considered.

speleology The study of caves and caverns.

spheroidal weathering Weathering of rock surfaces that creates a rounded or spherical shape as the corners of the rock mass are weathered faster than its flat faces.

spontaneous liquefaction Process whereby water-saturated sands, clays, or artificial fill suddenly become fluid upon shaking, as in an earthquake.

stack A pillarlike rocky island that has become detached from a sea cliff or headland by wave erosion.

stage The height of flood waters in feet or meters above an established datum plane.

stalactite Conical deposit of mineral matter, usually calcite, that has developed downward from the ceiling of a limestone cave.

stalagmite Conical deposit of mineral matter, usually calcite, developed upward from the floor of a cavern or cave.

steppe An extensive treeless grassland in the semiarid mid-latitudes of Asia and southeastern Europe. It is drier than the *prairie* that develops in the mid-latitudes of the United States.

storm surge The sudden rise of sea level on an open coast due to a storm. Storm surge is caused by strong onshore winds and results in water being piled up against the shore. It may be enhanced by high tides and low atmospheric pressure.

strain Deformation resulting from an applied stress (= force per unit area). It may be elastic (recoverable) or ductile (nonreversible).

stratabound Said of mineral deposits that are enclosed in sedimentary rocks.

stratification (bedding) The arrangement of sedimentary rocks in strata or layers.

stratigraphic trap An accumulation of oil that results from a change in the character (permeability) of the reservoir rock, rather than from structural deformation.

stress The force applied on an object per unit area, expressed in pounds per square foot (lb/ft^2) or kilograms per square meter (kg/m^2).

strip-mining Surficial mining, in which the resource is exposed by removing the overburden.

subduction The sinking of one lithospheric plate beneath another at a convergent plate margin. (*See* convergent boundary.)

subsidence Sinking or downward settling of the earth's surface due to solution, compaction, withdrawal of underground fluids, or to cooling of the hot lithosphere or to loading with sediment or ice.

subtropical highs The climatic belts at latitudes of approximately 30°–35° north and south of the equator characterized by semipermanent high-pressure air masses, moderate seasonal rainfall, summer heat, dryness, and generally clear skies.

supernova The explosive destruction of a massive star that occurs when the star's nuclear fuel has been depleted; the star catastrophically collapses inward.

surcharge Weight added to the top of a natural or artificial slope by sedimentation or by artificial fill.

surging glacier A glacier that alternates periodically between brief periods (usually 1–4 years) of very rapid flow, called *surges,* and longer periods (usually 10–100 years) of near-stagnation.

suspension (*a*) A mode of sediment transport in which the upward currents in eddies of turbulent flow are capable of supporting the weight of sediment particles and keeping them held indefinitely in the surrounding fluid (as silt in water or dust in air). (*b*) The state of a substance in such a mode of transport.

sustainable society A population capable of maintaining itself while leaving sufficient resources for future generations.

sustained yield (hydrology) The amount of water an aquifer can yield on a daily basis over a long period of time.

S-wave *See* transverse wave.

syncline A concave-upward fold that contains younger rocks in its core. (*Contrast with* anticline.)

synthetic fuels (synfuels) A manufactured fuel; for example, a fuel derived by liquefaction or gasification of coal.

tailings The worthless rock material discarded from mining operations.

talus Coarse angular rock fragments lying at the base of a cliff or steep slope from which they were derived.

tectonism (adj. *tectonic*) Deformation of the earth's crust by natural processes leading to the formation of ocean basins, continents, mountain systems, and other earth features.

tephra A collective term for all *pyroclastic* material ejected from a volcano.

terminal moraine *See* end moraine.

terranes (*See* exotic terannes.)

texture The size, shape, and arrangement of the component particles or crystals in a rock.

tidewater glacier A glacier that terminates in the sea, usually ending in a cliff of ice from which icebergs are discharged.

till Unsorted and unstratified drift, generally unconsolidated, that is deposited directly by and beneath a glacier without subsequent reworking by water from the glacier.

tombolo A sandbar that connects an island with the mainland or another island.

trace element An element that occurs in minute quantities in rocks or plant or animal tissue. Some are essential for human health.

trade winds Steady winds blowing from areas of high pressure at 30° north and south latitudes toward the area of lower pressure at the equator. This pressure differential produces the northeast and southeast trade winds in the Northern and Southern Hemispheres, respectively.

transform faults Plate boundaries with mostly horizontal (lateral) movement that connect spreading centers to each other or to subduction zones. (*See* conservative boundary.)

translational slide *See* block glide.

transported soils Soil developed on *regolith* transported and deposited by a geological agent. Hence the term *glacial* soils, alluvial soils, *eolian* soils, *volcanic* soils, and so forth.

transuranic elements (TRU) Unstable (radioactive) elements of atomic numbers greater than 92.

transverse dune A strongly asymmetrical sand dune that is elongated perpendicularly to the direction of the prevailing winds. It has a gentle windward slope and a steep leeward slope that stands at or near the angle of repose of sand.

transverse wave (S-wave) A seismic wave propagated by a shearing motion that involves oscillation perpendicular to the direction of travel. It travels only in solids.

tropical desert A hot, dry desert occurring between latitudes 15° and 30° north and south of the equator where subtropical air masses prevail.

troposphere That part of the atmosphere next to the earth's surface, in which temperature generally decreases rapidly with altitude and clouds and convection are active. Where the weather occurs.

tsunami Large sea wave produced by a submarine earthquake, a volcanic eruption, or a landslide. Incorrectly known as a tidal wave.

tuff A coherent rock formed from volcanic ash, cinders, or other *pyroclastic* material.

tundra A treeless plain characteristic of arctic regions with organic rich, poorly drained tundra soils and permanently frozen ground.

unconfined ground water Underground water that has a free (static) water table; that is, water that is not confined under pressure beneath an aquiclude.

upwelling A process whereby cold, nutrient-rich water is brought to the surface. Coastal upwelling occurs mostly in trade-wind belts.

U-shaped valley Straight, steep-walled and flat-floored valley eroded by a valley glacier.

valley glacier (alpine glacier) A glacier flowing down between the walls of a mountain valley in all or a part of its length, usually originating in a cirque. Any glacier in a mountain range except an ice sheet or an ice cap.

vein A thin, sheetlike igneous intrusion into a crevice.

ventifact Any stone or pebble that has been shaped, worn, faceted, cut, or polished by the abrasive or sandblast action of windblown sand, generally under arid conditions.

Volcanic Explosivity Index (VEI) A scale for rating volcanic eruptions according to the volume of the ejecta, the height to which the ejecta rise, and the duration of the eruption.

volcanogenic Of volcanic origin; e.g., volcanogenic sediments or ore deposits.

water cycle *See* hydrologic cycle.

water table The contact between the zone of aeration and the zone of saturation.

wave base The depth at which water waves no longer move sediment, equal to approximately half the wavelength.

wave frequency The number of waveforms that pass in a unit of time, expressed as cycles per second (cps). It is equal to $1/T$ (*see* wave period).

wavelength The distance between two equivalent points on two consecutive waves, for example, crest to crest or trough to trough.

wave of oscillation A water wave in which water particles follow a circular path with little forward motion.

wave of translation A water wave in which the individual particles of water move in the direction of wave travel, as in broken waves (surf).

wave period (T) The time (in seconds) between passage of equivalent points on two consecutive waveforms.

wave refraction The bending of wave crests as they move into shallow water.

weathering The physical and chemical breakdown of materials of the earth's crust by interaction with the atmosphere and biosphere.

welded tuff A glassy pyroclastic rock that has been made hard by the welding together of its particles under the action of retained heat.

wetted perimeter (stream) The length of stream bed and bank that are in contact with running water in the stream in a given cross-section.

wind The movement of air from a region of high pressure to a region of low pressure.

zero population growth A population growth rate of 0; that is, the number of deaths in a population is equal to the number of live births on average.

zone of aeration (vadose zone) The zone of soil or rock through which water infiltrates by gravity to the water table. Pore spaces between rock and mineral grains in this zone are filled with air; hence the name.

zone of saturation The ground-water zone where voids in rock or sediment are filled with water.

INDEX

Credits

Chapter 1
Opener, p.2, Breck P. Kent, JLM Visuals; **1.02 and 1.03** adapted from George A. Kiersch, *The Heritage of Engineering Geology: First Hundred Years, 1888–1988,* 1991, Geological Society of America, Vol. 3, p. 4 and 3 respectively; **1.05a and b** Dr. James C. Anderson, University of Southern California; **1.07 and 1.08** Blue Planet Group.

Chapter 2
Opener, p. 16, D. D. Trent; **2.08c and d** JLM Visuals; **Case Study 2.01, fig. 1b** Martin Land, Science Photo Library/Photo Researchers; **fig. 1c** Carl Frank, Photo Researchers; **fig. 1d** Martin Land, Science Photo Library/Photo Researchers; **Case Study 2.01, fig. 2** After Christopher B. Gunn, "A Descriptive Catalog of the Drift Diamonds of the Great Lakes Region, North America," *Gems and Gemology,* Vol. 12, Nos. 10–11, 1968, pp. 297–303 and 333–334, © Gemological Institute of America; **2.17, 2.18, 2.22** D. D. Trent.

Chapter 3
Opener, p. 42, Shawn Gerrits, JLM Visuals; **3.02a and b** Courtesy of Patricia G. Gensel, University of North Carolina; **3.03c** Courtesy of Scott Katz; **3.08** NGDC; **3.14b** Courtesy of H. E. Enlows, USGS.

Chapter 4
Opener, p. 60, USGS; **4.01** Frederick Larson, *San Francisco Chronicle;* **4.02f** B. Bradley, Department of Geology, University of Colorado; **4.06** Courtesy of the University of Southern California; **4.13b** Dorothy Stout; **4.13d** Dennis Fox; **4.14** Dr. James C. Anderson, University of Southern California; **4.15** C. J. Langer, USGS; **4.17b** B. Bradley, University of Colorado; **4.18a** Courtesy of NOAA, **4.18b** Courtesy of Richard Hilton; **4.19** Courtesy of NGDC; **4.21b** USGS, Menlo Park; **Case Study 4.03** Courtesy of Dave Johnson; **4.24a** Courtesy of NGDC; **4.26** Courtesy of the United States Navy; **4.28a, b, and c** Courtesy of Henry Helbush; **4.30a and b, and 4.32b** Dr. James C. Anderson, University of Southern California; **4.32c** E. V. Leyendecker, NBS, **4.32d** C. Arnold, Building Systems Dev., Inc.; **3.35 and 3.36** Dennis Fox; **4.37** E. V. Leyendecker, USGS; **4.38** EOS, American Geophysical Union; **4.43** Sykes, Lynn and Seeber, 1985, "Great Earthquakes and Great Asperities, San Andreas Fault, California"; *Geology,* Vol. 13, pp. 835–838, December; **4.44** Courtesy of Kerry Sieh, California Institute of Technology; **4.47** James Williams.

Chapter 5
Opener, p. 100, NOAA; **Case Study 5.01, fig. 1b** David M. Bailey, Grant County Airport, Moses Lake, WA; **fig. 2** NOAA; **5.08** EROS Data Center; **5.09a** USGS, **5.09b** J. D. Griggs; **5.10** J. B. Stokes, USGS; **5.14a** Björn Bölstad, Photo Researchers; **5.24** Jim Mori, USGS; **Case Study 5.02, fig. 2** Jim Mori, USGS; **fig. 3** NOAA; **5.27** USGS; **Case Study 5.03, figs. a and b** T. Orban, Sygma; **5.30** J. P. Lockwood, USGS; **5.33a** Tracy Knauer, Photo Researchers; **5.34** John Mead, Science Photo Library/Photo Researchers; **5.35** Van Bucher, Photo Researchers; **5.36** Richard Brooks, Photo Researchers.

Chapter 6
Opener, p. 134, John Minnich, JLM Visuals; **6.03a and b** John S. Shelton; **6.08** Bruce Hayes, Photo Researchers; **6.11** John S. Shelton; **Case Study 6.01, fig. 1,** Visuals Unlimited; **6.17** United States Department of Agriculture; **6.19** Doug Burbank, University of Southern California; **6.20** Visuals Unlimited; **Case Study 6.02, fig. 1,** UPI Bettmann, **fig. 3,** The Bettmann Archive; **6.21** Doug Moran; **6.25** 1957 Professor Troy Péwé, Arizona State University; **6.27a** R. A. Combellick, Alaska Division of Geological and Geophysical Surveys; **6.27b** Professor Troy Péwé, Arizona State University; **6.29a** Atlantic-Richfield Co., ARCO.

Chapter 7
Opener, p. 158, B. Pipkin; **7.04** USGS; **7.06** Courtesy of Earl W. Hart, California Division of Miners and Geology; **7.12c** D. D. Trent; **7.14a and b** Courtesy of Randall W. Jibson, USGS; **Case Study 7.01, fig. 4,** Courtesy of Professor Perry Ehlig; **7.18b** NGDC; **7.19** Canadian Air Force; **7.27a** Hollingsworth and Kovacs, California Geology; **Case Study 7.03, fig. 1** Dedman's Photo Shop, Skagway, Alaska; **7.27b** D. D. Trent.

Chapter 8
Opener, p. 188, Porterfield-Chickering, Photo Researchers; **Case Study 8.01, fig. a, b, c, d, and e** Los Angeles Times Photo; **8.04a, and 8.05a and b** City of Long Beach, Department of Oil Properties; **8.07** Anne Bryant, United States Air Force Photo; **8.09** Ronny Jacques, Photo Researchers; **8.12** R. V. Dietrich; **8.13b** James Ruhl Enterprises, Fullerton, CA; **Case Study 8.02, fig. 1** EROS Data Center; **8.14a** USGS, **8.14b** Frank Kujawa, University of Central Florida, GeoPhoto Publishing Company; **8.15** USGS; **8.19** E. W. Osterwald, USGS; **8.20** Doug Burbank; **8.21a** Scott Fee, American Speleological Society.

Chapter 9
Opener, p. 208, Les Stone, Sygma; **9.10** D. D. Trent; **9.14b** Victor R. Baker, University of Arizona; **9.17** Chris Stewart, Black Star; **Case Study 9.03, fig. 1 and 3** Security Pacific National Bank Historical Collection, Los Angeles County Library; **9.28** Landsat imagery courtesy of the Earth Observation Satellite Company, Lanham, Maryland; **9.29** Lisa Wigoda; **9.36** *Imprints of Time,* Bradford B. VanDiver, 1988, Mountain Press Publishing Company, Missoula, MT; **9.38** Amy Lauta.

Chapter 10
Opener, p. 242, Jim Richardson; **10.18a and b** Visuals Unlimited; **10.19** Martin Land, Science Photo Library, Photo Researchers.

Chapter 11
Opener, p. 266, Ruth Flanagan, © 1993, Kalmbach Publishing Co., *Earth* magazine; **11.05b and 11.06b** Tom Servais, *Surfer* magazine; **11.17** NOAA; **11.22a** Craig H. Everts; **11.28a and b** GeoPhoto Publishing Co.; **11.35** P. Godfrey; **11.42** Suzanne and Nick Geary, Tony Stone Worldwide.

Chapter 12
Opener, p. 296, Breck P. Kent, JLM Visuals; **12.07** Howard G. Wilshire, USGS; **12.08, 12.09, 12.10** D. D. Trent; **12.12** Rodman Snead, JLM Visuals; **12.14 and 12.15** D. D. Trent; **12.16, fig. a, b, and d** D. D. Trent, **fig. c** Joseph Allen; **12.22** Austin Post, USGS; **12.23, 12.24, 12.26, 12.28** D. D. Trent; **Case Study 12.03, fig. 2 and 3b** D. D. Trent; **12.30** after R. Dott and R. Batten, *Evolution of the Earth,* Third Edition, 1981, McGraw-Hill, Inc., Figure 18.6, p. 514; **12.31** Foundation for Glacier and Environmental Research; **12.33 and 12.34** D. D. Trent; **12.36** Baggeroer and Munk, "Heard Island Feasibility Test," *Physics Today,* V. 45, n. 9, September 1992, p. 24, fig. 2; **12.37** Nansen and Lebedeff, "Global Trends of Measured Surface Air Temperatures," *Journal of Geophysical Research,* V. 29, n. D11, Nov. 20, 1987, p. 13370, fig. 15; **12.39a** M. Elliot, **12.39b, c, and d** D. D. Trent.

Chapter 13
Opener, p. 334, Peter Menzel; **13.08** Steven Dutch; **Case Study 13.02, fig. 1** Drake Well Museum, Titusville; **Case Study 1303, fig. 2** Bill Nation, Sygma; **13.11** after *AAPG Explorer,* March 1991, American Association of Petroleum Geologists, p. 39; **13.17** Visuals Unlimited; **13.19** Peggy Foster; **13.23** D. D. Trent; **13.29** Visuals Unlimited.

Chapter 14
Opener, p. 370, Bill Bachman, Photo Researchers; **14.04** © 1985 by Wendell E. Wilson; **14.07** Courtesy Kennecott Corporation; **14.08** D. D. Trent; **14.09** John Delaney, University of Washington; **14.10** D. D. Trent; **14.11** Dr. Bruce Heezen, courtesy of Scripps Institution of Oceanography, University of California; **14.12** Precision Graphics, after "Alaska's Oil/Gas & Minerals," *Alaska Geographia,* V. 9, n. 4, 1982; **Case Study 14.01, fig. 2** Courtesy of the California State Library; **fig. 3** Photo from Donald B. Sayner Collection; **14.14** after "Alaska's Oil/Gas & Minerals," *Alaska Geographia,* V. 9, n. 4, 1982; **Case Study 14.02, fig. 1** D. D. Trent; **fig. 2** Precision Graphics, courtesy of the Jamestown Mine, Sonora Mining Corporation; **14.16** after Brian J. Skinner, "Mineral Resources of North America," in *Geology of North America,* Vol. 4; **14.17** Courtesy of Homestake Mining Company; **14.19** D. D. Trent; **14.21** Marc J. Plumley; **14.22** Courtesy of IMC Fertilizer, Inc.; **Case Study 14.05, fig. 1** courtesy of California Division of Mines and Geology; fig. 2 D. D. Trent; **14.23** Precision Graphics, from Kenneth A. Cole and Ann Kirkpatrick, "Cyanide Heap Leaching in California," *California Geology;* **14.24, 14.25, 14.26, and 14.27** D. D. Trent.

Chapter 15
Opener, p. 408, Dennis Capolongo, Greenpeace; **15.17** Courtesy Richard Proctor; **Case Study 15.04, fig. a and b** Michel J. Philippot, Sygma; **15.28** Courtesy of United States Department of Energy.